CIVIL ENGINEERING MATERIALS

WOODHEAD PUBLISHING SERIES IN CIVIL AND STRUCTURAL ENGINEERING

CIVIL ENGINEERING MATERIALS

From Theory to Practice

QIANG YUAN

ZANQUN LIU

KEREN ZHENG

CONG MA

ELSEVIER

中南大學出版社
www.csupress.com.cn
·长 沙·

Elsevier
Radarweg 29, PO Box 211, 1000 AE Amsterdam, Netherlands
The Boulevard, Langford Lane, Kidlington, Oxford OX5 1GB, United Kingdom
50 Hampshire Street, 5th Floor, Cambridge, MA 02139, United States

Library of Congress Cataloging-in-Publication Data
A catalog record for this book is available from the Library of Congress

British Library Cataloguing-in-Publication Data
A catalogue record for this book is available from the British Library

ISBN: 978-0-12-822865-4

For information on all Elsevier publications visit our website at https://www.elsevier.com/books-and-journals

Publisher: Glyn Jones
Editorial Project Manager: Naomi Robertson
Production Project Manager: Surya Narayanan Jayachandran
Cover Designer: Victoria Pearson

Typeset by TNQ Technologies

Contents

Preface *vii*

1. Fundamentals of materials **1**
1.1 Composition and structure 1
1.2 Physical properties 5
1.3 Mechanical properties 11
1.4 Durability 15
Exercises 16

2. Inorganic cementing materials **17**
2.1 Portland cement 17
2.2 Calcium sulfoaluminate cement 37
2.3 Calcium aluminate cements 41
2.4 Alkali-activated cement 47
2.5 Magnesium-based cements 52
Exercises 56

3. Portland cement concrete **59**
3.1 Introduction 59
3.2 Types of concrete 62
3.3 Raw materials 64
3.4 Concrete at fresh state 113
3.5 Mechanical properties 129
3.6 Deformation 137
3.7 Durability 146
3.8 Mix design 163
3.9 Self-compacting concrete and its application in high-speed rail 171
3.10 Steam-cured concrete 190
Exercises 202

4. Metal **205**
4.1 Introduction 205
4.2 Structural steel 206
4.3 Standards and selection of building steel 220
4.4 Corrosion and prevention of steel 232
4.5 Nonferrous metals 234
Exercises 238

5. Wood **239**
5.1 Introduction 239
5.2 Structure and composition 240
5.3 Engineering properties 241
5.4 Wood-based composites 250
5.5 Durability 257
Exercises 259

6. Polymers **261**
6.1 Engineering plastics 263
6.2 Sealants 271
6.3 Adhesive 272
6.4 Fiber reinforced polymer 276
Exercises 284

7. Asphalt **287**
7.1 Asphalt cement 287
7.2 Liquid asphalts 306
7.3 Asphalt concrete 308
Exercises 324

8. Cement-based composites **327**
8.1 Cement asphalt composite 327
8.2 Ultrahigh-performance concrete 362
Exercises 376

Index *377*

Preface

Civil engineering materials are the basics for construction. Human civilization is built on all kinds of buildings and infrastructures that are made up of various materials. The proper understanding and use of materials are of paramount significance to the engineers, which determine the quality of the buildings and infrastructure. For the past decades, governments and construction industries all around the world have made a huge investment in construction works, and massive amounts of civil engineering materials have been manufactured and consumed. In order to meet the requirements of new structures, traditional and newly developed civil engineering materials have been innovatively put into use, and new knowledge and experiences have been generated, which are invaluable to academia and industry. The use of new materials and new technologies promote the development of structure, and the development of structure encourages the use of new materials and technologies. It is the right time to write a textbook dedicated to civil engineering materials, which includes the new knowledge and experiences in this field, and the fundamental theory for materials as well. Since almost half of the global construction works happen in China, some Chinese experiences are introduced particularly.

This book covers a wide range of materials, from organic to inorganic, from metal to nonmetal, and from traditional to newly developed. Materials are introduced based on the relations among composition, structure, and properties. Firstly, fundamentals of materials are provided from the perspective of materials science, and then the materials described subsequently can be related to these basic theories. Specifically, seven types of civil engineering materials, i.e., inorganic binder, concrete, metal, asphalt, wood, polymer, and composite, are described in this book. Most importantly, the new knowledge and experiences obtained recently in China, especially in the field of high-speed railway, are incorporated in this book. For instance, ultrahigh-performance concrete and self-compacting concrete are newly developed materials, and have been widely used in the construction of infrastructure. Steam-cured concrete is a traditional way for fast production of concrete members. This has been widely used in China for precast box girders and other concrete members. The basic knowledge and innovative application of polymer, wood, steel, composite, and asphalt are also introduced.

The authors of this book are from Central South University and ShenZhen University in China. Qiang Yuan from Central South University is responsible for the plan of this book and the writing of Chapter 3 and part of Chapter 8. Keren Zheng from Central South University is responsible for the writing of Chapters 1 and 2. Zanqun Liu from Central South University is responsible for the writing of Chapters 4 and 5 and part of Chapter 8. Cong Ma from ShenZhen University is responsible for Chapters 6 and 7.

Many masters and PhD students helped with the editing and figure drawing of the manuscript during the preparation of this book. The authors would like to acknowledge Ms. Yuman Wang, Mr. Shenghao Zuo, Mr. Tsegaye Lakew Berihun, and Mr. Ghimire Prateek for their contributions to this book.

This book is intended for undergraduate and graduate students in civil engineering or material science. It can also be used as a general reference book for professional engineers and researchers, or a tool book for professional engineers and researchers.

Qiang Yuan, Zanqun Liu
Keren Zheng, Cong Ma

CHAPTER 1

Fundamentals of materials

1.1 Composition and structure

1.1.1 Composition

1.1.1.1 Chemical composition

The chemical composition of a material can be defined as the distribution of the individual components that constitute the material.

The material can be a pure substance, which contains only one chemical component; in this case, the chemical composition corresponds to the relative amounts of the elements constituting the substance. Normally, it can be expressed with a chemical formula. For an example, the chemical formula for water is H_2O, thus the chemical composition of water may be interpreted as a 2:1 ratio of hydrogen atoms to oxygen atoms.

For a mixture, the chemical is equivalent to quantifying the concentration of each component. Component responds to chemically recognizable species (Fe and C in carbon steel, H_2O and NaCl in salted water). There are different ways to define the concentration of a component, and there are also different ways to define the composition of a mixture. It may be expressed as molar fraction, volume fraction, mass fraction, molality or normality or mixing ratio.

1.1.1.2 Phase composition

Phase, in thermodynamics, refers to chemically and physically uniform or homogeneous quantity of a matter that can be separated mechanically from a nonhomogeneous mixture, and that may consist of a single substance or a mixture of substances. The concept of phase is also introduced to characterize the composition of materials containing more than one component. A phase in a material has uniform physical and chemical characteristics, and different phases in a material are separated from one another by distinct boundaries. In materials, a phase may contain one or more components. In other words, a multicomponent material can exist as a single phase if the

Civil Engineering Materials
ISBN 978-0-12-822865-4
https://doi.org/10.1016/B978-0-12-822865-4.00001-5

different chemical components are intimately mixed at the atomic length scale. In the solid state, such mixtures are called solid solution.

The components or phases in inorganic materials can be minerals. Minerals are naturally occurring, inorganic substances with quantifiable chemical composition and a crystalline structure. Portland cement clinker is man-made, and contains mainly four phases: Alite, belite, aluminate, and ferrite; however, we also called these phases as minerals.

1.1.2 Structure

Generally, the term structure for materials refers to the arrangement of internal components of materials. The structure of materials can be classified by the general magnitude of various features being considered. The three most common major classifications of structure are as follows: ①Atomic structure, which includes features such as the types of bonding between the atoms, and the way the atoms are arranged; ②Microstructure, which includes features that can be seen using a microscope, but seldom with the naked eye; ③Macrostructure, which includes features that can be seen with the naked eye.

Actually, most properties are highly structure sensitive and the structure virtually determines everything about a material: its properties, its potential applications, and its performance within those applications. Therefore, it is very important to understand the basis for the structure of materials to be able to control the properties and reliability of engineering materials.

1.1.2.1 Atomic structure

All materials are made of atoms. There are only about 100 different kinds of atoms in the entire universe. However, these same 100 atoms form thousands of different substances ranging from the air we breathe to the metal used to support tall buildings. It is the interaction between atoms and atomic bonding, to hold these atoms together and form different substances. According to their nature, the bonds can be categorized into two classes based on the bond energy. The primary bonds (>100 kJ/mol) are ionic, covalent, and metallic. The secondary bonds are of the van der Waals, or hydrogen.

Ionic bonding occurs between metal atoms and nonmetal atoms. To become stable, the metal atom tends to lose one or more electrons in its outer shell, thus becoming a positively charged ion (aka cations). Since electrons have a negative charge, the atom that gains electrons becomes a negatively charged ion (aka anion). As a result, the atoms in an ionic compound are held together since oppositely charged atoms are attracted to one another.

Where a compound only contains nonmetal atoms, a covalent bond is formed by atoms sharing two or more electrons. Nonmetals have four or more electrons in their outer shells (except boron). With these many electrons in the outer shell, it would require more energy to remove the electrons than would be gained by making new bonds. Therefore, both the atoms involved share a pair of electrons. Each atom gives one of its outer electrons to the electron pair, which then spends some time with each atom. Consequently, both atoms are held near each other since both atoms have a share in the electrons.

Metallic bonding is a type of chemical bonding that rises from the electrostatic attractive force between conduction electrons (in the form of an electron cloud of delocalized electrons) and positively charged metal ions. It may be described as the sharing of free electrons among a structure of positively charged ions (cations). Metallic bonding accounts for many physical properties of metals, such as strength, ductility, thermal and electrical resistivity and conductivity, opacity, and luster.

van der Waals bonding includes attraction and repulsions between atoms, molecules, and surfaces, as well as other intermolecular forces. van der Waals bonding differs from covalent and ionic bonding in that they are caused by correlations in the fluctuating polarizations of nearby particles. van der Waals force is a distance-dependent interaction between atoms or molecules and comparatively weak.

When the atoms, ions, or molecules have an opportunity to organize themselves into regular arrangements, or lattices by the bonds mentioned above in a solid, the solid is classified as a crystalline material. Hence, a crystalline solid possesses long range, regularly repeating units.

If there is no long-range structural order throughout the solid, the material is best described as amorphous. Quite often, these materials possess considerable short-range order over distances of 1−10 nm or so. However, the lack of long-range translational order (periodicity) separates this class of materials from their crystalline counterparts. Examples of amorphous solids are glass and some types of plastic. They are sometimes described as supercooled liquids because their molecules are arranged in a random manner somewhat as in the liquid state. As shown in Fig. 1.1, silicon and oxygen are bonded by covalent bond to form regular unit, silicon−oxygen tetrahedron, when the silicon−oxygen tetrahedrons are arranged in regular way, the solid is called quartz, crystalline SiO_2, whereas silicon−oxygen tetrahedrons are arranged in a random way, they form glass (amorphous SiO_2).

The atomic structure primarily affects the chemical, physical, thermal, electrical, magnetic, and optical properties.

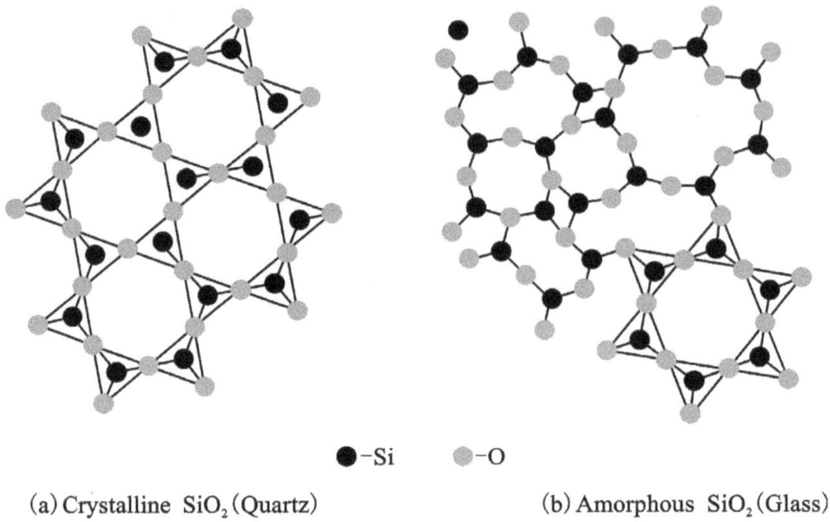

Figure 1.1 Schematic comparison between crystalline SiO_2(quartz) and amorphous SiO_2(glass).

1.1.2.2 Microstructure

The term "microstructure" is used to describe the arrangement of phases and defects within a material, the appearance of the material on the nm—μm length scale. A complete description of microstructures involves describing the size, shape, and distribution of grains and second-phase particles and their composition.

Microstructure can be observed using a range of microscopy techniques. The microstructural features of a given material may vary greatly when observed at different length scales. For this reason, it is crucial to consider the length scale of the observations you are making when describing the microstructure of a material.

Microstructures determine the mechanical, physical, and chemical properties of materials. For example, the strength and hardness of materials are determined by the number of phases and their grain sizes. The electrical and magnetic properties and also the chemical behavior (corrosion) are determined by the grain size and defects (vacancies, dislocations, grain boundaries, etc.) presented in the material. As a consequence, the behavior of such multiphase material is determined by the properties of the individual phases and the fashion in which these phases interact. As a general rule, the mechanical properties such as ductility, strength, resistance to creep and fatigue of engineering materials are determined by their (micro)structure at different geometric scales.

Figure 1.2 The BSE and particles packing images of cement-based materials.

The microstructure of cement-based materials is controlled by their constituents, the mixture proportions, processing (e.g., mixing, consolidation, and curing), and degree of hydration. The properties of the hardened cement-based materials are dependent on their microstructure; the capillary pore structure (black areas in Fig. 1.2), which includes the interface transition zone between the cement paste and aggregates usually governs the transport properties of concrete, while larger voids reduce the strength of concrete.

1.1.2.3 Macrostructure

Macrostructure describes the appearance of a material in the scale millimeters to meters, it is the structure of the material as seen with the naked eye. The term macrostructure is sometimes used to refer to the largest components of the internal structure. Grain flow, cracks, and porosity are all examples of macrostructure features of materials. Macrostructure also determines properties of materials, especially the mechanical properties.

1.2 Physical properties

Properties of a material refer to the features we can sense, measure, or test. For example, if we have a sample of metal in front of us, we can identify that the material is gray, hard, or shiny. Testing shows that the material is able to conduct heat and electricity and it will react with an acid. These are some of the metal's properties.

Physical properties are those that can be observed without changing the identity of the substance. The general properties of matter such as density, specific gravity, fineness, thermal conductivity, heat capacity, etc., are examples of physical properties.

1.2.1 Density and specific gravity

Mass (m) is a fundamental measure of the amount of matter. The space the mass occupies is its volume, and the mass per unit of volume is its density. Hence, it is simple to calculate density of an object by dividing its mass by its volume. However, this is pretty complicated in the case of building materials.

A lot of building materials, such as wood, cementitious materials, and ceramics, are porous. For porous particles, the mass is a finite value, but how about the volume? As shown in Fig. 1.3, a stack of porous particles contains a lot of pores, and these pores can be divided into two groups, i.e., open pores and closed pores. When the particles are immersed in water, water can enter open pores, but it cannot enter closed pores. Hence, different volumes of the porous particles can be defined. For a particulate solid, it additionally includes the space left void between particles.

Envelope volume: The volumes of the solid and the voids within the particle, that is, within close-fitting imaginary envelopes completely surrounding the particle.

Apparent volume or skeletal volume: The volumes of the solid in the particles and closed (or blind) pores within the particle. This volume definition excludes volumes of open pores.

True or absolute volume: The volume of the solid in the particle, which excludes volumes of all pores.

Accordingly, we can define different densities for porous materials as follows:

Apparent density: The mass of a particle divided by its apparent (skeletal) volume.

Envelope density: The ratio of the mass of a particle to the envelope volume of the particle.

True density: The mass of a particle divided by its true (absolute) volume.

For a collection of discrete particles of solid porous material, the bulk density is the ratio of the mass of the collection of discrete pieces of solid material to the sum of the volumes of the solids in each piece, the voids within the pieces, and the voids among the pieces of the particular collection. For powder materials, bulk density is also called bulk powder density.

Weight (w) is a measure of the force exerted by a mass and this force is produced by the acceleration of gravity. Therefore, on the surface of the earth, the mass of an object is determined by dividing the weight of an

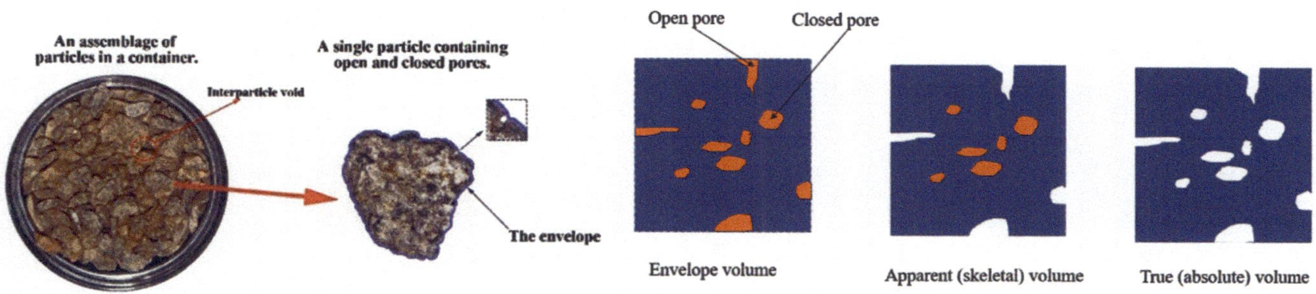

Figure 1.3 A schematic picture well illustrates the physical meaning of these density definitions.

object by 9.8 m/s^2 (the acceleration of gravity on the surface of the earth). Since we are typically comparing things on the surface of the earth, the weight of an object is commonly used rather than calculating its mass. Specific gravity is the ratio of the density of a substance compared to the density of freshwater at 4°C. At this temperature, the density of water is at its greatest value and equals 1 g/cm^3. Since specific gravity is a ratio, it has no units. Specific gravity values for a few common substances are as follows: Au, 19.3; mercury, 13.6; alcohol, 0.7893; benzene, 0.8786. Note that since water has a density of 1 g/cm^3, the specific gravity is the same as the density of the material measured in g/cm^3.

1.2.2 Fineness

Fineness indicates the fineness or coarseness degree of powdery materials. It is often expressed as standard sieve percentage or specific surface area.

Fineness can also be expressed by percentage of particles of various sizes or average value of unit weight material. The population of particles of various sizes is termed as particle size distribution. D_{50} is usually used to represent the particle size of group of particles, which characterizes the median diameter or medium value of particle size distribution. For instance, if $D_{50} = 5.8$ μm, then 50% of the particles in the sample are larger than 5.8 μm and 50% smaller than 5.8 μm.

The specific surface area is the surface area of the powdery material per unit weight. There are many methods to determine the specific surface area, such as gas adsorption, organic molecular adsorption, and air permeability. Blaine's air permeability apparatus is commonly used for cementitious materials, which consists essentially of a means of drawing a definite quantity of air through a prepared bed of cement of definite porosity.

Fineness, PSD, and specific surface area are fundamental characteristics of cementitious materials, they affect the properties of building materials in many important ways. Taking cement for an example, the finesses affects its hydration rate, water demand, workability of fresh concrete prepared with the material.

1.2.3 Thermal conductivity and heat capacity

Thermal conductivity is the ability of a material to transfer heat. Thermal conductivity is quantified using the unit of $W/(m \cdot K)$, and is the reciprocal of thermal resistivity, which measures the ability of materials to resist heat transfer. Thermal conductivity can be calculated as the following equation:

$$k = Q \times L/A(T_2 - T_1) \tag{1.1}$$

where Q is heat flow, W; L is length or thickness of the material, mm; A is surface area of material, m^2; $T_2 - T_1$ is temperature gradient, K.

The thermal conductivity of a specific material is highly dependent on a number of factors, including the temperature gradient, the properties of the material, and the path length that the heat follows. The thermal conductivity of the materials around us varies substantially, from those with low conductivities such as air with a value of 0.024 W/(m·K) at 0°C to highly conductive metals like copper, 385 W/(m·K).

The thermal conductivity of materials determines how we use them, for example, those with low thermal conductivities are excellent at insulating our homes and businesses, while high thermal conductivity materials are ideal for applications where heat needs to be moved quickly and efficiently from one area to another, as in cooking utensils and cooling systems in electronic devices. By selecting materials with the thermal conductivity appropriate for the application, we can achieve the best performance possible.

Heat capacity describes how much heat must be added to a substance to raise its temperature by 1°C:

$$C = Q/\Delta T \tag{1.2}$$

where C is heat capacity; Q is energy (usually expressed in joules); ΔT is the change in temperature (Celsius or in Kelvin).

Specific heat and heat capacity are related by mass:

$$C = m \times S \tag{1.3}$$

where C is heat capacity; m is mass of material; S is specific heat.

1.2.4 Linear coefficient of thermal expansion

The average amplitude of the atoms' vibration within the material increases when heat is added to most of the materials. This, in turn, increases the separation between atoms and causes materials to expand. It is usually expressed as a fractional change in length or volume per unit temperature change; a linear expansion coefficient is usually used in describing the expansion of a solid. The linear coefficient of thermal expansion (α) describes the relative change in length of a material per degree temperature change.

$$\alpha = \frac{\Delta l}{l_i \cdot \Delta T} \tag{1.4}$$

where l_i is initial length; Δl is the change in length; ΔT is change in temperature.

Thermal expansion (and contraction) must be taken into account when designing structures. The phenomena of thermal expansion can be challenging when designing bridges, buildings, aircraft, and spacecraft, but it can be put to beneficial uses.

1.2.5 Wetting and capillarity

Wetting is the ability of liquid to form interfaces with solid surfaces, or refers to describe how a liquid deposited on a solid (or liquid) substrate spreads out. To determine the degree of wetting, the contact angle (q) that is formed between the liquid and the solid surface is measured. The smaller the contact angle and the smaller the surface tension, the greater the degree of wetting.

As shown in Fig. 1.4, a wetting liquid is a liquid that forms a contact angle with the solid which is smaller than 90 degrees. A nonwetting liquid creates a contact angle between 90 and 180 degrees with the solid. Assuming that there are no other factors involved (e.g., roughness), when the contact angle formed between water and a solid surface is smaller than 90 degrees, the solid is hydrophilic. On the contrary, water creates a contact angle between 90 and 180 degrees with a solid, which means that water cannot spread on the solid surface autogenously, then the solid is hydrophobic.

Capillarity is the ability of a substance to draw another substance into it. It occurs when the adhesive intermolecular forces between the liquid and a substance are stronger than the cohesive intermolecular forces inside the liquid. The effect forms a concave meniscus where the substance is touching a vertical surface. The same effect is what causes porous materials to soak up liquids. Capillary forces pull a wetting liquid toward a low contact angle with the surface and wets the surface. A completely wetting liquid forms a zero-contact angle into a capillary by creating a curved meniscus at the rising liquid front. This phenomenon can be described with the Young–Laplace equation and the Laplace pressure inside a capillary.

Figure 1.4 Schematic illustration of contact angle of both hydrophobic surface and hydrophilic surface.

1.3 Mechanical properties

1.3.1 Loading and strength

The application of a force to an object is known as loading. Materials can be subjected to many different loading scenarios and a material's performance is dependent on the loading conditions. There are five fundamental loading conditions: tension, compression, bending, shear, and torsion (Fig. 1.5).

If material is subjected to a constant force, it is called static loading. If the loading of the material is not constant but instead fluctuates, it is called dynamic or cyclic loading. The way material is loaded greatly affects its mechanical properties and largely determines how, or if, a component will fail; and whether it will show warning signs before the failure actually occurs.

In mechanics of materials, the strength of a material is its ability to withstand an applied load without failure or plastic deformation. According to different loading conditions, the strength includes tensile strength, compressive strength, flexible strength, shear strength, and others.

1.3.2 Elasticity and plasticity

Elasticity is the property of solid materials to return to their original shape and size after the forces deforming them have been removed. When a force is applied to a certain cross-sectional area of an object, that object will develop both stress and strain as a result of the force.

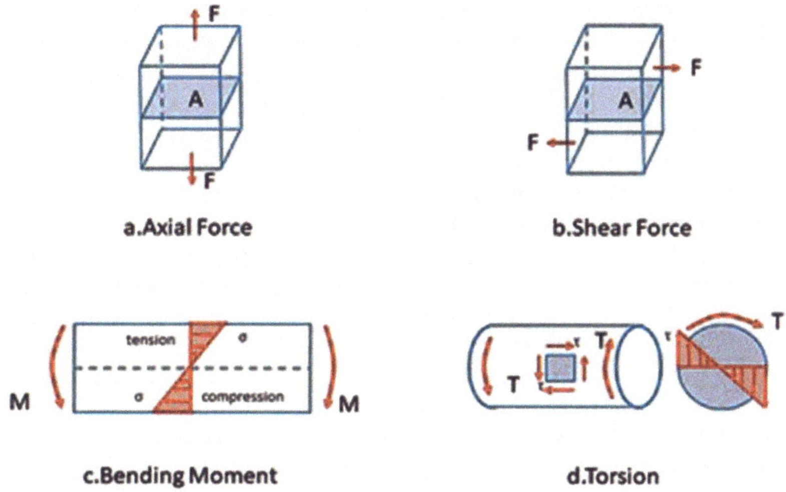

Figure 1.5 The fundamental loading conditions and illustration.

Stress is the force carried by the member per unit area;

$$\varepsilon = \frac{L - L_0}{L_0} = \frac{\delta}{L_0} \tag{1.5}$$

where F is the applied force; A is the cross-sectional area over which the force acts.

Strain is the ratio of the deformation to the original length of the part:

$$\varepsilon = \frac{L - L_0}{L_0} = \frac{\delta}{L_0} \tag{1.6}$$

where L is the deformed length; L_0 is the original undeformed length; δ is the deformation (the difference between the two).

Stress is proportional to strain in the elastic region of the material's stress—strain curve (below the proportionality limit, where the curve is linear). The coefficient that relates a particular type of stress to the resulted strain is called an elastic modulus (plural, moduli).

$$E = \sigma/\varepsilon \tag{1.7}$$

Elastic moduli are properties of materials, not objects. The elastic modulus, also known as the modulus of elasticity, or Young's modulus, is essentially a measurement of the stiffness of a material. As a result, it is commonly used in design and engineering applications.

Plasticity, ability of solid material to flow or to change shape permanently when subjected to stresses of intermediate magnitude between those producing temporary deformation, or elastic behavior, and those causing failure of the material, or rupture. Plasticity enables a solid under the action of external forces to undergo permanent deformation without rupture. Plastic deformation is a property of ductile and malleable solids.

Most of the building materials are not pure elastic materials. Some materials only have elastic deformation if the stress is not large, but plastic deformation will happen to them when the stress is beyond a limit, such as low-carbon steel. Under external forces, some materials will have elastic deformation and plastic deformation at the same time, but elastic deformation will disappear and plastic deformation still maintains when the stress is removed, such as concrete.

1.3.3 Brittleness and toughness

Brittleness is a property of materials which enables it to withstand permanent deformation. Cast iron and glass are examples of brittle materials. They will break rather than bend under shock or impact. Generally, the brittle materials have high compressive strength but low in tensile strength.

Toughness means the ability of a material to deform plastically and to absorb energy in the process before fracture occurs. The emphasis of this definition should be placed on the ability to absorb energy before fracture. Ductility is a measure of how much something deforms plastically before fracture, but note that a material is ductile does not make it tough. The key to toughness is a good combination of strength and ductility. A material with high strength and high ductility will have more toughness than a material with low strength and high ductility. Therefore, one way to measure toughness is by calculating the area under the stress–strain curve from a tensile test. This value is simply called "material toughness" and it has the unit of energy per volume. Material toughness equates to slow absorption of energy by the material.

It is the property of a material which enables it to withstand shock or impact. Toughness is the opposite condition of brittleness. The toughness may be considering the combination of strength and plasticity. Manganese steel, wrought iron, mild steel, etc., are examples of toughness materials. There are several variables that have a profound influence on the toughness of a material. These variables are strain rate (rate of loading), temperature, and notch effect.

1.3.4 Hardness

Hardness is the resistance of a material to localized deformation. The term can apply to deformation from indentation, scratching, cutting, or bending. In metals, ceramics, and most polymers, the deformation considered is plastic deformation of the surface. For elastomers and some polymers, hardness is defined as the resistance to elastic deformation of the surface. Hardness measurements are widely used for the quality control of materials because they are quick and considered to be nondestructive tests when the marks or indentations produced by the test are in low-stress areas. There are a large variety of methods used for determining the hardness of a substance. Historically, it was measured on an empirical scale, determined by the ability of a material to scratch another, diamond being the hardest and talc the softer. There are a few different hardness tests: Mohs, Rockwell, Brinell, Vickers, etc. They are popular because they are easy and nondestructive (except for the small dent).

1.3.5 Dynamic mechanical properties

A lot of structures are subjected to dynamic load during their service time such as bridges, rails. Dynamic mechanical properties refer to the response of a material to a periodic force. These properties may be expressed in terms of a dynamic modulus, a dynamic loss modulus, and a mechanical damping term. Polymers, and particularly rubbers, are often deliberately selected for products which are to be subjected to dynamic mechanical loading.

Stress analysis involves the use of the frequency-dependent dynamic moduli of the polymers. Assume, for example, that the polymer is subjected to a sinusoidal stress σ of amplitude σ_o and frequency ω, i.e., $\sigma = \sigma_0 \sin\omega t$. Stress analysis concerned with the dynamic mechanical properties normally assumes that polymers are linearly viscoelastic. Hence, the strain response ε to the imposed sinusoidal stress can be described as $\varepsilon = \varepsilon_0 \sin(wt - \delta)$ where δ is the phase angle. This is shown diagrammatically in Fig. 1.6. The imposed stress and the material response do not coincide, and the phase angle δ is the difference between the two curves.

Note that the strain response lags behind the stress by the phase angle—owing to the viscous component of the material. Some, but not all, of the energy stored during the deformation of the material is dissipated. Since the material is assumed to be linear, the stress is proportional to the strain at all times, i.e., $\sigma = E_e$, but E is a function of the frequency ω. Because the stress and strain are not in phase, E must be treated as a complex function:

$$E^* = E' + iE'' \tag{1.8}$$

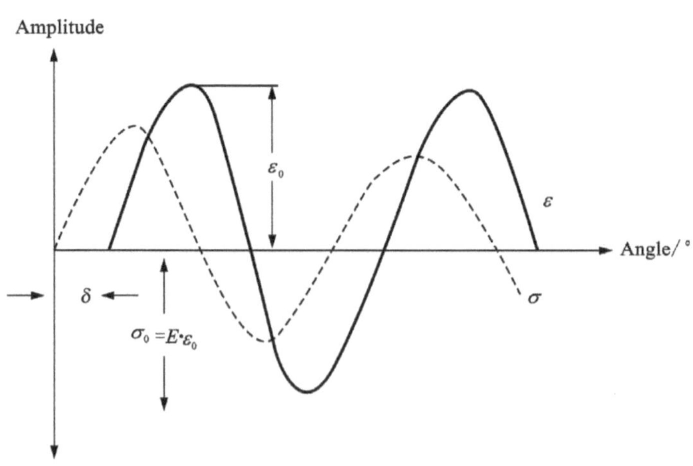

Figure 1.6 The sinusoidal stress σ and corresponding strain ε response for a linear viscoelastic material.

where and E' and E'' are the in-phase and out-of-phase components of the modulus.

From the above definitions of the dynamic moduli and by manipulation of the linear relationship between the sinusoidal stress and the corresponding strain response, the phase angle δ can be expressed as follows:

$$\tan \delta = E''/E' \tag{1.9}$$

where tan δ is commonly called the loss tangent or damping factor; E'' and E' are the most commonly measured dynamic properties of rubbers, representing the elastic stiffness and damping or hysteresis properties, respectively. Sometimes the "argument" of the complex modulus $|E|$ is used instead of E^*, and is given by the following equation:

$$|E| = \sqrt{(E')^2 + (E'')^2} \tag{1.10}$$

At very high frequencies ($\omega = 10^4 - 10^8$ cycles/s or Hz) rubber is very stiff with a glass-like modulus. At these frequencies the polymer molecules do not have time to react in response to the forcing oscillations. The damping factor is then small but it increases to a maximum value in the "leathery" transition region between the glassy modulus and the usual (low) modulus which is characteristic of rubbers that are deformed slowly ($\omega < 1$ cycle/s or Hz).

1.4 Durability

For materials, durability is the ability to service for the long term without significant deterioration by resisting the effects of the heavy use, drying, wetting, heating, freezing, thawing, corrosion, oxidation, volatilization, etc.

According to the deterioration mechanisms, the deleterious factors mainly consist of physical actions, chemical reactions. Physical actions include wetting and drying, change in temperatures, and freeze-and-thaw cycle.

For cement-based materials, chemical reactions leading to degradation include acid attack, salt attack, alkali-aggregate reaction, carbonation, and reinforcement corrosion. Two types of corrosion can be distinguished for steel and other metals: direct reaction of the corrosive compound with the metal and corrosion that occurs through the water present at the metal surface. For asphalt, plastic, rubber, and other organic materials will be damaged due to aging.

Durability is one of the major requirements to be considered in the use of building materials. More knowledge about deterioration mechanisms and the measures to counteract these are to be provided in other chapters in this book.

Exercises

1. Please summarize the factors which influence the durability of materials for civil engineering.
2. Water is easy to spread over the surface of concrete and transport in concrete because it is a hydrophilic material, how to improve the impermeability of concrete without changing its porosity and pore structure?
3. When designing buildings such as airport terminals, why the linear coefficient of thermal expansion of used materials should be considered ?
4. State the general relationship between the composition, structure, and properties of materials.
5. Complete the following form to describe the change in properties of a materials as its porosity increases. (using ↑when increasing; ↓ when decreasing, – when unchanged, and ? for unclear)

Porosity	Density	Envelope density	Strength	Water sorption	Resistance to frost	Thermal conductivity
↑						

CHAPTER 2

Inorganic cementing materials

2.1 Portland cement

Portland cement was invented in 1824, and its name is derived from its similarity to Portland stone, a type of building stone quarried on the Isle of Portland in Dorset, England. Nowadays, Portland cement is the most widely used man-made material because of low cost, easy availability of raw materials, good workability of fresh cement-based materials, and versatility.

Portland cement is obtained by grinding clinker together with adequate gypsum. Clinker is a hydraulic material which shall consist of at least two-thirds by mass of calcium silicates. Clinkers are nodules of a sintered material that is produced when a raw mixture of predetermined composition is heated to high temperature. The main mineral phases contained in a Portland clinker are calcium silicates (Ca_3SiO_5, Ca_2SiO_4), aluminate ($Ca_3Al_2O_6$) and alumino-ferrite ($Ca_4(Al_xFe_{x-1})_4O_{10}$), briefly denoted as C_3S, C_2S, C_3A, and C_4AF.

According to GB-175, clinker makes up more than 90% of the cement, along with a limited amount of calcium sulfate ($CaSO_4$, which controls the set time), and up to 5% minor constituents (or fillers) as allowed by various standards.

Blended cement can be defined as a uniform mix of Portland cement and blending materials such as fly ash, limestone, and slag to enhance its properties for different uses. Blended cement can improve workability, strength, durability, and chemical resistance of concrete. Moreover, the replacement of clinker with blending materials can reduce the CO_2 footprint of cement and concrete, and use solid industrial byproducts in an eco-efficient way.

The composition of Portland cement is defined by the mass percentages and composition of the raw sources of lime, iron, silica, and alumina as well as the temperature and duration of heating. It is this variation in raw materials source and the plant-specific characteristics, as well as the finishing processes (i.e., grinding and possible blending with gypsum, limestone, or

Civil Engineering Materials
ISBN 978-0-12-822865-4
https://doi.org/10.1016/B978-0-12-822865-4.00002-7

supplementary cementing materials), that define the cement produced. Based on this, we can produce Portland cement with special properties such as rapid setting, high early strength, and low hydration heat. The commonly used special cement based on Portland cement include white Portland cement, expansive cement, sulfate resisting cement, etc.

2.1.1 Manufacture

Cement is manufactured through a closely controlled chemical combination of calcium, silicon, aluminum, iron, and other ingredients. Common materials used to manufacture cement include limestone, shells, and chalk or marl combined with shale, clay, slate, blast furnace slag, silica sand, and iron ore. The most common way to manufacture Portland cement is through a dry method.

The first step is to quarry the principal raw materials, mainly limestone, clay, and other materials. After quarrying the rock is crushed. This involves several stages. The first crushing reduces the rock to a maximum size of about 6 inches. The rock then goes to secondary crushers or hammer mills for reduction to about 3 inches or smaller. The crushed rock is combined with other ingredients such as iron ore or fly ash and ground, mixed, and fed to a cement kiln. The cement kiln heats all the ingredients to about 1450°C in huge cylindrical steel rotary kilns lined with special firebrick. The finely ground raw material (i.e., raw meal) is fed into the higher end. At the lower end is a roaring blast of flame, produced by precisely controlled burning of powdered coal, oil, alternative fuels, or gas under forced draft. Manufacture of cement by dry process is schematically shown in Fig. 2.1.

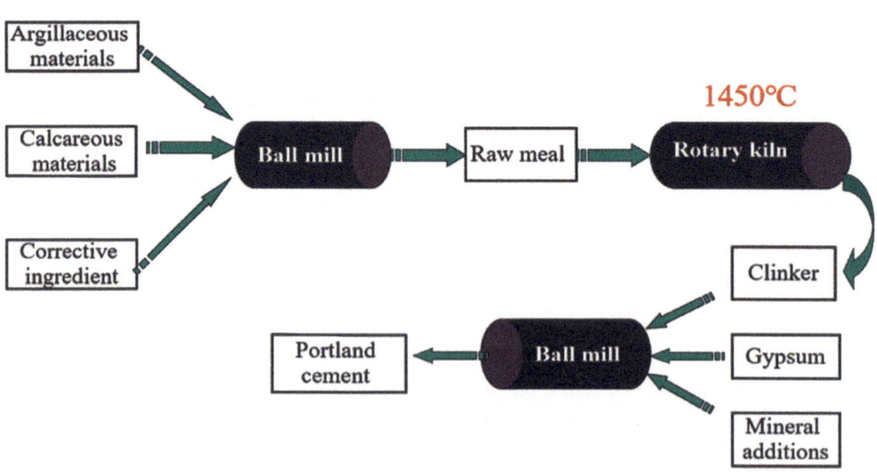

Figure 2.1 Manufacture of cement by dry process.

As the material moves through the kiln, certain elements are driven off in the form of gases. The remaining elements unite to form a new substance called clinker. Clinker comes out of the kiln as gray balls, about the size of marbles.

During the sintering of clinker, the following reactions take place:

$$CaCO_3 = CaO + CO_2 \quad (2.1)$$

$$2CaO + SiO_2 = Ca_2SiO_4 \text{ (dicalcium silicate (}C_2S\text{))} \quad (2.2)$$

$$3CaO + SiO_2 = Ca_3SiO_5 \text{ (tricalcium silicate (}C_3S\text{))} \quad (2.3)$$

$$3CaO + Al_2O_3 = Ca_3Al_2O_6 \text{ (tricalcium aluminate (}C_3A\text{))} \quad (2.4)$$

$$4CaO + Al_2O_3 + Fe_2O_3 = Ca_4Al_2Fe_2O_{10} \text{ (alumino-ferrite(}C_4AF\text{))} \quad (2.5)$$

Clinker is discharged red-hot from the lower end of the kiln and generally is brought down to handling temperature in various types of coolers. The heated air from the coolers is returned to the kilns, a process that saves fuel and increases burning efficiency. After the clinker is cooled, cement plants grind it and mix it with small amounts of gypsum. The cement is now ready for transport to ready-mix concrete companies to be used in a variety of construction projects.

Although the dry process is the most modern and popular way to manufacture cement, some kilns use a wet process. The two processes are essentially alike except in the wet process, the raw materials are ground with water before being fed into the kiln.

2.1.2 Composition

For cement, chemical analyses are normally given in the oxide form. From the chemical analysis, the quantity of each of the four above-mentioned main minerals can be calculated using the "Bogue" calculation. A typical example of mineral composition is shown in Table 2.1.

Fig. 2.2 presents the image of polished cement clinker derived from optical microscope, from which four minerals can be discerned.

Table 2.1 Typical composition of ordinary Portland cement.

Mineral	Phase name	Abbreviation	Typical amount
Tricalcium silicate	Alite	C_3S	50%–70%
Dicalcium silicate	Belite	C_2S	10%–20%
Tricalcium aluminate	Aluminate	C_3A	5%–10%
Tetra-calcium aluminoferrite	Ferrite	C_4AF	5%–15%

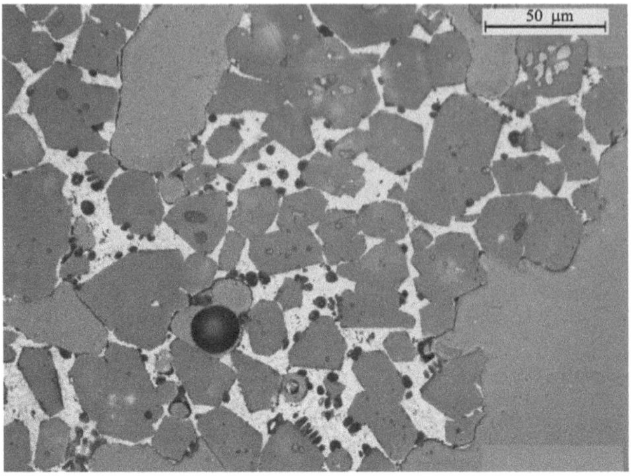

Figure 2.2 The SE image of cement clinker (Chemistry of Cement and Concrete, Fifth Edition, Elsevier).

Small amounts of clinker sulfate (sulfates of sodium, potassium, and calcium), free-lime(f-Cao), and periclase (f-MgO) also present in Portland cement clinker.

2.1.3 Hydration

The reaction of Portland cement with water is termed as "hydration." This involves many different reactions, often occurring simultaneously. As the reactions proceed, the products of the hydration gradually bond together with the individual sand and gravel particles, as well as other components, to form a solid mass (concrete).

2.1.3.1 The hydration process: reaction

When water is added, the following reactions are mostly exothermic, that is, the reactions generate heat. We can get an indication of the rate at which the minerals are reacting by monitoring the rate at which heat is evolved using a technique called isothermal calorimetry. An illustrative example of the heat evolution curve produced is shown below.

Fig. 2.3 shows typical heat evolution plotted against time from mixing to 60 h. During the hydration, three principal reactions occur:

①Almost immediately on adding water, some of the clinker sulfates and gypsum dissolve, producing an alkaline, sulfate-rich solution. ② Soon after mixing, C_3A (the most reactive of the four main clinker minerals) reacts with the water and sulfate in solution to form small rod-like crystals of ettringite (Stage I on the heat evolution curve mentioned above). The

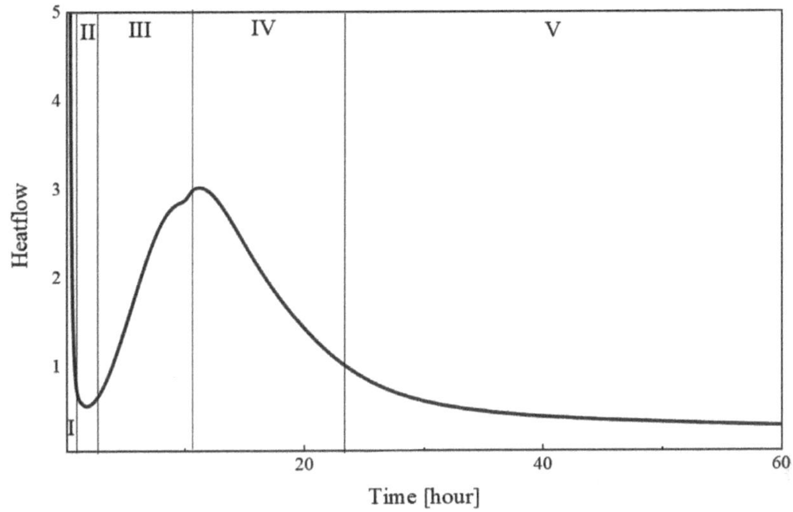

Figure 2.3 Schematic representation of a calorimetry curve of OPC.

reaction of C_3A with water is strongly exothermic but does not last long, typically only a few minutes, that is because the presence of gypsum can inhibit the dissolution of C_3A. Then a period of a few hours of relatively low heat evolution occurs, which is called the dormant, or induction period (Stage II). C_3S also dissolve into water quickly, releasing a lot of Ca^{2+} into the alkaline, sulfate-rich solution. However, the reaction of C_3S slows down as the high concentration of Ca in the solution resist the further dissolution of the phase. The first part of the dormant period, up to perhaps half-way through, corresponds to when concrete can be placed. As the dormant period progresses, the paste becomes too stiff to be workable.

At the end of the dormant period, the reaction of alite speeds up, leading to massive precipitation of calcium silicate hydrate(C—S—H) and calcium hydroxide (CH). This corresponds to the main period of hydration (Stage III), during which concrete strength increases. The individual grains react from the surface inwards, and the anhydrous particles become smaller. C_3A hydration also continues, as fresh crystals become accessible to water.

The period of maximum heat evolution occurs typically between about 10 and 20 h after mixing and then gradually tails off. In a mix containing PC only, most of the strength gain has occurred within about a month. Where PC has been partly replaced by other materials, such as fly ash, strength growth may occur more slowly and continue for several months or even a year.

At an early stage, the reaction of C_2S is insignificant because its solubility is much lower than that of C_3S. It is reported that the reaction of C_2S becomes significant at 10 days or so. As a result, C_2S contributes the strength development mainly at later ages.

Ferrite reaction also starts quickly as water is added, but then slows down, probably because of a layer of iron hydroxide gel forms, coating the ferrite and acting as a barrier, preventing further reaction. Fig. 2.4 presents the reaction degree of four clinker minerals at various hydration time.

2.1.3.2 Hydration products

The products of the reaction between cement and water are termed as "hydration products." There are typically four main types:

2.1.3.2.1 Calcium silicate hydrate

This is the main reaction product and is the main source of concrete strength. It is often abbreviated, using cement chemists' notation, to "C—S—H," the dashes indicating that no strict ratio of SiO_2 to CaO is inferred. The Si/Ca ratio is somewhat variable but typically approximately

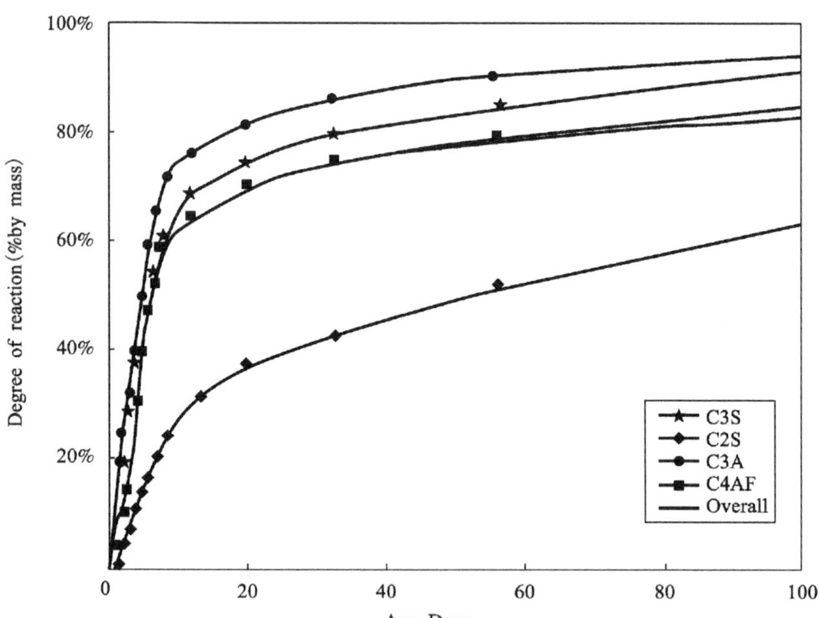

Figure 2.4 The reaction degree of different clinker phases as a function of time. *(Adapted from CCR, 30(6), 2000, Pages 855—863.)*

0.45−0.50 in hydrated Portland cement, the ratio may be up to perhaps about 0.6 if slag or fly ash or micro silica is present, depending on the proportions. C−S−H that occupies the boundary region of the anhydrous cement grain is called an "inner product" and that forms in the originally water-filled pore spaces is called an "outer product." Those two types form at different stages of hydration and exhibit distinct morphologies. Outer product C−S−H has a fibrillar morphology, the inner product exhibits a dense, homogeneous morphology with the significantly decreased porosity compared to the outer product (Fig. 2.5).

2.1.3.2.2 Calcium hydroxide or portlandite

$Ca(OH)_2$, often abbreviated to "CH," has a layered structure at atomic level (Fig. 2.6). CH is formed mainly from alite hydration. Alite has a Ca/Si ratio of 3:1 and C−S−H has a Ca/Si ratio of approximately 2:1, so excess lime is available to produce CH.

2.1.3.2.3 AFm and AFt phases

These are two groups of minerals that occur in cement, and elsewhere. They are products derived from C_3A and C_4AF. The general definitions of these phases are somewhat technical, but, for example, ettringite is an AFt phase because it contains three (t-tri) molecules of anhydrite and monosulfate is an AFm phase because it contains one (m-mono) molecule of anhydrite. The most common AFt and AFm phases in hydrated cement are as follows:

Figure 2.5 The morphology of C−S−H observed by TEM. *(From I.G. Richardson, CCR, 34(9), 2004:1733−77.)*

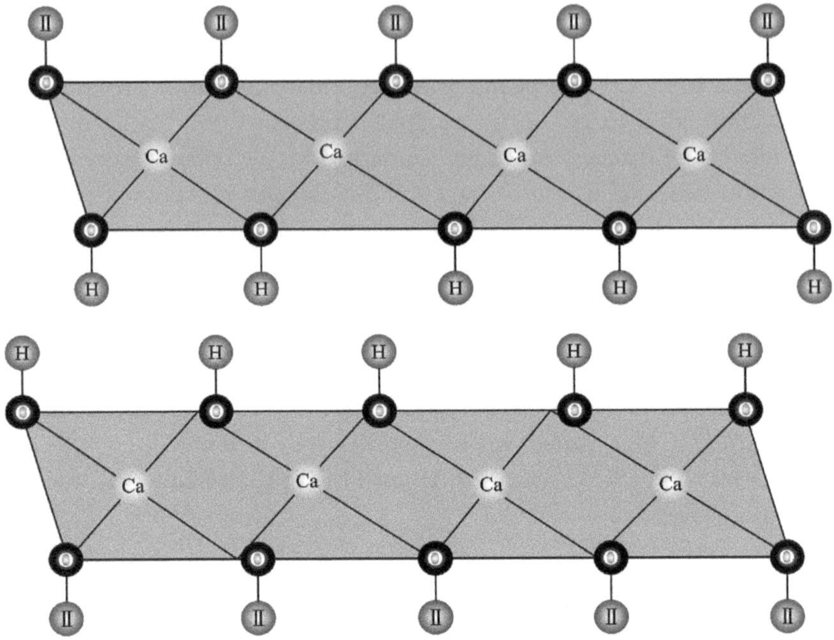

Figure 2.6 The atomic structure of portlandite.

2.1.3.2.4 Ettringite

Ettringite is present as rod-like crystals in the early stages of reaction or sometimes as massive growths filling pores or cracks in mature concrete or mortar. The chemical formula for ettringite is $[Ca_3Al(OH)_6.12H_2O]_{2.2}$ H_2O or, mixing notations, $C_3A.3CaSO_4.32H_2O$. Monosulfate: monosulfate tends to occur in the later stages of hydration, a day or two after mixing. The chemical formula for monosulfate is $C_3A.CaSO_4.12H_2O$. Note that both ettringite and monosulfate are compounds of C_3A, $CaSO_4$ (anhydrite), and water, in different proportions. Monocarbonate: the presence of fine limestone, whether interground with the cement or present as fine limestone aggregate, is likely to produce monocarbonate ($C_3A.CaCO_3.11H_2O$) as some of the limestone reacts with the cement pore solution. Other AFm phases that may be present are hemicarbonate, hydroxy-AFm, and Friedel's salt. Hydrogarnet forms mainly as the result of C_4AF or C_3A hydration. Hydrogarnets have a range of compositions, of which C_3AH_6 is the most common phase forming from normal cement hydration and then only in small amounts.

2.1.3.3 Setting and hardening

The process of hydration can be thought of as the progressive conversion of free (liquid) water in the capillary pores into bound water in the solid hydration products. The binding of water ensures that the hydration products occupy a greater volume than the solid reactants (i.e., the cement minerals) that they replace. The greater the volume fraction of the paste occupied by solid phases, the stronger and stiffer the cement paste or concrete.

2.1.3.3.1 The hydration leads to setting and hardening

Setting is a process where a fresh cement paste of freely flowing or plastic consistency is converted into a set material which has lost its unlimited deformability and crumbles under the effect of a sufficiently great external force. It is followed by the hardening of the paste in which the hardness, strength, and modulus of elasticity increase until an ultimate value of these parameters is attained.

After mixing a Portland cement with adequate amounts of water ($w/c = 0.3-0.7$), cement grains are initially evenly distributed in the liquid phase. A fresh cement paste is formed whose rheological properties (also called consistency) depend on the water/cement ratio, the fineness of the cement, and its composition. Within minutes following mixing, flocculation of the cement particles takes place, associated with an increase in the viscosity of the paste. This initial flocculation of cement particles is brought about by opposite zeta potentials and by weak van der Waals forces. The use of plasticizer, superplasticizer, can break down the initial flocculation, thus enhancing the flowability of the paste.

As the hydration proceeds, more solid hydration products fill the spaces occupied initially by water, contributing to the microstructural development of cement paste as shown in Fig. 2.7.

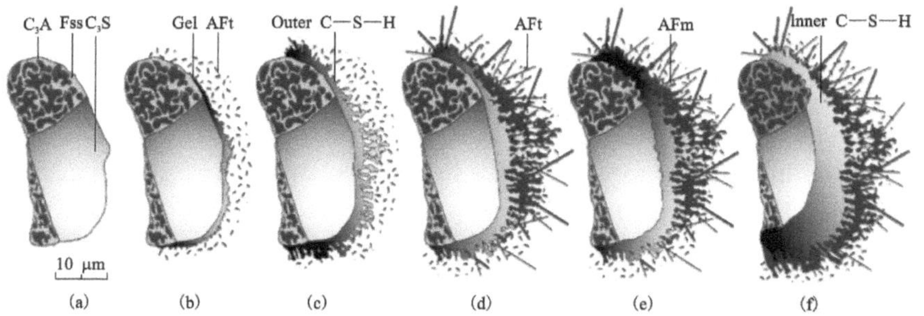

Figure 2.7 The process of microstructural development for a grain of cement. *(Adapted from Karen, the microstructure of concrete, 1984.)*

In addition to solids, the hydrated cement paste contains several types of voids which have an important influence on its properties. Figure 2.8 shows the volume evolution of solid phase, porosity (assumed to be filled with pore solution), and chemical shrinkage.

2.1.3.3.2 Interlayer space in C—S—H

Powers assumes the width of the interlayer space within the C—S—H structure to be 18 Å and determines that it accounts for 28% porosity in solid C—S—H. This void size is too small to have an adverse effect on the strength and permeability of the hydrated cement paste. However, as discussed below, water in these small voids can be held by hydrogen bonding, and its removal under certain conditions may contribute to drying shrinkage and creep.

2.1.3.3.3 Capillary voids

Capillary voids represent the space not filled by the solid components of the hydrated cement paste. The total volume of a typical cement—water mixture remains essentially unchanged during the hydration process. The average bulk density of the hydration products is considerably lower than the density of anhydrous Portland cement; it is estimated that 1 cm^3 of cement, on complete hydration, requires about 2 cm^3 of space to accommodate the products of hydration. Thus, cement hydration may be looked upon as a process during which space originally occupied by cement and water is being filled more and more by hydration products. Space not taken up by the cement or the hydration products consists of capillary voids. The volume and size of the capillary voids are determined by the original distance between the anhydrous cement particles in the freshly mixed cement paste (i.e., water/cement ratio), and the degree of cement hydration. A method of calculating the total volume of capillary voids, popularly known as porosity, cement pastes having either different water—cement ratios or different degrees of hydration will be described later.

In well-hydrated, low water—cement ratio pastes, the capillary voids may range from 10 to 50 nm; in high water—cement ratio pastes, at early ages of hydration, the capillary voids may be as large as 3—5 μm (Fig. 2.8).

It has been suggested that the pore size distribution, not the total capillary porosity, is a better criterion for evaluating the characteristics of a hydrated cement paste. Capillary voids larger than 50 nm, referred to as

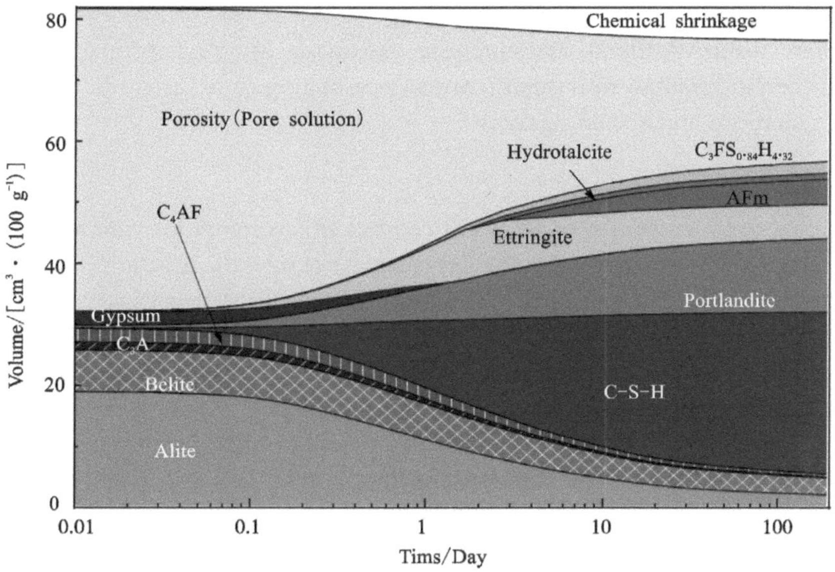

Figure 2.8 The volume of various hydration products and the porosity as a function of time.

macropores in modern literature, are probably more influential in determining the strength and impermeability characteristics, whereas voids smaller than 50 nm, referred to as micropores, play an important part in drying shrinkage and creep. Whereas capillary voids are irregular in shape, air voids are generally spherical. A small amount of air usually gets trapped in the cement paste during concrete mixing. For various reasons, as discussed in Chapter 8, admixtures may be added to concrete to entrain purposely tiny air voids. Entrapped air voids may be as large as 3 mm; entrained air voids usually range from 50 to 200 μm. Therefore, both the entrapped and entrained air voids in the hydrated cement paste are much bigger than the capillary voids, and are capable of adversely affecting the strength.

2.1.4 Properties

Cement properties can be categorized into two types: chemical properties and physical properties. Physical properties are important for qualities controlling purpose of concrete or cementitious materials, while chemical properties are closely related to the durability of cement-based materials.

2.1.4.1 Physical properties

When using Portland cement, the following physical properties are considered: fineness, soundness, consistency, setting time, strength, the heat of hydration, and specific gravity.

2.1.4.1.1 Fineness

Fineness or particle size of Portland cement affects hydration rate and thus the rate of strength gain. The smaller the particle size, the greater the surface area-to-volume ratio, and thus, the more area available for water—cement interaction per unit volume. The effects of greater fineness on strength are generally seen during the first 7 days. When the cement particles are coarser, the coarser particles may not be completely hydrated. This causes low strength and low durability. For rapid development of strength, a high fineness is necessary. The fineness also has an effect on the water demand for normal consistency, which will be mentioned below.

There are various methods for determining the fineness of cement particles. The Blaine air-permeability method is the most commonly used method for Portland cement. In the Blaine air-permeability method, given volume of air is passed through a prepared sample of definite density. The number and size of the pores in a sample of given density is a function of the particles and their size distribution and determines the rate of air flow through the sample. Calculations are made and the fineness is expressed in terms of cm^2/g or m^2/kg.

2.1.4.1.2 Soundness

When referring to Portland cement, "soundness" refers to the volume stability of the cement paste. The cement paste should not undergo large changes in volume after it has set. However, when excessive amounts of free CaO or MgO are present in the cement, these oxides can slowly hydrate and cause expansion of the hardened cement paste. Most Portland cement specifications limit magnesia content and expansion. The addition of excessive of gypsum during grinding can also lead to unsoundness. Soundness test determines the soundness of the cement, and is performed with the help of Le-Chatelier apparatus.

2.1.4.1.3 Consistency

Consistency means the required water to produce plastic cement paste for particular cement. According to consistency, one can know the water—cement ratio for better workability of the mix. It is measured by Vicat Test.

In Vicat Test cement paste of normal consistency is taken in the Vicat Apparatus. The plunger of the apparatus is brought down to touch the top surface of the cement. The plunger will penetrate the cement up to a certain depth depending on the consistency. A cement is said to have a normal consistency when the plunger penetrates 10 ± 1 mm.

2.1.4.1.4 Setting time

As soon as the water is mixed with Portland cement, the hydration process starts and it begins to set. The activity of changing from a fluid state to a solid state is known as the setting. The knowledge of the setting time of the cement is always helpful in deciding the time duration to mix, transport, place, and compact the concrete effectively. Setting is divided into two different categories: Initial setting time and final setting time. Initial setting time: Initial setting time can be defined as the time when the cement paste starts losing its plasticity. As per IS specification, the minimum initial setting time is 30 min for ordinary Portland cement and 60 min for low heat cement. Final setting time: The final setting is defined as the time taken to reach the cement paste to become into a hardened mass. In construction, initial setting time should not be too early and final setting time should not be too late. Normally, initial setting time is 30—45 min and final setting time is below 10 h. Standard test method for the determination of the initial and final setting time of the hydraulic cement is measured by Vicat needle apparatus.

2.1.4.1.5 Strength

Cement paste strength is typically defined in three ways: compressive, tensile, and flexural. These strengths can be affected by various factors, such as water—cement ratio, cement—fine aggregate ratio, curing conditions, size and shape of a specimen, the manner of molding and mixing, loading conditions, and age. While testing the strength, the following should be considered: Cement mortar strength and cement concrete strength are not directly related. Cement strength is merely a quality control measure. The tests of strength are performed on cement mortar mix, not on cement paste. Cement gains strength over time, so the specific time of performing the test should be mentioned.

2.1.4.1.6 Heat of hydration

Hydration generates heat, which can affect the quality of the cement and also be beneficial in maintaining curing temperature during cold weather.

On the other hand, when heat generation is high, especially in large structures, it may cause undesired stress. The heat of hydration is affected most by C_3S and C_3A present in cement, and also by water—cement ratio, fineness, and curing temperature.

2.1.4.1.7 Bulk density

When cement is mixed with water, the water replaces spaces where there would normally be air. Because of that, the bulk density of cement is not very important. Cement has a varying range of density depending on the cement composition percentage. The density of cement of different types may be anywhere from 0.9 to 3.15 g/cm^3.

2.1.4.1.8 Specific gravity (relative density)

Specific gravity is generally used in mixture proportioning calculations. Portland cement has a specific gravity of 3.15, but other types of cement (for example, Portland-blast-furnace-slag and Portland-pozzolan cement) may have specific gravities of about 2.90.

2.1.4.2 Chemical properties

Durability and some physical properties are related to the chemical properties of Portland cement. For general application purposes, the following chemical properties are considered: loss of ignition (L.O.I), insoluble residue, total chloride content, alkali content, and sulfur content.

2.1.4.2.1 Loss of ignition

Heating a cement sample at 900—1000°C (that is, until a constant weight is obtained) causes weight loss. This loss of weight upon heating is calculated as loss of ignition. Improper and prolonged storage, adulteration during transport, prehydration, and carbonation, all of which could make the loss of ignition increased.

2.1.4.2.2 Insoluble residue

Referring to the residue after treated by alkalis or acids, insoluble residue reflects the impurity of the tested cement. The insoluble residue is limited by various specifications. As per GB175, max of 5.0% is required.

2.1.4.2.3 Total chloride content

Chloride content contained in cement may lead to corrosion of reinforcing bars, it is generally limited by specifications, for GB175, the total chloride content is required to be less than the loss of ignition 0.10%.

2.1.4.2.4 Alkali

Referring to potassium oxide (K_2O) and sodium oxide (Na_2O) content contained in cement. Cement containing large amounts of alkali can cause some difficulty in regulating the setting time of cement. Low alkali cement, when used with calcium chloride in concrete, can cause discoloration. In slag-lime cement, ground granulated blast furnace slag is not hydraulic on its own but is "activated" by addition of alkalis. At high levels of alkali content, a reaction called alkali-aggregate reaction may take place, leading to a deleterious expansion in concrete. There is an optional limit in total alkali content of 0.60%, calculated by the equation $Na_2O + 0.658\ K_2O$.

Because the excessive amounts of free-MgO and SO_3 may result in unsoundness, the magnesia (MgO) and sulfur content should be controlled. An excess amount of magnesia may make the cement unsound and expansive, but a little amount of it can add strength to the cement. Production of MgO-based cement also causes less CO_2 emission. All cement is limited to a content of 6% MgO. Free lime, which is sometimes present in cement, may cause expansion.

2.1.5 Corrosion and prevention of hardened cement

2.1.5.1 Corrosion of hardened cement

In normal conditions, Portland cement—based materials are durable. But when exposed to particular environmental conditions, corrosion may take place in the hydrated Portland cement, leading to strength degradation, expansion, etc. Soft water, acidic water, and sulfates can result in corrosion of hydrated Portland cement.

2.1.5.1.1 Soft water corrosion (dissolving corrosion)

Soft water includes rainwater, snow water, pure water, industrial condensation water, and water from rivers and lakes containing less bicarbonate. After long-term contacting with water, calcium hydroxide in cement could be dissolved out (each liter of water can dissolve more than

1.3 g of calcium hydroxide). The dissolution of portlandite results in mass loss, increases porosity, leading to degradation of the concrete structure. The dissolution of portlandite further leads to the decomposition of C—S—H, promoting the degradation of the concrete structure. This kind of corrosion is also known as leaching corrosion.

2.1.5.1.2 Acid corrosion

Carbonic acid corrosion: Usually there is a lot of carbon dioxide dissolved in industrial sewage and groundwater. Carbon dioxide in water reacts with calcium hydroxide in hardened cement to produce calcium carbonate; if calcium carbonate continues to react with carbonic water, it will produce calcium bicarbonate, which is easy to dissolve in water. With the dissolution loss of calcium bicarbonate and the dissolution of other hydration products in cement, hardened cement may suffer structural breakage.

$$Ca(OH)_2 + CO_2 + H_2O = CaCO_3 + 2H_2O \quad (2.6)$$

$$CaCO_3 + CO_2 + H_2O + Ca(HCO_3)_2 \quad (2.7)$$

General acid corrosion: In industrial wastewater, groundwater, and swamp water, there is a certain amount of inorganic and organic acids which have different corrosion influences on hardened cement. The chemical compounds formed when the acid reacts with calcium hydroxide in hardened cement either dissolves in water or leads to volume expansion, which may result in the breakage of hardened cement. Moreover, due to the serious loss of calcium hydroxide, the alkalinity of the hardened cement declines and other hydrates decompose greatly. Therefore, the cement's strength decreases sharply.

2.1.5.1.3 Strong alkali corrosion

Generally, the corrosion is very slight when alkali concentration of the solution is not high and is deemed harmless. However, the exposure of Portland cement with higher aluminate content to strong alkali (NaOH, KOH), hydrated aluminate will be dissolved and bring about damage to hardened Portland cement.

2.1.5.1.4 Sulfate attack caused corrosion

Sulfate attack occurs when sulfates react with compounds in the cement paste such as monosulfate, portlandite, and C—S—H gel. The reaction products may include ettringite, gypsum, and thaumasite. Destabilization of the C—S—H gel may be another consequence of the reaction. External sulfate attack occurs when the sulfates enter the concrete from the surrounding environment. Internal sulfate attack occurs when there is an excess of sulfate in the original mixture. Delayed ettringite formation, or DEF, is a form of internal sulfate attack that can happen with steam-cured and mass concrete when internal concrete temperatures exceed 70°C.

These changes may vary in type or severity but commonly include extensive cracking, expansion, loss of bond between the cement paste and aggregate, alteration of paste composition, with monosulfate phase converting to ettringite and, in later stages, gypsum formation. The necessary additional calcium is provided by the calcium hydroxide and calcium silicate hydrate in the cement paste. The effect of these changes is an overall loss of concrete strength.

Solutions containing magnesium sulfate are generally more aggressive, for the same concentration. This is because magnesium also takes part in the reactions, replacing calcium in the solid phases with the formation of brucite (magnesium hydroxide) and magnesium silicate hydrates. The displaced calcium precipitates mainly as gypsum.

$$2H_2O + Ca(OH)_2 + MgCl_2 = CaCl_2 + Mg(OH)_2Ca(OH)_2 + MgSO_4 \\ = CaSO_4 \cdot 2H_2O + Mg(OH)_2 \quad (2.8)$$

Other sources of sulfate which can cause sulfate attack include seawater, oxidation of sulfide minerals in clay adjacent to the concrete (sulfuric acid is produced which reacts with the concrete), bacterial action in sewers (anaerobic bacterial produce sulfur dioxide, which dissolves in water and then oxidizes to form sulfuric acid). In masonry, sulfates are present in bricks and can be gradually released over a long period of time, causing sulfate attack of mortar, especially where sulfates are concentrated due to moisture movement.

2.1.5.2 Prevention of the corrosion of hardened cement

From the above six types of corrosion to hardened cement, we can conclude that it is mainly because of the internal compositions (such as calcium hydroxide and calcium aluminate hydrate) and hardened cement is not dense and compacted enough that corrosive media intrudes and creates corrosion. To prevent such cases, it is necessary to take the following approaches.

2.1.5.2.1 Using appropriate cement

Selecting suitable types of cement according to corrosive environments. If hardened cement is to suffer soft water corrosion, cement with hydrates containing less calcium hydroxide should be used (e.g., cement with lower content of tri-calcium silicate); if it is to suffer sulfate salt attack, sulfate-resistant cement with less tri-calcium aluminate is preferable; cement with admixtures has its corrosion resistance enhanced.

2.1.5.2.2 Increase the impermeability

Improving the impermeability of cement is an important measure to prevent corrosion to hardened cement. Hardened cement with lower porosity has stronger impermeability, and corrosive media is hard to intrude in. Decreasing water to cement ratio, improving aggregate size gradation, and incorporation of mineral additives can refine the pore structure of cement-based materials, improve the impermeability, hence enhancing the corrosion resistance of the resulting materials.

2.1.5.2.3 Surface protective covering

In case that the corrosion is so strong that the above approaches fail to work, it may help to cover the hardened cement with a layer made of corrosion-resistant material, such as acid-resistant stone, glass, ceramic, asphalt, paint and plastic, etc.

2.1.6 Application

Portland cement is used for general construction purposes where special properties are not required. It is normally used for the reinforced concrete buildings, bridges, pavements, and where soil conditions are normal. It is also used for most concrete masonry units and for all uses where the concrete is not subject to special sulfate hazard or where the heat generated by the hydration of cement is not objectionable. It has great resistance to cracking and shrinkage but has less resistance to chemical attacks.

It is favorable to use Portland cement for cold weather concreting. When for marine structures, the C_3A contained in Portland cement should be between 5% and 8%.

2.1.7 Special Portland-based cements

2.1.7.1 White Portland cement

White Portland cement has essentially the same properties as gray cement, except for color. The color of white cement is determined by its raw materials and the manufacturing process. Metal oxides, primarily iron and manganese, influence the whiteness and undertone of the material. Obtaining this color requires high purity raw materials (low Fe_2O_3 content), and some modification to the method of manufacture (a higher kiln temperature required to sinter the clinker in the absence of ferric oxides as a flux). The main requirement is to have a low iron content which should be less than 0.5 wt.% expressed as Fe_2O_3 for white cement, and less than 0.9 wt.% for off-white cement. Other metallic oxides such as Cr_2O_3 (green), MnO (pink), TiO_2 (white), etc., in trace content, can also give color tinges, so for a given project, it is best to use cement from a single batch.

After adding pigments, white cement produces clean, bright colors, especially for light pastels. Many different colors of concrete can be created, and just like paint, two or more pigments can be combined to achieve a wide range of colors.

White Portland cement is ideal for decorative concrete and architectural concrete applications such as swimming pools and spas, colored mortars, ornamental statuary, reflective floors, floor tiles and pavers, cast stone, terrazzo, tile grout, glass fiber reinforced concrete products, concrete countertops, concrete roof tiles, traffic calming and delineation, median barriers, bridge parapets, sound walls, retaining walls, and reflective concrete paving.

2.1.7.2 Sulfate resistance cement

Sulfate resisting Portland cement is a type of Portland cement in which the amount of tricalcium aluminate (C_3A) is restricted to lower than 5% and $2C_3A + C_4AF$ lower than 25%. The SRC can be used for structural concrete wherever OPC or PPC or slag cement are useable under normal conditions. The use of SRC is particularly beneficial in such conditions where the concrete is exposed to the risk of deterioration due to sulfate attack, for example, in contact with soils and groundwaters containing excessive amounts of sulfates as well as for concrete in seawater or exposed directly to the sea coast.

2.1.7.3 Expansive cement

Expansive cement is usually based upon Portland cement with an expansive component. When mixed with water, expansive cement forms a paste that tends to increase in volume to a significantly greater degree than Portland cement paste after setting.

There are several types of expansive cement depending upon the type of expansive compound used in the cement. Expansive cement is divided into three types: ①Type K expansive cement—a mixture of Portland cement, anhydrous tetra-calcium trialuminate sulfate (C_4A_3S), calcium sulfate ($CaSO_4$), and lime (CaO); ②Type M expansive cement—interground or blended mixtures of Portland cement and calcium sulfate suitably proportioned; ③Type S expansive cement—a Portland cement containing a high computed tricalcium aluminate (C_3A) content and an amount of calcium sulfate above the usual amount found in Portland cement. Other components such as calcined limestone and lightly burned MgO can also be used to produce expansive cement.

This cement is used in large, continuous floor slabs without joints. It works well to fill holes in foundations and to create self-stressed concrete that is stronger than conventional Portland cement concrete. Prestressed concrete components for bridges and buildings are made using this material. It also can be used for the construction of water retaining structures, repairing the damaged concrete surfaces, and grouting of anchor bolts.

2.1.8 Blended cement

The classification of blended cement is based on the type of blending material used in the cement. As per GB-175, blended cement is classified into four types.

2.1.8.1 Portland-slag cement

Blends containing up to 70% slag, cement are used for general construction. It gives low heat of hydration. The slag should be more than 50% and up to 70%. It is used for marine and offshore structures (very high chloride and sulfate resistant); sewage disposal treatment works; water treatment plants. Constructions which are expected to be attacked by dissolved chlorides and sulfate ions should mainly use Portland-slag cement.

2.1.8.2 Portland-pozzolan cement

Used for general construction, this blend can contain up to 50% pozzolan. Fly ash is the most common pozzolan used in blended cement. It gives low heat of hydration and reduces the leaching of calcium hydroxide. This cement should be used only after proper evaluation. It is used for hydraulic structures-dams, retaining walls; marine structures; mass concrete works like bridge footings under aggressive conditions; masonry mortar; and plastering.

2.1.8.3 Portland–limestone cement

Relatively new to the US market, Portland–limestone cement contains between 5% and 15% percent interground limestone.

2.1.8.4 Ternary blended cement

Ternary cements are blends of two complementary supplementary materials such as fly ash, slag cement, or silica fume.

2.1.8.5 Advantages of blended cement

It provides a finer texture than PC when mixed and placed, so it can be used for finishing and elevation works. Water consumption is less, which makes it easy to work with and shape. The strength gained after 28 days is significantly stronger than PC in both compressive and flexural stress. The permeability of blended concrete is low, due to which the life of concrete is extended by reducing the penetration of aggressive water run-off compounds such as sulfates and chlorides. Cracks occurred due to thermal stress by variation of temperature are reduced by the use of blended cement. Problems related to alkali–silica reaction were reduced by using a mix of blended cement as either silica fume and slag or silica fume and fly ash.

2.2 Calcium sulfoaluminate cement

Calcium sulfoaluminate cement (CSA cement) was developed in China in the 1970s. Designed by the China Building Materials Academy (CBMA), they were intended for the manufacturing of self-stress concrete pipes due to their swelling properties. This special cement distinguishes itself from Portland cement by a high-speed setting, fast strength development, and a lower shrinkage.

2.2.1 Manufacture and composition

CSA cement clinkers are usually produced by mixing limestone, clay, and/or bauxite and gypsum, as sources of calcium, silicon/aluminum, and sulfur, respectively. Besides the raw materials mentioned above, various industrial by-products or waste materials like fly ash, blast furnace slag, phosphogypsum, baghouse dust, or scrubber sludge can be used for the manufacturing of calcium sulfoaluminate—based clinkers.

Calcium sulfoaluminate cements are made from clinkers that include ye'elimite ($Ca_4(AlO_2)_6SO_4$ or C_4A_3S) as a primary phase. Depending on the raw meal composition, calcium sulfoaluminate—based clinkers may contain various minor phases such as belite, calcium aluminate ferrate, excess anhydrite, gehlenite, or mayenite. Usually, about 15—25 wt.% of gypsum is interground with the clinker for optimum setting time, strength development, and volume stability.

2.2.2 Hydration

The hydration of the calcium sulfoaluminate cements (CSA) depends mainly on the amount and reactivity of the added calcium sulfate as well as on the kind and amount of minor phases present. The water demand for complete hydration is determined by the amount of calcium sulfate added which is at a maximum around the addition of 30%. The required water/cement ratio for complete hydration is higher compared to an OPC, e.g., 0.78 for pure ye'elimite reacting with 2 mol of anhydrite.

The main hydration product of CSA is ettringite, which precipitates together with amorphous $Al(OH)_3$ until the calcium sulfate is consumed after around 1—2 days of hydration. Afterward, monosulfate is formed. In the presence of belite, strätlingite occurs as an additional hydration product. Immediately after wetting, the following reaction takes place:

$$C_4A_3S + 2CS \cdot H_2 + 34H \rightarrow C_3A \cdot 3CS \cdot 32H + 2AH_3 \tag{2.9}$$

This reaction also takes place with anhydrite or bassanite although at different rates. Formation of ettringite will continue while calcium sulfates are present. If there are not enough sulfates to react with ye'elimite phase, the reaction will start:

$$C_4A_3S + 18H \rightarrow C_3A \cdot CS \cdot 12H + 2AH_3 \tag{2.10}$$

This corresponds to the hydration of ye'elimite to form an AFm type phase with different water contents, depending on the time of hydration, water-to-cement (W/C) ratio, and initial mineral phase assemblage.

Depending on the minor phases present in CSA cements, various other hydration products may occur such as C—S—H phases, strätlingite, monocarboaluminate, gibbsite, or hydrogarnet. Minor phases also react with water or calcium sulfate and somewhat contribute to rapid hardening. All these reactions yield ettringite and take place during the first hours of hydration. Depending on the availability of calcium sulfates again, there are alternative reactions for the direct hydration of these phases, and also yield aluminum or iron hydroxides (AH_3 or FH_3) which are initially amorphous.

When the CSA clinker contains belite, this phase reacts with water in the same way as it in Portland cement to produce C—S—H and portlandite.

$$C_2S + (x + 2 - y)\ H \rightarrow C_ySH_x + (2 - y)\ CH \tag{2.11}$$

However, belite coexists with aluminum-rich amorphous hydrates, promoting the formation of strätlingite

$$C_2S + AH_3 + 5H \rightarrow C_2ASH_8 \tag{2.12}$$

Microstructural investigations revealed mainly the formation of large space-filling ettringite needles, together with monosulfate, aluminum hydroxide, and calcium silicate hydrates, leading to a very dense, low-porosity microstructure.

2.2.3 Properties

2.2.3.1 Rapid strength gain

The primary advantage of CSA cements is that concrete made with CSA instead of Portland cement often achieves compressive strengths in excess of 35 MPa in 24 h. It's possible to achieve 28 days strength in 24 h. This is the main reason that CSAs are used in place of ordinary Portland cement for certain applications.

2.2.3.2 Lower carbon

Another key advantage is that CSA cements are also significantly greener. Portland cement is fired in kilns at temperatures of around 1500°C, whereas CSA cements only need to be fired at temperatures of around 1250°C. The resulting CSA clinker is softer than OPC clinker, requiring less energy to grind. The production of 1000 kg CSA cement emits 216 kg of CO_2, which shows a 62% reduction compared to the production of Portland. The production of CSA cement has the lowest emission of CO_2 compared to other alternative cements such as magnesium cement (Sorel), sodium metasilicate (water glass), and calcium aluminate cement.

2.2.3.3 Lower alkalinity

The pH value is only 10.5–11; the pH of ordinary Portland cement (OPC) is around 13, which is 100–300 times more alkaline than CSA cement. The low alkalinity naturally minimizes the chance for alkali-aggregate reaction.

2.2.3.4 Lower shrinkage

CSA cements demonstrate very low shrinkage characteristics. This is due in part for two reasons. The first is that CSAs require about 50% more water than Portland cement for proper hydration. The minimum recommended water to cement ratio (w/c) is 0.35, whereas with OPC it's around 0.22–0.25. Because of the higher water of hydration requirements, most of the mix water is consumed for hydration, and less excess water is available to cause problems with shrinkage. The second reason is that the very rapid strength gain can prevent shrinkage cracks because the concrete strength increases more rapidly than do the concrete's shrinkage stresses.

2.2.3.5 Shorter curing time

Curing with CSA is important, but wet curing durations are often measured in hours, not days or weeks. Optimal hydration and slab stability are achieved when the CSA concrete is kept wet for at least 3–4 h after casting. During the initial hydration phase, the concrete demands moisture, and the rapid reaction generates significant heat. If sufficient moisture is not provided during curing, cracking and curling are possible.

2.2.4 Application

The advantages mentioned such as rapid set, high early strength, and shrinkage reduction make it a very versatile addition or replacement of Portland cement.

Applications where fast setting and high early strength are necessary: an airport runway, a bridge repair, or a damaged freeway that must be returned to service in a very short amount of time, production of precast concrete. Likewise, it is suitable for acid treatment, polishing, or blasting even after 6 h.

Applications where low shrinkage is needed: shrinkage compensated (low shrinkage) concretes, grouts and mortars, self-levelers, tile adhesives, repair mortars, water plug.

Special attentions: The quick set and the water demand require special attention to ensure a positive end result. A higher water, material, or ambient temperature may accelerate binding times as well as increasing the mixing speed. Superplasticizers, especially polycarboxylates, and viscosity modifiers work the same with CSAs as with Portland cement.

It is not recommended to use pozzolans with CSA, because CSAs do not react with pozzolans. The use of pozzolans with CSA may weaken the resulting concrete.

2.3 Calcium aluminate cements

Calcium aluminate cements (CACs) are made from limestone (or lime) with bauxite or other aluminous material low in SiO_2. CACs have been also termed aluminous or high-alumina cements, but the former term is less specific, and the latter is more appropriately used for white CACs made using purified alumina. Ciment fondu is the name of CACs in French.

2.3.1 Manufacture and composition

There are two main methods of manufacturing calcium aluminate cement on an industrial scale. The first is by fusion in which a raw feedstock of bauxite and limestone is melted together in a reverberatory furnace at 1450–1600°C. In the second, high purity limestone and high purity alumina are sintered together in a rotary kiln, which produces a much purer grade of cement known as high alumina cement.

In order to produce a cement with the desired rapid-hardening properties, both raw materials must be low in SiO_2. The molten clinker is tapped off continuously from the furnace, solidifies, and is typically crushed and ground to a fineness of about 300 m^2/kg. The color of cements produced from bauxite can vary from yellow-brown to black, but is commonly grayish black. White CACs are usually made by sintering calcined alumina with quicklime (calcium oxide) or high-purity limestone. Most of the commercial white CACs contain CA and $C_{12}A_7$; those high in $A1_2O_3$ may also contain CA_2, CA_6 or α-Al_2O_3.

2.3.2 Hydration

In commercial CACs, phases that hydrate significantly at early ages are normally CA and, if present, $C_{12}A_7$. Some sintered cements made at relatively low temperatures also contain free lime or CH, which are very active in the hydration process. The reaction of C_2A, present in many white cements high in $A1_2O_3$, is very slow, possibly because of gel formation.

Hydration of CA, or in fact any calcium aluminate compound, can be thought of as a three-stage process, of which the primary stage is the dissolution step. For CA, the anhydrous grains react immediately upon their addition to water, dissolving congruently to yield calcium ions and aluminate ions. As a consequence of this, the resulting solution will increase in both conductivity and pH until a point of supersaturation is attained. The dissolution process is typically thought of as being congruent, yet there is often more lime than alumina present in solution.

2.3.2.1 The initial stage

This initial stage is exothermic and associated heat evolution can be detected easily by calorimetric methods. Once a state of supersaturation of Ca^{2+} and $Al(OH)_4^-$(aq) species is reached, the reaction reaches a dormant or nucleation stage.

2.3.2.2 The second stage

In the hydration reaction (nucleation), the solution will remain supersaturated with ions. Dissolution and hydrate formation occurs at a very slow rate, which maintains a very high concentration of Ca^{2+} and $Al(OH)_4^-$(aq) species. A clear reflection of the fact is that the pH and conductivity of the solution will be constant.

2.3.2.3 The final stage

At the end of the nucleation phase of the reaction, massive precipitation and growth of the hydrate species will occur. This immediately results in a reduction in the amounts of ions present in the solution, so if any reacted CA is present it will quickly be dissolved to attain supersaturation once again. However, this will precipitate almost instantaneous to yield hydrated species. Dissolution and hydrate precipitation will now proceed simultaneously though eventually, the rate of the reaction will become infinitely large as the anhydrous CA diminishes. At this point, the conductivity will drop sharply corresponding to the drop of ionic species present in solution. Mass precipitation of species of this nature in the hydration reaction is accompanied by a considerably large exotherm (which can be detected to determine the end point of the hardening reaction).

The hydration of the CA phase is temperature dependent. At lower temperatures, CAH_{10} will form (typically below 20°C). In the intermediate temperature ranges between 21 and 30°C, C_2AH_8 will form. Under conditions of elevated temperature C_3AH_6 will form, which is the most thermodynamically stable and the least soluble of the calcium aluminate hydrates. In addition to the hydrates formed at intermediate temperatures, crystalline gibbsite (AH_3) will form.

Due to the metastable nature of CAH_{10} and C_2AH_8 over long periods of time or at elevated temperatures, both phases will undergo a transformation into the hydrogarnet phase (C_3AH_6) as shown in Fig. 2.9. This process is called conversion. Conversion is initiated by the nucleation of C_3AH_6 and takes place in solution. Because of the loss of bound water, conversion causes an increase in porosity, which can be accompanied by a decrease in strength (Fig. 2.10). This is a problem in structural concrete provided that a sufficiently high cement content and a sufficiently low water/cement ratio are employed.

CACs harden rapidly as soon as the massive precipitation of hydrates begins. Because of their relatively high cost, calcium aluminate cements are used in a number of restricted applications where performance achieved justifies costs.

2.3.3 Properties

2.3.3.1 Strength

The high alumina cement is black in color and its rate of strength development is very high. High alumina cement gains sufficient strength for the removal of the form work at 8 h. About 80% of its ultimate strength is developed at the age of 24 h. The high rate of gain of strength is due to its rapid hydration. Its ultimate strength is also higher than that of ordinary cement.

Its rate of heat development is also very high. However, the total heat of hydration is of the same order with ordinary Portland cement. The high rate of heat development makes this cement unfit for mass concrete work, but rapid rate of heat development is of great advantage when concrete is placed in freezing weather.

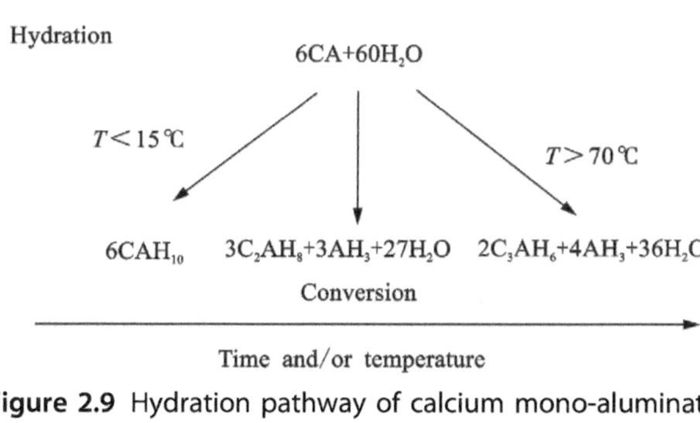

Figure 2.9 Hydration pathway of calcium mono-aluminate.

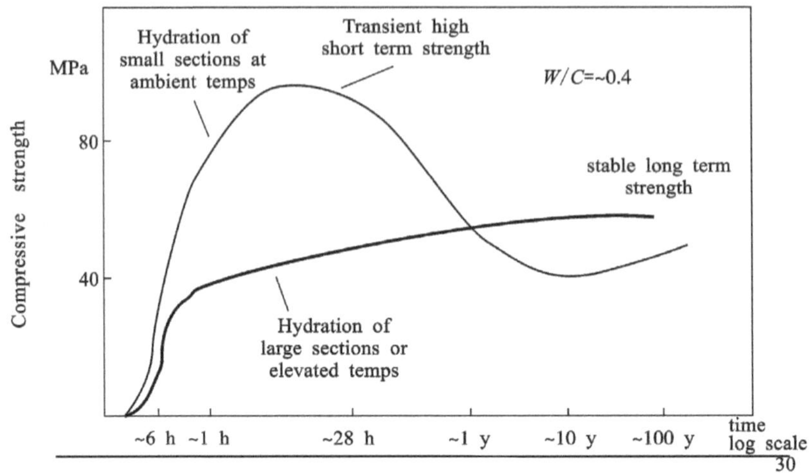

Figure 2.10 The strength development of CAC Concretes. *(Adapted from Peter C. Hewlett, Chemistry ofCement and Concrete fifth Edition.)*

2.3.3.2 Workability and setting time

High alumina cement is slow setting cement. Its initial setting time is 4–5 h and final setting time is about 30 min later after initial setting. No additives should be used because the setting time of high alumina cement is greatly affected by the addition of plaster, lime, Portland cement, and organic matter. To lower down the setting time of this cement 1%–2% hydrated lime may be added to it. When rapid setting is of vital importance, i.e., stopping the ingress of water or temporary construction between the tides, etc., mixtures of ordinary Portland cement and high alumina cement in suitable proportions are used, but the ultimate strength of such pastes is quite low.

In normal concrete construction, the two cements should not be allowed to come in contact with each other as in such cases flash set will occur. The flash set or accelerated setting time is due to the formation of a hydrate of C_4A by the addition of lime from the ordinary Portland cement and to calcium aluminate from the high alumina cement. Also, gypsum contained in Portland cement may react with hydrated calcium aluminates and cause flash set.

For the same water/cement ratio, equal mix proportions concrete made with high alumina cement exhibits more workability than Portland cement concrete due to the fact that particles of high alumina cement have smoother surface than Portland cement particles, as the raw materials of high alumina cement fuse fully in the hearth. Secondly, high alumina cement has lower total surface area 2500−3200 cm^2/g.

If the layers of two cements are to be placed, then they should be laid at different times. If the first layer is laid of high alumina cement, then the layer of Portland cement should be laid at least 24 h later. In case first layer is made with Portland cement, then the concrete made with high alumina cement should be laid after 3−7 days. Contamination through plant or tools should also be avoided.

2.3.3.3 Durability

It is highly resistant to sulfate attack. This is due to the absence of $Ca(OH)_2$ in hydrated high alumina cement and secondly due to the protective influence of the relatively inert gel formed during hydration. Meanwhile, calcium aluminate cement is free from the attack of CO_2 dissolved in pure water. However, this cement is not acid resistant, and can be attacked readily by nitric, hydrochloric, or hydrofluoric acids. But it can withstand well very dilute solutions of acids (pH value greater than 3.5−4.0) found in industrial effluents. Caustic alkalis even in dilute solutions attack this cement with great vigor by dissolving the alumina gel.

Though this cement stands extremely well to seawater, seawater should never be used as mixing water. By the use of seawater as mixing water, the setting and hardening of the cement are affected very badly. Similarly, calcium chloride should never be added to this cement.

2.3.3.4 Refractory properties

High alumina cement concrete is one of the foremost refractory materials but its performance varies with the range of temperature. Between room temperature and about 500°C, concrete made with high alumina cement loses strength more than concrete made with Portland cement, then up to 800°C the two concretes are comparable but above 1000°C, high alumina cement gives excellent performance.

Between 700 and 1000°C, solid reactions between the cement and fine aggregate take place. This reaction is known as ceramic bond, which is responsible for the increase in the strength of high alumina cement concrete between 800 and 1000°C. The rate of reaction increases with the rise of temperature.

Thus, concrete made with high alumina cement and crushed fired brick as aggregate can withstand very high temperatures up to 1350°C. Temperatures above 1350°C up to 1500°C can be withstood by using special aggregates such as fused alumina or carborundum. Temperature up to 1800°C can be withstood over a prolonged period of time by concrete made from special white calcium aluminate cement with fused alumina aggregate. This cement contains 70%–80% Al_2O_3 (Alumina), 20%–25% lime, and about 1.0% iron and silica. Its composition reaches to C_3A_5. However, its price is very high.

Refractory concrete made with high alumina cement has a good resistance to acid attack. The chemical resistance increases by firing at 900–1000°C. As soon as the concrete hardens, it can be put to use, that is, it should not be pre-fired. The refractory brick work expands on heating, thus it needs provision of expansion joints, while high alumina cement concrete can be cast monolithically or with joints according to the exactly required size and shape. Thus, refractory high alumina cement concrete can withstand considerable thermal shocks. Refractory linings can be made by creating high alumina cement mortar.

For insulation purposes when temperature rise is expected up to 950°C, lightweight concrete made with high alumina cement and lightweight aggregates can be used with advantage. The density and thermal conductivity of this concrete are of the order 500–1000 kg/m^3 and 0.21–0.29 W/(m·K), respectively.

2.3.4 Application

2.3.4.1 Heat-resistant and refractory concretes

Although CACs were originally developed for improved sulfate resistance, they are now mainly used as binders in heat-resistant concrete, dense refractory concrete, and thermally insulating concrete. Cement with higher alumina content is often used, because it has the ability to combine precision-graded aggregates together in the green state. When these materials heat up consequently, cement sinters with the aggregate, thereby creating a refractory matrix.

2.3.4.2 Rapid repair and construction

CACs are used in construction concretes, where rapid strength development is required, even at low temperatures. Typical applications of rapid concrete are overnight repair work on highway slabs con, repairs of airport runways and ramps, industrial infrastructure repairs with minimum disruption to operations, repairs of bridge joint closures, works in tunnels and mines with rapid turnaround, and cold-weather repairs.

Being used as a component in blended cement formulations, various properties such as ultra-rapid strength development and controlled expansion are required.

2.3.4.3 Building chemistry products

Floor leveling compounds, sealers, rapid floor screeds, tile adhesives, bedding mortars, tile grouts, and repair mortars are different types of building chemistry products. The mineral base of this wide range of products is a mixture of CAC and Portland cement. Additionally, the blend may include slag, admixtures, gypsum, polymers, lime, and fine calcareous material.

2.3.4.4 Sewer applications

Hardened CACs can be used as a protective layer against microbial corrosion such as in sewer infrastructure for their high resistance to biogenic sulfide corrosion and improved resistance to abrasion.

2.3.4.5 Chemical-resistant concretes

CACs were originally developed for improved sulfate resistance and early CAC concrete was successfully used in tunneling works through gypsum deposits. The porosity matrix of concretes prepared by mixing calcium aluminate cement is lower. They also exhibit higher resistance to both abrasion and sulfate attack. Consequently, calcium aluminate cement is desirable for making chemical-resistant concretes, which are mainly used in products such as industrial floorings, for example, cast house floors.

2.4 Alkali-activated cement

Alkali-activated cement, also termed as alkali activated -materials (AAM), is still not clearly defined in the literature. In this book, alkali-activated cement is defined as a class of cementitious materials derived by the reaction of an alkali metal source (solid or dissolved) with a solid aluminosilicate powder. This definition excludes lime-pozzolana—based systems, other materials where an elevated pH is generated through the supply of alkaline earth compounds. In the book, geopolymers are viewed as a subset of AAMs, where the binding phase is almost exclusively aluminosilicate and highly coordinated.

According to the composition of the cementitious component(s), alkali-activated cements can be classified into two categories: ① high-calcium alkali-activated cements; ② low-calcium alkali-activated cements.

2.4.1 Manufacture

There are two main ways to produce alkali-activated cements: one-part mix (dry powder combined with water) and two-part mix (liquid activator) pathway. The majority of the products that are already in the market are produced in manner of two-part mix. The production of a one-part mix alkali-activated binder system has been demonstrated via co-calcination or intergrinding of various aluminosilicate powder precursors and solid activators.

Variety of industrial by-products and waste as well as a number of aluminosilicate raw materials can be used as the cementitious components in alkali-activated cements. These materials include granulated blast-furnace slag, granulated phosphorus slag, steel slag, coal fly ash, volcanic glass, zeolite, metakaolin, silica fume, and nonferrous slag. Blast-furnace slag requires quenching (granulation or pelletization) and grinding to yield a reactive material.

The most used activators are alkali silicates (water glass) and hydroxides. Na_2CO_3 and Na_2SO_4 can also be used for activation of high-calcium precursors such as blast-furnace slag. Sulfate activators can be used in production of low-Ca binders. Some by-products such as red mud, maize stalk, cob ash, and concentrated sodium aluminate solution can also be used as alternative activator in production of AAM.

2.4.2 Alkali activation process and products

Alkali-activated cements gain their strength, and other properties, via chemical reaction between a source of alkali and aluminate-rich materials. The exact reaction mechanism of alkali-activated binders is not yet quite understood, which is dependent on the prime material as well as on the alkaline activator.

Alkali activation is composed of conjoined reactions of destruction-condensation, that include the destruction of the prime material into low stable structural units, their interaction with coagulation structures, and the creation of condensation structures. The process includes the following steps: ① dissolution of the glassy precursor particles, ② nucleation and growth of the initial solid phases, ③ interactions and mechanical binding at the boundaries of the phases formed, and ④ ongoing reaction via dynamic

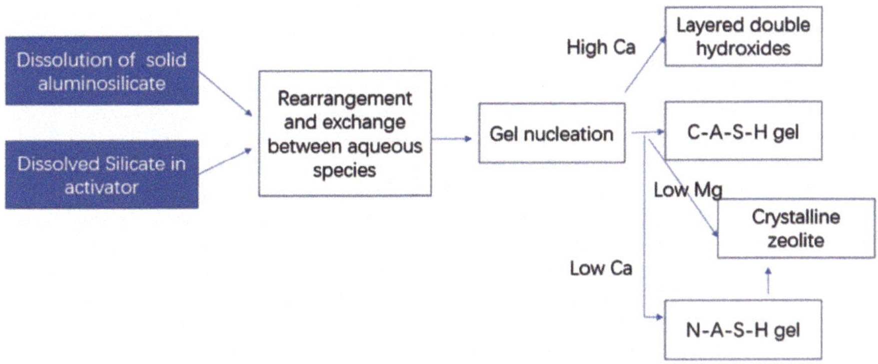

Figure 2.11 Schematic illustration of the process and reaction of products of alkaline activation under different conditions and at different compositions. *(Adapted from I. Garcia-Lodeiro, A. Palomo, A. Fernández-Jiménez. Handbook of Alkali-Activated Cements, Mortars and Concretes (2015), Woodhead Publishing).)*

chemical equilibria and diffusion of reactive species through the reaction products formed at advanced times of curing. Fig. 2.11 schematically shows the process and products of alkali activation under different conditions and various compositions.

In general, two types of reaction mechanisms of alkali-activated binders are established based on the composition of precursors (Fig. 2.11). The first system involves the activation of calcium-rich raw materials like blast-furnace slag, with high content of SiO_2, Al_2O_3, and CaO. The activation is realized using moderate alkaline solutions leading to calcium silicate hydrates-like phases as reaction products. The second mechanism involves the alkali activation of low-calcium and calcium-free prime materials using medium to high alkaline solutions, leading to a polymeric network with the formation of amorphous zeolite-like phases with high mechanical strength similar to OPC.

It is difficult to determine the exact phases formed because of the high chemical complexity of the systems due to varying parameters like the chemical compositions of the prime materials, the alkaline activating solutions, the used liquid/solid ratio, and the curing conditions. Fig. 2.12 shows the influence of composition of raw materials on product types in alkali-activated binder systems.

In high-Ca alkali activated systems, secondary reaction products exist as AFm type phases (mainly in NaOH-activated binders, and also strätlingite in silicate-activated binders), hydrotalcite (relatively high contents of MgO), and zeolites such as gismondine and garronite (formed in with high Al_2O_3 and low (<5%) MgO).

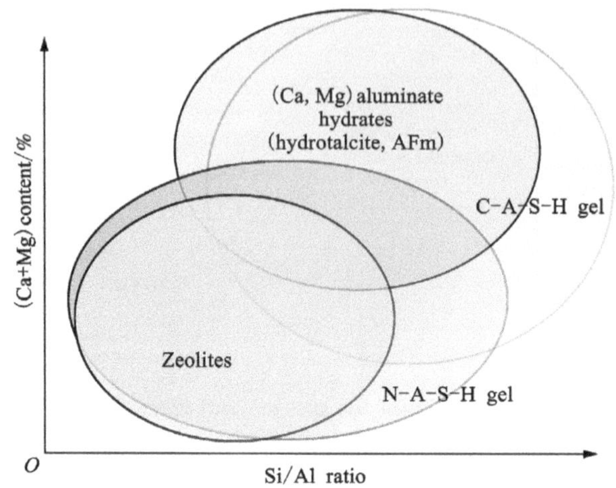

Figure 2.12 Schematic representation of phase formation within alkali-activated binders; ordered products are shown in green and disordered products in blue. Zeolites form as the main crystalline secondary products in systems with low Ca content and/or low Mg content, while hydrotalcite forms from Mg-rich slags, and AFm-type phases form in most blast furnace-slag based binders in the presence of sufficient Al. *(Adapted from John L. Provis, Materials and Structures (2014) 47:11–25.)*

2.4.3 Properties

The physical and chemical properties of AAM depend largely on the raw materials used and the curing condition.

Setting times of alkali-activated materials are in general similar to those of Portland cement-based materials. Alkali-activated slag cements tend to harden rather rapidly, and in some cases, retarders are used to regulate the setting rates. The workability of AAM is sensitive to the composition and fineness of raw material, alkali concentration, and modulus of alkali activator.

AAM has better resistance to acid attack, sulfate attack when compared with OPC, but there remain questions over carbonation and freeze-thaw resistance which are not yet fully understood.

Although the alkali content of the majority of alkali-activated concretes is very high, alkali-aggregate reaction has not proven to be the case either in field or laboratory investigations.

These materials, and in particular the low-calcium systems described as "geopolymers," have been shown in many instances to have superior high temperature properties when compared to Portland cement—based materials.

Performance with respect to chloride ingress (diffusion or migration) appears in general to be comparable to that of materials based on Portland cement, but depends fundamentally on achieving a satisfactory microstructure in the binder through judicious mix design.

Efflorescence, the white "bloom" which can appear on the surfaces of some alkali-activated mortars or concretes, is usually related to the carbonation of mobile alkalis from the pore solution. This problem can be addressed through implementation of curing conditions or addition of pore-refining additives, to reduce the rate of moisture movement through the hardened cements.

The drying and autogenous shrinkage of alkali-activated slag binders is more serious than that of OPC and other alkali -activated binders. Albeit, reinforcing the plain compositions with fibers, using expanding admixtures, curing at elevated temperatures, using a combination of slag with fly ash or metakaolin reduces successfully the drying shrinkage.

In general, mechanical performance of alkali-activated binders fulfill or even outperform the standard strength requirements for OPC.

2.4.4 Application

Alkali-activated binders can be used in a wide variety of ways, in precasting as well as in situ concreting, in the construction of buildings of significant scale, and other civil infrastructure components.

① Structural alkali-activated BFS concretes, including reinforced concrete, plain concrete, precast concrete components (both reinforced and unreinforced, pipes), self-leveling concretes applied in road construction.

② Mortars, grouts and renders, masonry blocks.

③ Foamed and lightweight concretes.

④ High temperature applications.

⑤ Fire-resistant products, including tunnel linings, high rise buildings, lift doors, and marine structures/coatings. Structures can be protected by a fire-resistant coating, which may be sprayed (as shotcrete) or precast during the construction phase, with particular value in protecting metal beams from deformation during a fire.

⑥ Matrices for the immobilization of toxic and nuclear wastes, both organic and inorganic: Alkali-activated binders have been used in practice for the immobilization of a wide variety of wastes. The leachability of Pb, Cr, Cd, and Zn species incorporated into alkali-activated BFS pastes (either individually or as components of complex mixed industrial wastes) has been noted to be low.

2.5 Magnesium-based cements

Before the widespread use of Portland cement, magnesium oxide—based cements were among the world's most popular cement products. Magnesium-based cement is a big family, which use MgO as a building block. Today, there are three commonly used magnesium-based cements, i.e., magnesium oxychloride cement (MOC), magnesium oxysulfate cement (MOS), and magnesium phosphate cement (MPC). Magnesium oxychloride cement (MOC) is also known as Sorel cement, because it was first discovered by Sorel in 1867.

2.5.1 Manufacture and composition

The manufacture of magnesium-based cement involves the production of MgO. The bulk production of MgO is achieved by calcination of crushed magnesite, through the process

$$MgCO_3 \xrightarrow{>600^\circ C} MgO + CO_2 \tag{2.13}$$

Magnesium oxychloride cement is manufactured by mixing magnesia with magnesium chloride solutions in well-defined proportions to produce the magnesium oxychloride, which is the bonding phase. The magnesium oxysulfate cements formed can be regarded as variants of Sorel cements.

Likely, magnesium oxysulfate cements can be produced by adding magnesium chloride solutions to calcium sulfates or calcium phosphate—sulfate mixtures.

Magnesium phosphate cement is derived from reactions between a soluble phosphate (typically an ammonium or potassium phosphate) and dead burnt (above 1400°C) magnesium oxide.

Magnesium deposits exist in abundance in every corner of the Earth and cover roughly 8% of its surface. Depending on where they are mined, magnesium oxide cements require only 20%—40% of the energy used to produce Portland cement.

2.5.2 Hydration

Magnesium oxychloride cements are based on the aqueous reaction between MgO and $MgCl_2$, the main phases of MOC cement include MgO, brucite ($Mg(OH)_2$), and crystal phases (phase 3 and phase 5). The reactions to produce phase 3, phase 5, and $Mg(OH)_2$ during hydration are as follows:

$$3MgO + MgCl_2 + 11H_2O \rightarrow 3Mg(OH)_2 \cdot MgCl_2 \cdot 8H_2O \tag{2.14}$$

$$5MgO + MgCl_2 + 13H_2O \rightarrow 5Mg(OH)_2 \cdot MgCl_2 \cdot 8H_2O \tag{2.15}$$

A parallel or competitive reaction is the hydration of magnesium oxide due to the presence of excess water according to the chemical equation.

$$MgO + H_2O \rightarrow Mg\ (OH)_2 \quad (2.16)$$

Table 2.2 summarizes the hydration products of magnesium phosphate cements depending on the starting material.

Magnesium oxysulfate (MOS) cements are similar in concept to MOC cements, except that $MgSO_4$ is used instead of $MgCl_2$. A similar nomenclature scheme is applied to the $xMg(OH)_2{\cdot}yMgSO_4{\cdot}zH_2O$ phases. The phase transformation and stability of hydrates in the MOS cement paste are primarily associated with two molar ratios, i.e., $MgO/MgSO_4$ and $MgSO_4/H_2O$. It is found that as the molar ratio of MgO/MgSO4 increases from 3 to 9, the main hydrated products are 3-1-8 phase, 5-1-7 phase, and MH. Other phases: 5-1-3 (or 5-1-2) phase ($5Mg(OH)_2{\cdot}MgSO_4{\cdot}3H_2O$), 1-1-5 phase ($Mg(OH)_2{\cdot}MgSO_4{\cdot}5H2O$), 1-2-3 phase ($Mg(OH)_2{\cdot}2MgSO_4{\cdot}3H_2O$) may also occur within the temperature range from 30 to 120°C: Typically, formation of hydrates of MOS follow the equations. It is believed that the formation of 5-1-7 phase will result in high strength and good water resistance.

$$3MgO + MgSO_4 + 11H_2O \rightarrow 3Mg(OH)_2{\cdot}MgSO_4{\cdot}8H_2O \quad (2.17)$$

$$5MgO + MgSO_4 + 8H_2O \rightarrow 5Mg(OH)_2{\cdot}MgSO_4{\cdot}3H_2O \quad (2.18)$$

2.5.3 Properties

Magnesium-based cements share some common properties including fast setting and rapid strength gain, high strength, low electrical and thermal conductivity, good abrasion resistance, superior bonding strength, noncombustible, resisting fire, and lightweight.

Table 2.2 Hydration products of different types of MPC in the literature.

Raw materials	Hydration products
H_3PO_4; MgO	$Mg(H_2PO_4)_2{\cdot}2H_2O$ and minor $Mg(H_2PO_4)_2{\cdot}4H_2O$
$NH_4H_2PO_4$; MgO	$NH_4MgPO_4{\cdot}6H_2O$ and $(NH_4)_2Mg(HPO_4)_2{\cdot}4H_2O$
$(NH_4)_2HPO_4$; MgO	$Mg_3(PO_4)_2{\cdot}4H_2O$ $Mg(OH)_2$ and $NH_4MgPO_4{\cdot}6H_2O$
$(NH_4PO_4)_n$; MgO	$NH_4MgPO_4{\cdot}6H_2O$
$Al(H_2PO_4)_3$; MgO	$Mg(HPO_4){\cdot}3H_2O$ and amorphous $AlPO_4$
KH_2PO_4	$MgKPO_4{\cdot}6H_2O$

2.5.3.1 Fast setting and rapid strength gain

1d strength of magnesium oxychloride cements can reach about 80% of the ultimate strength. For MPC compressive strength can reach 28 MPa in 1 h and above 40 MPa in 3h. MPC can quickly set in a low temperature environment (−20°C) without heat treatment.

2.5.3.2 High strength

For MOC the compressive strength is generally more than 50 MPa after curing 28d, which can even break through 200 MPa after adding modifier, in which case, the rupture strength is more than 10 MPa, and the bending stress performance will be better.

2.5.3.3 High bonding strength

These cements bind exceptionally well with a wide range of ingredients and substrates. A wide variety of indigenous rock, plant and wood fibers, and other cellulose granules can be added to it as aggregate or reinforcement.

2.5.3.4 Low electrical and thermal conductivity

Magnesium-based cements are completely nonconductive of electricity, as well as heat and cold, enabling being used for flooring, radar stations, and hospital operating rooms. Heat transfer is lower than most of inorganic materials 0.218 W/(m·K), Sorel cement floors exist with low thermal conductivity close to wooden floors.

2.5.3.5 Flame retardant

The main component of MOC has a fire resistance of 2800, thus MOC products are generally high temperature resistant. Refractory property can reach above 300°C even put some glass fiber in it. Because of the fire resistance of magnesium oxychloride cement, it is widely used in the production of fire prevention board.

2.5.3.6 Good abrasion resistance

The abrasion resistance of magnesium oxychloride cement is 3 times of ordinary Portland cement.

However, MOC are poor in water resistance and poor freeze-thaw resistance. Crystals in the MOC cement are easy to hydrolyze when they meet water, forming flaky brucite, which has a high porosity and loose structure. And loose structure will allow water to contact with internal crystals, further promoting crystal hydrolysis, and eventually leading to serious deterioration of mechanical properties. Usually, 2 months later, the magnesium oxychloride cement specimen has serious quality loss, with its compressive strength decreased by more than 80%.

MOCs are significantly water soluble and release corrosive solutions. In the process of storage and use, if the environmental humidity is higher, the surface of the magnesium oxychloride cement products will produce some water and white powdery substance, commonly known as "white frost" In contrast, MPC show high dimension stability, nearly no shrinkage occurs during hydration of MPC.

2.5.4 Application

2.5.4.1 MOC

The very good properties of magnesium oxychloride phases in association with the properties of the aggregates and other raw materials allow the production of Sorel cement with unique properties for various applications.

Sorel cements are most commonly used for industrial flooring, due to their elastic properties and resistance to accumulated static loads.

Due to its very high bonding strength, MOC is used to produce wood packaging materials, particle boards, and wall materials with high strength, good machinability, refractory, heat insulation, and acoustic insulation. Additionally, MOC is also largely used in the manufacture of grinding and polishing stones, to bond SiC grains with the main end-users of marble, granite, and mosaic.

Because of the fire resistance of magnesium oxychloride cement, products with MOC are fire resistant. It is widely used in the production of fire prevention board, and as inorganic flame-retardant agents.

Precautions and limitations: its major limitation comes from its nonresistance to prolonged contact with water. Consequently, it cannot be installed as exterior constructions and also in areas which need repeated water flow cleaning. MOC is corrosive so metallic parts have to be protected when they adjoin with it.

2.5.4.2 MOS

The principal industrial application of magnesium oxysulfate cement is in the manufacture of lightweight insulating panels.

2.5.4.3 MPC

MPC is widely used for the rapid repair of concrete pavements, roads, bridge surfaces, runways, grounds of industrial plants, slab in ballast-free railway, and shotcrete applications, due to its rapid high strength growth, good durability, high bonding strength, low dry shrinkage, approximate color with old concrete, and good abrasive resistance.

MPC is also used for production of a magnesia-based refractory concrete for fireproofing coatings applications, particle boards. MPC foam concrete with high fluidity and low cement usage is specially applied to sandwich structures, earth-remaining walls, and running tracks or playgrounds. MPC has been developed to obtain porous material—based foamed concrete for cast-in-situ construction, high-temperature resistance, and acoustic insulation.

Hazardous waste/nuclear waste encapsulation. MPCs are most promising in the waste management area and especially in chemically solidification/stabilization (S/S) of various mixed toxic wastes as well as low level nuclear waste. MPCs provide an active chemical immobilization and physical encapsulation of the contaminants in the matrix. The main mechanism of contaminants entrapment inside the microstructural bond of waste forms is that phosphate reactions of MPC.

Other applications include carbon fiber reinforced mortar, corrosion protective overlays, coatings, and bioceramic. Sometimes, it is used for improving the water resistance of Sorel cement because of its much higher ability to resist water attack than that of MOC.

Exercises

1. What do you understand by the following terms: alite, belite, periclase, ferrite phase, portlandite, C—S—H, ettringite, and AFm?
2. Why is gypsum added to the cement clinker? Typically, how much is the amount of added gypsum?
3. Approximately what is the combined percentage of calcium silicates in Portland cement? What are the typical amounts of C_3A and C_4AF in Portland cement (P·I type, GB175, similar to ASTM Type III)?
4. Which chemical reactions take place during the hydration of Portland cement?

5. Which one of the four major mineral phases of Portland cement contributes most to the strength development during the first few weeks of hydration? Which mineral phase or phases are responsible for rapid stiffening and early setting problems of the cement paste?
6. Assuming the chemical composition of the calcium silicate hydrate formed on hydration of C_3S or C_2S corresponds to $C_3S_2H_3$, make calculations to show the proportion of calcium hydroxide in the final products and the amount of water needed for full hydration.
7. When does the monosulfate form? Summarize possible AFm phase occurring in cementitious materials AFm phase.
8. Please choose one or more appropriate types of cement to prepare concrete used in the following structures or elements, and explain the reasons: (1) prefabricated concrete sleeper; (2) piers of a marine bridge exposed to seawater; (3) refractory concrete; (4) repairs of bridge joint closures; (5) matrix for glass fiber referenced concrete panel.
9. Compare the differences of properties between Portland cement and blended cements.
10. What is the biggest impact of production of cement and concrete on our environment, and how to reduce it?
11. After hydration, 1unit volume of Portland cement produces 2 units volume of hydration products, given density of water $= 1.0\ g/cm^3$, density of Portland cement $= 3.1\ g/cm^3$, please calculate the minimum water to cement ratio (w/c) for complete hydration.

CHAPTER 3

Portland cement concrete

3.1 Introduction

Concrete refers to as a class of composite materials consisting of binder and particles, where particles are bound together by the binders. There are many types of concrete, which mainly depend on the type of binders, such as Portland cement concrete, asphalt concrete, and polymer concrete. Portland cement concrete is the most widely used construction material. In this chapter, the term concrete refers to Portland cement concrete unless stated otherwise.

Compared with other building materials, concrete in general and reinforced concrete, in particular, is indeed heavily loaded with dichotomies. Stretched between liquid/air and solid, gel and crystalline, smooth and rough, compact and porous, metal and inorganic nonmetal, compression and tension, brittle and ductile, material and process, material and structure, experimentation and computation, engineers and architects, technicality and art, worthless and precious, historical and unhistorical, concrete is permanently moving or transgressing the frame of taxonomy.

Human civilization is built on concrete, which is the world's most widely used construction material, due to its versatility, durability, sustainability, and economy, as explained as follows:

3.1.1 Versatility

① It can be used for various infrastructures, such as building, dam, bridge, road, nuclear station, man-made island.
② It can be shaped into any shape as long as the mold can be fabricated.
③ It can be used for different purposes, such as load-bearing, fire resistance, partition walls, even sculpture.

3.1.2 Durability

① For concrete itself, it can last hundreds of years without significant loss of its mechanical properties in the normal environment.

Civil Engineering Materials
ISBN 978-0-12-822865-4
https://doi.org/10.1016/B978-0-12-822865-4.00003-9

② With an embedded steel bar, the durability of steel-reinforced concrete is greatly reduced. However, the durability of reinforced concrete still can meet the needs of humankind.
③ Normal concrete can deteriorate very quickly under some severe environments, such as the marine environment, frosting zone, sulfate-bearing area. However, with special design and measures, concrete can be durable even under a severe environment.

3.1.3 Sustainability

① On the one hand, the production of Portland cement is energy-consuming and brings a large portion of carbon footprint to the environment.
② On the other hand, the concrete can be recycled, and many industrial by-products can be used in concrete. These make concrete more sustainable.

3.1.4 Economy

① In comparison with steel, the raw materials of concrete are much cheaper, such as cement, stone, and sand.
② The domestic cement industry is regional in nature. The logistics of shipping cement limits distribution over long distances. As a result, customers traditionally purchase cement from local sources. Sand and stone are definitely locally available materials in most areas, which can be natural or artificial.

In China, 2.298 billion cubic meters of concrete were produced in 2017. It is the most important and most widely used building material for municipal infrastructure, transportation infrastructure, hydraulic infrastructure, office buildings, and houses (Fig. 3.1).

It is generally accepted that Portland cement is first invented by Joseph Aspdin, who is an English mason. In 1824, he patented a special binder, which was named as Portland cement. This is because the color of the product of Aspdin resembled the color of the natural limestone quarried on the Isle of Portland in the English Channel. The name has endured and is now used throughout the world, with many manufacturers adding their own trade or brand names.

In 1845, I. C. Johnson claimed to have burned the cement raw materials with unusually strong heat until the mass was nearly vitrified, producing a Portland cement as we now know about it. The first cement plant on

Figure 3.1 Cement plant.

Polish Land was built in Grodziec and started production in 1857. However, the cement had low quality at that time. This cement was exported from England out the world in the middle of the 19th century. The first recorded shipment of Portland cement to the United States was in 1868. The first Portland cement plant in the United States was opened in Coplay, Pennsylvania, in 1871. The first Portland cement Plant in China, Qixing cement Corp. located in Tangshan, Hebei province, was established in 1889.

As the cement technology is diffused very quickly, it soon became the most important and widely used building materials. The invention of Portland cement is a milestone in the history of human civilization. It provides human the basic and economical construction materials, and lays the physical foundation for the entire human.

It is worth to mention that cement and concrete is a ductile material with very low tensile strength. If the ductility of concrete cannot be modified, concrete cannot be used in many situations. Luckily, a steel bar was used in combination with concrete, i.e., reinforced concrete.

The idea to use iron or steel reinforcement to overcome the tensile weakness of concrete emerged almost concomitantly with the rise of the cement industry and in parallel with the use of plain concrete. It seems that the real discoverer of reinforcement of cement paste or mortars with metal is Joseph Louis Lambot, a farmer who established a structure in the south of France. However, another inventor, Joseph Monier, obtained more recognition. His first patent was filed in 1867, but many more patents followed from 1861 to 1891; the patents were an application of reinforced cement—based materials for larger water reservoirs, staircases, fences, beams, floors, sewage pipes, aqueducts, beams, footbridges, and railway sleepers, etc. Monier was describing "a frame of round or square iron rods of any size and thickness, according to the strength that I want to give to them," which is much closer to modern reinforcement than his initial trusses of iron wire. However, Monier did not make much progress in reinforced concrete because of a lack of understanding of the mechanism of reinforcement correctly.

In 1874, an American, William E. Ward, a mechanical engineer, went one step further by realizing the importance of the strong adhesion of cement on steel in order to allow for the smooth load transmission from the concrete to the bar. At the same time, another American, Thaddeus Hyatt, made an important contribution by showing that the thermal expansion coefficient of iron and plain concrete is very close, proving that reinforced concrete beams would not collapse in case of fire. Then, a steel bar and concrete can collectively work perfectly together, and perfect for the infrastructures of humankind.

In 1889, the first concrete reinforced bridge was built. Although the first true reinforced concrete civil engineering work was the bridge built in Wiggen, Switzerland, by Hennebique's company in 1892, the most iconic work is probably the 16 stories Ingalls Building, the first reinforced concrete high-rise building with 16-story and 64 m high, in Cincinnati. It symbols that humankind entered a new era of reinforced concrete.

3.2 Types of concrete

Basically, concrete is a material in which the particles with different sizes are bound together by inorganic binders. However, many types of concrete are commercially available. There is no strict classification of concrete. The types of concrete can be classified based on various ways, such as differences in bulk density, application, and construction methods. It is worth to mention that there is no essential difference among these concretes. The authors try to classify concrete into different types, as shown in Fig. 3.2.

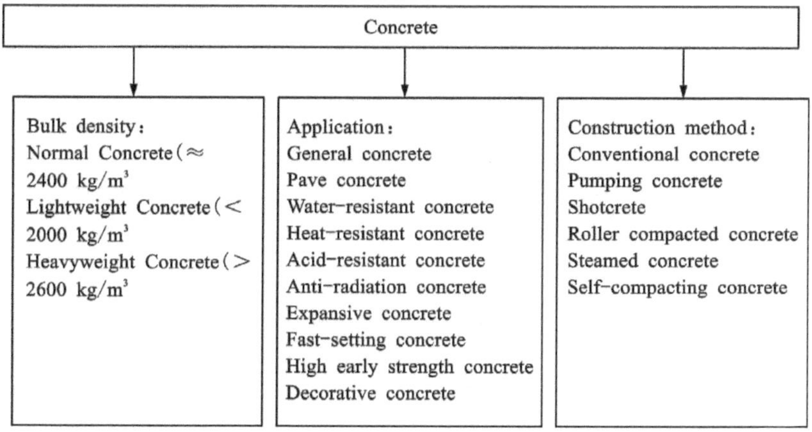

Figure 3.2 Types of concrete.

3.2.1 Based on bulk density

Normal concrete has a bulk density of about 2400 kg/m^3. In order to decrease the self-weight of concrete structure, lightweight concretes with a bulk density lower than 2000 kg/m^3 are often used. For some special applications, such as radiation insulator walls, military facilities, heavyweight concrete may be used, whose bulk density is higher than 2600 kg/m^3.

3.2.2 Based on application

For the concretes used in different environments, special functions may be required. For example, in the industrial areas in contact with an acid solution that is aggressive to concrete, acid-resistant concrete may be needed. In case of repair of traffic infrastructure, fast-setting concrete may be needed. In the case of fast removal of formwork, the early strength of concrete is required to develop fast, and high early strength concrete is needed.

3.2.3 Based on the construction method

Vibration is often necessary for conventional concrete to ensure the consolidation of concrete. However, self-compacting concrete can fill in the mold without vibration, just under the action of gravity. Shotcrete is a type of concrete conveyed through a hose and pneumatically projected at high velocity onto a surface, as a construction technique.

Figure 3.3 Construction methods of concrete.

Roller compacted concrete is a very dry concrete and nearly has no slump. It is placed in a manner similar to paving. The material is delivered by dump trucks or conveyors, spread by small bulldozers or special pavers, and then compacted by vibratory rollers. It is mainly used for the construction of the dam, as shown in Fig. 3.3.

3.3 Raw materials

Concrete is a complex composite with many types of raw materials. Traditionally, cement, water, and sand aggregate are the main ingredients for concrete. Except for water, aggregate is the most inexpensive component of Portland cement concrete. Conversely, cement is the most expensive one. With the concrete technology advancing, chemical admixtures and mineral admixtures were rapidly developed and widely applied in concrete. Admixtures have become indispensable ingredients for concrete, especially for some special concrete, like self-compacting concrete, ultrahigh-performance concrete. Fig. 3.4 gives the specific surface area and particle size of raw materials. It can be seen that concrete is a particulate material consisting of particles with various sizes ranging from mm to nm.

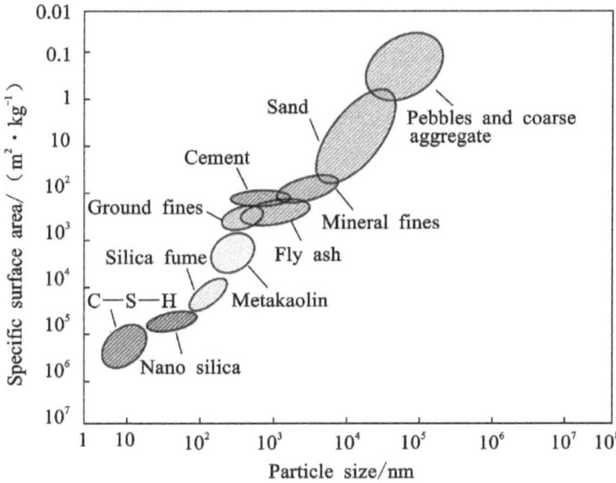

Figure 3.4 A surface area versus particle size plot for aggregates, fillers, cement, and hydrates.

3.3.1 Mixing water

Water plays a very important role in batching concrete. Water is mixed with cementitious materials to form cement paste, which contributes to the flowability of concrete in a fresh state and binds the aggregate together in a hardened state. In addition, water is an indispensable ingredient for the hydration of cementitious materials. Water can also be used to wash aggregate and as a curing agent for cement and concrete.

Water is available everywhere on the earth. However, not any water can be used as the mixing water for concrete. The water can be used for mixing concrete include tap water, natural surface water, underground water, urban recycling water, etc. In Chinese standard JGJ 63-2006, the ionic types and concentrations of mixing water are strictly required, as well as the insoluble impurity content, as illustrated in Table 3.1. It can be seen that for different concrete elements, the requirements for the matters in the water are different. For concrete with a steel bar, the requirement for the dissolved and undissolved matters are stricter than that of plain concrete. Mixing water for prestressed concrete is purer than the water for reinforced concrete.

The typical chemical compositions of tap water and seawater are given in Table 3.2. Obviously, the dissolved matter of tap water shown in Table 3.2 is far lower than 2000×10^{-6}, which is the upper limit in Table 3.1.

Table 3.1 Requirements for mixing water for concrete.

Items	Prestressed concrete	Reinforced concrete	Plain concrete
pH	≥5.0	≥4.5	≥4.5
Insoluble substance/(mg/L)	≤2000	≤2000	≤5000
Soluble substance/(mg/L)	≤2000	≤5000	≤10,000
Cl^-/(mg/L)	≤500	≤1000	≤3500
SO_4^{2-}/(mg/L)	≤600	≤2000	≤2700
Alkali content/(rag/L)	≤1500	≤1500	≤1500

Although higher concentrations are not always detrimental to concrete, some types of cement may react adversely. Thus, the upper limit of 2000×10^{-6} of dissolved matter should not exceed. In order to evaluate the qualification of unknown water, a reference mortar sample should be made with distilled water or known city water. If the 7 and 28 d strengths equal to at least 90% of the reference sample, the water can be accepted as mixing water for concrete, according to ASTM C94.

Impurities in water may affect the hydration, setting time, flowability, strength development, and volume stability, and may also cause efflorescence, discoloration, and premature corrosion of steel bar. It was reported that sodium carbonate might cause fast hydration of cement, and bicarbonate may either accelerate or deaccelerate the hydration, depending on the concentration of bicarbonate. Sodium chloride or sodium sulfate are also found to be an accelerator for the hydration of cement. However, chloride is the potential to initiate the corrosion of steel. This greatly limits the concentration of chloride ions. Iron salts in natural groundwater usually do not exceed $20\text{–}30 \times 10^{-6}$. This has no effect on the hydration of cement.

Seawater has $35,000 \times 10^{-6}$ dissolved solids, which is far higher than 2000×10^{-6}. It should never be used for reinforced concrete, since the chloride may greatly increase the corrosion risk of a steel bar. Normally, seawater is not used as mixing water for plain concrete, since it may cause a high early strength, and low later strength. It is worth mentioning that due to the lack of freshwater, there are many scientists who are studying the possibility of making concrete with seawater. However, this is still not a successful technology, and cannot be used in practice.

After proper treatment, industrial wastewater and sanitary sewage can be used in concretes. For the sewage water passing through a good disposal system, the concentrations of solids are usually too low to have any remarkable effect on concrete. Wastewaters from tanneries, paint factories,

Table 3.2 Typical chemical compositions of tap water and seawater (parts per million).

Chemical	Analysis number						
	1	2	3	4	5	6	Sea water[a]
Silica(SiO_2)	2.4	0.0	6.5	9.4	22.0	3.0	—
Iron(Fe)	0.1	0.0	0.0	0.2	0.1	0.0	—
Calcium(Ca)	5.8	15.3	29.5	96.0	3.0	1.3	50–480
Magnesium(Mg)	1.4	5.5	7.6	27.0	2.4	0.3	260–1410
Sodium(Na)	1.7	16.1	2.3	183.0	215.0	1.4	2190–12,200
Potassium(K)	0.7	0.0	1.6	18.0	9.8	0.2	70–550
Bicarbonate(HCO_3)	14.0	35.8	122.0	334.0	549.0	4.1	—
Sulfate(SO_4)	9.7	59.9	5.3	121.0	11.0	2.6	580–2810
Chloride(Cl)	2.0	3.0	1.4	280.0	22.0	1.0	3960–20,000
Nitrate(NO_3)	0.5	0.0	1.6	0.2	0.5	0.0	—
Total dissolved solids	31.0	250.0	125.0	983.0	564.0	19.0	35,000

[a]Different seas contain different amounts of dissolved salts.

coke plants, chemical plants, etc., may contain harmful impurities. Before using this water as mixing water for concrete, a comparative strength test should be run.

Generally speaking, mixing water containing inorganic acid concentrations as high as $10{,}000 \times 10^{-6}$ has no adverse effects on concrete strength. It is reported that the acceptance of mixing waters containing acid should be based on concentrations of acids rather than the pH value, since the latter is an intensity index.

Sugar is a good retarder for the hydration of cement at its low concentration; 7 d strength may be reduced, and 28 d strength may be increased. When the concentration of sugar increases to 0.2% by mass of cement, the hydration of cement may be accelerated. When the concentration of sugar increases to 0.25%, there may be a rapid setting.

Clay or fine powders may exist in industrial or natural water. The existence of clay or fine powders mainly affects the action of superplasticizer (SP), since the molecule of SP may adsorb onto the surface of clay or fine powders. Besides, clay or fine powders have little effect on cement and concrete. Mineral oils have less effect on the strength development of cement than vegetable or animal oils.

Whenever using water as mixing water for concrete, the testing reports of the water should be required. The dissolved and undissolved matters in water should meet the requirements in Table 3.1. If the water is questionable, a comparative test can be run to check the effect of water on the strength development of mortar. However, for reinforced concrete and prestressed concrete, any ions, mainly chloride, may initiate the corrosion of steel bar and should be strictly controlled.

3.3.2 Cement

Cement, most commonly Portland cement, is associated with the general term "concrete." A range of other materials can be used as the cement in concrete as well. One of the most familiar of these alternative types of cement is asphalt and polymer. Other cementitious materials such as fly ash, slag, and limestone powder are sometimes added as mineral admixtures—either preblended with the cement or directly as a concrete component—and become a part of the binder for the aggregate.

To produce concrete from most types of cement (excluding asphalt and polymer), water is mixed with the dry powder, fine and coarse aggregates, which produces a semiliquid slurry that can be poured into formwork to form elements with any shapes. The concrete solidifies and hardens through a chemical process called hydration. The water reacts with the cement, which bonds the other components together, creating a robust man-made stone.

In modern cement kilns, many advanced technologies are used to lower the fuel consumption per ton of clinker produced. Cement kilns are extremely large, complex, and inherently dusty industrial installations, and have emissions which must be controlled. It is reported that the cement industry accounts for 7% greenhouse gas emission of the total. Therefore, great efforts are devoted to reduce the energy consumption and lower the carbon emission caused by the production of cement. Even the most complex and efficient kilns require 3.3—3.6 GJ of energy to produce a ton of clinker, and a large amount of energy is used to grind it into cement. Many kilns can be fueled with difficult-to-dispose-of wastes, the most common being used tires. The extremely high temperatures and long periods of time at those temperatures allow cement kilns to efficiently and completely burn even difficult-to-use fuels.

3.3.3 Aggregate

3.3.3.1 Significance of aggregate

Aggregates are particles of random shape. They can be either in nature as sand, gravel, stones, or rock that can be crushed into particles or obtained from by-products or waste materials from an industrial process. Aggregate sizes vary from several cm to several mm. Aggregates generally occupy 60%—80% of the concrete volume. Traditionally, the aggregates are often thought of as being inert fillers. They do not have chemical reactions with cement, do not hydrate, and do not swell or shrink. They should have sufficient integrity to maintain their shape during the concrete mixing, and be sufficiently strong to withstand the stresses imposed on the concrete. The strength becomes a particular consideration with high-strength concrete. The mineral constituents are not generally of great importance, the notable exceptions being those that can participate in alkali—silica reaction, which will be discussed later.

However, since the increasing awareness of the importance of aggregates has been recognized, it is inappropriate to treat the aggregate as an inert filler. The characteristics of aggregate have a great influence on the performance of fresh and hardened concrete, including porosity, grading or size distribution, moisture absorption, shape and surface texture, crushing strength, elastic modulus, and the type and content of deleterious substances. These characteristics depend on the mineralogical composition of the parent rock, exposure conditions, and the type of equipment used for producing the aggregate.

The proper selection of aggregates for use in concrete mixtures is critical to fresh and hardened properties of concrete, as well as the economy. Since aggregates are cheap and stable, one of the objectives of mix design is to use as much aggregate as possible. The concrete can be thought of as two phases, in which aggregates are dispersed phase and cement paste is a continuous phase, the former disperse in the later. Aggregates function as a skeleton in hardened concrete, and cement paste contributes to the flowability of concrete. In order to maximize the content of aggregates, continuous particle sizes need to be used. This minimizes the void content of the aggregate mixture, and requires less cement paste to obtain the same flowability.

3.3.3.2 Classification of aggregates

There are many ways to categorize the aggregate. For instance, it can be classified based on the density, the origins, the shape, and the size. The classification of aggregates is given as follows.

3.3.3.2.1 Based on density

Normal weight aggregate: The most commonly used normal weight aggregates often have relative densities within a limited range of approximately 2.55–2.75, and therefore all produce concretes with similar densities normally in the range of 2200–2450 kg/m^3, depending on the mix proportions.

Lightweight aggregate: Lightweight aggregate is often defined as any aggregate with a dry loose bulk density of less than 1200 kg/m^3. Lightweight aggregates are used to produce lower density concretes, which are advantageous in reducing the self-weight of structures and also have better thermal insulation than normal weight concrete. The lightweight of the aggregate is due to the cellular or highly porous microstructure. Pumice, a naturally occurring volcanic rock of low density, has been used since Roman times, but it is only available at a few locations, and artificial lightweight aggregates are now widely available, as follows:

① Sintered pulverized fuel ash, formed by heating the pelletized ash from pulverized coal used in power stations until partial fusion and hence binding occurs.

② Expanded clay or shale, formed by heating suitable sources of clay or shale until the gas is given off and trapped in the semimolten.

③ Foamed slag, formed by directing jets of water, steam, and compressed air on to the molten slag from blast furnaces.

In each case, both fine and coarse aggregates can be produced, and various products are available in many countries. Because they all achieve lower specific gravity by increased porosity, they are all weaker than the normal density aggregates, and the application of lightweight aggregate generally results in an overall lowering in the concrete strength. The quality and properties of different aggregates vary considerably, and therefore produce different strength/density ratios. Lightweight aggregates are not as rigid as normal weight aggregates, and therefore produce concrete with a lower elastic modulus and higher creep and shrinkage. Some very porous aggregates are generally weak and are therefore more suitable for making nonstructural insulating concretes, instead of a structural member. The strength of lightweight concrete depends on the lightweight aggregate type and source, and also whether lightweight fine aggregates and/or natural sands are used.

Heavyweight aggregate: Where concrete of high density is required, for example, in radiation shielding, heavyweight aggregates can be used, such as barite, limonite, magnetite, ilmenite, hematite, iron, and steel punchings or shot. Densities of 3500–4500 kg/m^3 are obtained by using baryte, and about 7000 kg/m^3 by using steel shot.

3.3.3.2.2 Based on sizes

Fine aggregate (sand): the particles with a size smaller than 4.75 mm are called fine aggregate or sand.

Coarse aggregate: particles with a size greater than 4.75 mm are called coarse aggregate. Normally, the size of the aggregate is smaller than 50 mm.

3.3.3.2.3 Based on origins

Natural aggregate: Weathering and erosion of rocks produce particles of stone, gravel, sand, silt, and clay. Natural gravel and sand are usually dug or dredged from a pit, river, lake, or seabed. Some natural aggregate deposits of gravel and sand can be readily used in concrete with minimal processing. The quality of natural aggregate mainly depends on its parent rock. For instance, sand and gravel derived from igneous and metamorphic rocks tend to be sound, while sand and gravel derived from rocks rich in shale and siltstone are more likely to be unsound. Natural aggregate deposited at higher elevations from glaciers may be superior to deposits in low areas. Sand and gravel that have been smoothed by prolonged agitation in water are usually considered higher quality because they are harder and have a more rounded shape than less abraded sand and gravel. However, the smooth surface of natural gravels can reduce the bond strength with the cement paste and reduce overall concrete strength.

Manufactured aggregate: Manufactured coarse and fine aggregates are produced by crushing sound parent rocks by stone crushing machines. Manufactured aggregates differ from gravel and sand in their grading, shape, and texture. As a result of the crushing operation, manufactured aggregates often have a rough surface texture and angular shape, tend to be cubical or elongated in shape, and bad in size distribution. By using appropriate crushing techniques and machines, better shape and size distribution of aggregates can be obtained. The particle elongation and layered flakes of manufactured sands can be reduced through appropriate crushing techniques. Manufactured aggregates are less likely than gravel and sand to be contaminated by deleterious substances such as clay minerals or organic matter. However, large crushed powder content is the major concern for crushed sand. The crushed powder content may affect the water required and the workability of concrete. It is worth to mention that manufactured sand has become more and more important due to the lack of natural sand.

Recycled-concrete aggregate: The urbanization and the economic development of developing countries have remarkably increased the pace of development of the construction industry. As a result of these activities, old constructions are being demolished to make new buildings. Due to these large-scale demolitions, a huge amount of debris is generated all over the world, which is causing serious environmental pollution, including a disposal problem. Therefore, there is an urgent need to make use of the demolished waste. Using construction demolished waste as recycled aggregate can reduce the use of natural aggregates and the problem of mining them. However, in comparison to natural aggregate, the quality of recycled aggregate is poor, which restricts its use in varieties of construction applications. The cost of concrete production containing recycled aggregate also needs to be considered in terms of its large-scale application in the construction sector. The recent implementation of stringent rules and regulations on the disposal of several types of wastes, including construction demolished waste all over the world, however, can also help in the large-scale application of recycled aggregate in productions of various types of concrete productions. Recycling construction demolished waste in the making of new construction also increases the life cycle of construction materials. To improve the recycling amount of construction demolished waste, some countries ratified some governmental laws and specific regulations. Due to these initiatives, the recycling level of this material in some countries reached about 90% of the total generated amount. As construction demolished waste can be used as aggregate in new concrete preparation, this type of aggregate is characterized using methods similar to those used for natural aggregate.

For the preparation of aggregates, in general, waste concrete elements are mechanically broken into small-sized pieces. The small pieces are further crushed into small-sized pieces using crushers. After crushing, different sized fractions are screened using a sieving device and used as aggregates. Rubble from demolished concrete buildings is generally contaminated with mortar paste, gypsum, and minor quantities of other substances such as wood, plastics, metals, and glass. These impurities have several deleterious effects and therefore are unsuitable for concrete production. Consequently, in most of the cases, the impurities present in construction demolished waste must be separated during the process.

The procedure involves demolishing and removing the existing concrete, crushing the material in crushing machines, removing reinforcing steel and other embedded items, grading and washing, and stockpiling the resulting coarse and fine aggregate. Dirt, gypsum board, wood, and other foreign materials should be prevented from contaminating the final product. The recycled fine aggregate contains large amounts of hydrated cement and gypsum, and it is unsuitable for batching fresh concrete mixtures. However, the size fraction that corresponds to coarse aggregate, although coated with cement paste, has been used successfully in practices.

The recycled-concrete aggregate has been satisfactorily used as aggregate in granular subbases, lean-concrete subbases, soil-cement, and in new concrete as the primary source of aggregate or as a partial replacement of natural aggregate. Recycled-concrete aggregate generally has higher absorption and lower specific gravity than natural aggregate. This results from the high absorption of the more porous hardened cement paste within the recycled concrete aggregate. Absorption values typically range from 3% to 10%, depending on the concrete being recycled. This absorption rate lies between those for natural and lightweight aggregates. Absorption rates increase as coarse particle size decreases. The high absorption of the recycled aggregate makes it necessary to add more water to achieve the same workability and slump as concrete with conventional aggregates. This makes concrete made with recycled aggregate have a higher *w/c* ratio and thus lower strength. Dry recycled aggregate absorbs water during and after mixing. Recycled aggregate stockpiles should be prewetted or kept moist in practice, just like lightweight aggregates. The particle shape of recycled-concrete aggregate is similar to crushed rock, as shown in Fig. 3.5. The relative density decreases progressively as particle size decreases. The sulfate content of recycled-concrete aggregate should be determined to assess the possibility of delayed sulfate attack. The chloride content should also be determined where applicable.

(a) Gravel (b) Manufactured aggregate

(c) Scree (d) Recycled aggregate

Figure 3.5 Different types of aggregates.

It has been reported that, compared with concrete mixtures containing natural aggregate, the concrete mixtures containing recycled-concrete aggregate generally give at least two-thirds of the compressive strength and modulus of elasticity, and show satisfactory workability, and the carbonation, permeability, and resistance to freeze—thaw (FT) action have been found to be the same or even better than concrete produced with natural aggregates. However, drying shrinkage and creep of concrete made with recycled-concrete aggregates may be up to 100% higher than concrete made with a corresponding natural aggregate. A large amount of existing cement paste and mortar contributes to the additional shrinkage and creep. Therefore, recycled-concrete coarse aggregate is more often used in concrete than recycled-concrete fine aggregate, and recycled-concrete aggregate is often used in combination with natural aggregates. It is worth mentioning that the variability in the properties of the old concrete may, in turn, affect the properties of the new concrete.

Technically, the side effect caused by using recycled-concrete aggregate can be well controlled by appropriate mix design. Although the performances of concrete made with recycled-concrete aggregate maybe, to some extent, lower than that concrete with natural aggregate, it is acceptable for many practices. The major problem of recycled-concrete aggregate is the cost of crushing, grading, dust control, and separation of undesirable constituents. Due to the lack of natural aggregate and environmental protection needs, many authorities are making regulations to promote the use of recycled-concrete aggregate. The technical and economic benefits of using recycled-concrete aggregate can be achieved by then.

3.3.3.2.4 Based on mother rock

According to the origins of the rock, they can be classified into three major groups: igneous, sedimentary, and metamorphic; these groups are further subdivided according to mineralogical and chemical composition, texture or grain size, and crystal structure.

Igneous rocks are formed by the cooling of the magma. It is worth to mention that the grain size and degree of crystallinity has a paramount effect on the characteristics of a rock. Rocks with the same chemical composition but different grain size and degree of crystallinity may present completely different properties. The degree of crystallinity and the grain size of igneous rocks are determined by the formation conditions of rock. The cooling rate is the major condition for the formation of rock. Magma intruded at great depths cools at a slow rate and forms completely crystalline minerals with coarse grains (more than 5 mm grain size). Rocks of this type are called intrusive or plutonic. The rocks formed near the surface of the earth contain minerals with smaller crystals, because of the quick cooling rate. These fine-grained rocks (1−5 mm grain size) may also contain some glass and are called shallow-intrusive. Rapidly cooled magma, for instance, the rocks formed by volcanic eruptions, contains mostly noncrystalline or glassy matter; the glass may be dense or cellular, and the rock type is called extrusive or volcanic. From the perspective of silica content, magma may be supersaturated, saturated, or undersaturated. For a supersaturated magma, the free or uncombined silica crystallizes out as quartz after the formation of minerals such as feldspars and mica. In saturated or unsaturated magma, the silica content is insufficient to form quartz. Thus, based on the content of SiO_2, rocks can be classified into acid, intermediate, and basic rocks. They contain more than 65% SiO_2, 55%−65% SiO_2, and less than 55% SiO_2, respectively. The content of SiO_2 is especially important to the bonding strength between binder and aggregate, especially for asphalt concrete.

Sedimentary rocks are formed by the accumulation of sediments, are usually laid down underwater, or maybe accumulated by wind and glacial action. Sedimentary rocks may be divided into three basic categories: ① Clastic (detrital) sedimentary rocks are composed of the solid products of weathering (gravel, sand, silt, and clay) cemented together by the dissolved weathering products. ② Biogenic (biochemical) sedimentary rocks are those composed of materials formed by the activity of living organisms such as coal (compacted undecayed plant matter) and many limestones which are made up of the shells or other skeletal fragments from marine organisms. ③ Chemically precipitated (chemical) sedimentary rocks are those such as halite and gypsum, and some limestones, which form direct precipitation (crystallization) of the dissolved ions in the water.

Gravel, sand, silt, and clay are the main sources of unconsolidated sediments. Gravel and coarse sands usually consist of rock fragments; fine sands and silt consist predominately of mineral grains, and clays consist exclusively of mineral grains. Sandstone, quartzite, and graywacke are in the secondary category. Sandstones and quartzite consist of rock particles in the sand-size range. The rock breaking around the sand grains is called sandstone, and the rock breaks through the grains are called quartzite. The cementing or interstitial materials of sandstone may be opal, calcite, dolomite, clay, or iron hydroxide. Graywacke is a special class of sandstone, which contains angular and sand-size rock fragments in an abundant matrix of clay, shale, or slate. Chert and flint also belong to siliceous sedimentary rocks. Chert consists of poorly crystalline quartz, chalcedony, and opal. Limestones are the most widespread of carbonate rocks. They range from pure limestone consisting of the mineral calcite to pure dolomite. Usually, they contain both the calcium and magnesium carbonate minerals in various proportions, and significant amounts of noncarbonate impurities, such as clay and sand. In comparison with igneous rocks, the aggregates produced from stratified sediments can vary widely in characteristics, such as the shape, texture, porosity, strength, and soundness. This is because the conditions under which they are consolidated vary significantly. The rocks tend to be porous and weak when formed under relatively low pressure. They are dense and strong if formed under high pressure. Sedimentary rocks often contain impurities, which jeopardize their use as aggregate. For instance, limestone, dolomite, and sandstone may contain opal or clay minerals, which are detrimental to the behavior of aggregate.

Metamorphic rocks are igneous or sedimentary rocks that have changed their original texture, crystal structure, or mineralogical composition, which is caused by one or both of the conditions of elevated temperatures and high pressures. There are four types of metamorphisms. Regional Metamorphism is the result of high pressures and elevated temperatures associated with deep burial in an orogenic belt. Platy minerals (micas) and elongate minerals (hornblende) recrystallize and/or rotate into a new orientation perpendicular to the applied stress, while other minerals recrystallize into new crystals which are stable at the higher pressures and temperatures. Contact Metamorphism is the result of baking the surrounding country rocks by an igneous intrusion. The metamorphic aureole surrounding an igneous body maybe only 2 cm wide adjacent to a small dike, or it maybe 2 km wide at the contact with a large, slow-cooling granite pluton. Contact metamorphosed rocks may be bleached out looking and nondescript fine-grained. The hydrothermal alteration, sometimes considered a form of metamorphism, is related to the circulation of hot, mineral-laden fluids through rock bodies. This is particularly important in the alteration of ocean crust in the high heat flow regime near the mid-ocean ridges. Serpentinites form from the hydration of peridotites, olivine-rich rocks at the base of the oceanic crust. The hydrothermal alteration also occurs as a result of hot fluids escaping from a cooling pluton, in addition to the high-temperature contact metamorphism occurring there. Cataclastic metamorphic rocks form where rocks are being faulted and sheared. Cataclasite or fault breccias form in brittle fault zones and consist of larger angular rock fragments dispersed in a fine-grained matrix. Mylonites are foliated, actually sheared, stretched, and streaked rocks, formed in plastic shear zones, at depths and pressures too great for a rock to break. The rock becomes drawn out like modeling clay or bubble gum. Common metamorphic rocks are marble, schist, phyllites, gneiss, etc. The rocks are dense but frequently foliated. Some phyllites are reactive with the alkalis present in Portland cement.

It should be noted that the earth's crust consists of 95% igneous and 5% sedimentary rocks. However, sedimentary rocks cover 75% of the earth's landed area. Most of the natural mineral aggregates used in concrete, namely sand, gravel, and crushed rocks, are derived from sedimentary rocks. Common rocks in aggregates are given in Table 3.3.

3.3.3.3 Characteristics of aggregate

Since up to approximately 80% of the total volume of concrete consists of aggregate, aggregate characteristics significantly affect the performance of fresh and hardened concrete and have an impact on the cost-effectiveness of concrete. Aggregate characteristics of shape, texture, and grading influence workability, finishability, bleeding, pumpability, and segregation of fresh concrete and affect strength, stiffness, shrinkage, creep, density, permeability, and durability of hardened concrete. Construction and durability problems have been reported due to poor mixture proportioning and variation on

Table 3.3 Rock in aggregates.

Igneous rock	Metamorphic rock	Sedimentary rock
Granite	Marble	Conglomerate
Syenite	Metaquartzite	Sandstone
Diorite	Slate	Quartzite
Gabbro	Phyllite	Graywacke
Peridotite	Schist	Subgraywacke
Pegmatite	Amphibolite	Arkose
Volcanic glass	Hornfels	Claystone siltstone argillite and shale
Obsidian	Gneiss	Carbonate
Pumice	Serpentinite	Limestone
Tuff		Dolomite
Scorla		Marl
Perlite		Chalk
Pitchstone		Chert
Felsite		
Basalt		

grading. The understanding of certain aggregate characteristics is necessary for proportioning concrete mixtures. Since the characteristic of aggregate affects the properties of concrete differently, it is more appropriate to divide the study of aggregate properties into three categories that are based on physical, chemical, and mechanical properties:

① Physical properties: density, moisture absorption, soundness, particle size distribution, particle shape, and surface texture
② Mechanical properties: strength, hardness, and elastic modulus
③ Chemical properties: chemical mineral and deleterious substances present

Identification of the constituents of an aggregate cannot alone provide a basis for predicting the behavior of aggregates in service. A visual inspection will often disclose weaknesses in coarse aggregates. In the absence of a performance record, the aggregates should be tested before they are used in concrete. The details of the characteristics of aggregates are discussed as follows.

3.3.3.4 Particle shape and surface texture

The shapes of aggregates are often irregular. The shape and surface texture of aggregate particles influence the properties of fresh and hardened concrete. Shape refers to geometrical characteristics such as round, angular, elongated, or flaky. Compared to smooth and rounded particles, rough-textured, angular, and elongated particles require more cement paste to produce workable concrete mixtures. Aggregate particles that are angular require more cement to maintain the same water-cementing materials ratio. Angular or poorly graded aggregates may also be more difficult to pump. The bond between cement paste and a given aggregate generally increases as particles change from smooth and rounded to rough and angular. The increase in the bond is a consideration in selecting aggregates for concrete where flexural strength is important or where high compressive strength is needed.

Natural aggregates, such as sand and gravel from seashore or riverbeds, often have a well-rounded shape, because different forces may round the edges and corners of the aggregates. In contrast, crushed aggregates often possess very sharp edges and corners and are called angular. They generally produce equidimensional particles. Laminated limestones, sandstones, and shale tend to produce elongated and flaky fragments, especially when jaw crushers are used for crushing. The thickness of those particles is relatively smaller than two other dimensions which are referred to as flat or flaky, while those who are considerably bigger in length than the other two dimensions are called elongated. Flat and elongated particles content in aggregates should be strictly controlled. For instance, in American standards, elongated, blade-shaped aggregate particles should be limited to a maximum of 15% by mass of the total aggregate. This requirement is important not only for coarse aggregate but also for manufactured sands, containing elongated grains, which produce very harsh concrete. By choosing appropriate crushing machines, the elongated and flaky can be lowered to some extent.

Another term sometimes used to describe the shape of coarse aggregate is sphericity, which is defined as a ratio of surface area to volume. Spherical or well-rounded particles have high sphericity, but elongated and flaky particles possess low sphericity.

Surface texture, which is defined as the degree to which the aggregate surface is smooth or rough, is based on visual judgment. The surface texture of aggregate depends on the hardness, grain size, and porosity of the parent rock and its subsequent exposure to forces of attrition. Obsidian, flint, and dense slags show a smooth, glassy texture. Sand, gravels, and chert are smooth in their natural state. Crushed rocks such as granite, basalt, and limestone show a rough texture. Pumice, expanded slag, and sintered fly ash show a honeycombed texture with visible pores. A rougher texture is beneficial to the formation of a stronger physical bond between the cement paste and aggregate.

3.3.3.5 Gradation and size

Grading or particle size distribution significantly affects some characteristics of concrete, like packing density, voids content, and, consequently, workability, segregation, durability, and some other characteristics of concrete. However, since aggregate is irregular, and the size of the aggregate is difficult to measure. Thus, many definitions of the size of aggregate are defined, as shown in Fig. 3.6. The size of aggregates is defined as the smallest hole that the particles can pass through. The size of single particles has no practical meaning. The important features of aggregates are the range of sizes, the largest particles, and gradation, which is the distribution of particles of a granular material among various size ranges, usually expressed in terms of cumulative percentage larger or smaller than each of a series of sizes of sieve openings, or the percentage between a certain range of sieve openings.

A set of sieves fitting tightly one on top of the other is used to determine the size and gradation of aggregate. A sample of the aggregate to be analyzed is placed in the top sieve, which has the largest holes. The second sieve has smaller holes, and each succeeding sieve has holes smaller than the

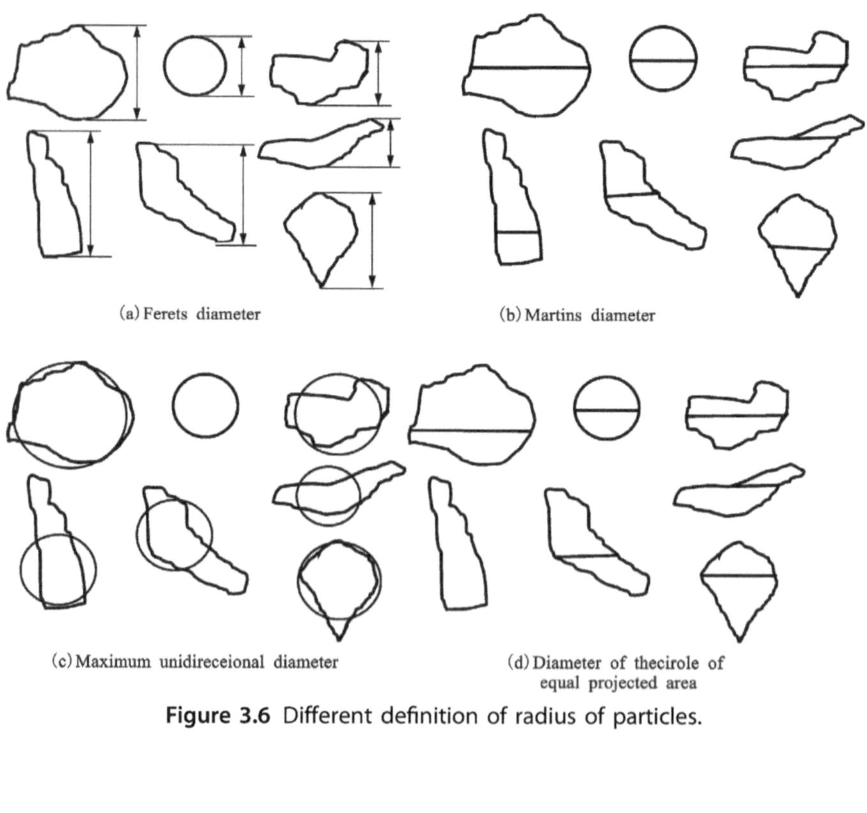

Figure 3.6 Different definition of radius of particles.

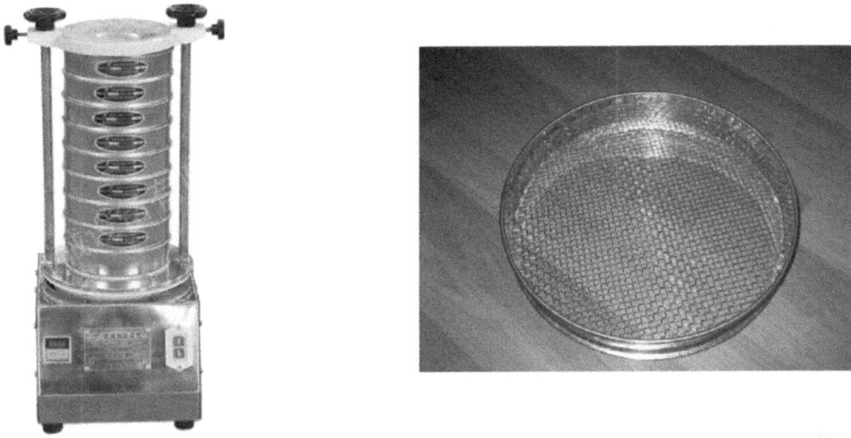

(a) Vibration machine (b) Sieve

Figure 3.7 Testing setup for aggregate gradation.

sieve above it. At the bottom is a solid pan. The pan collects all particles smaller than the openings in the smallest sieve, as shown in Fig. 3.7.

The sieve holes are often square. Particles are considered to be the size of the holes in the sieve on which they are caught. Whether or not flat particles and long, narrow ones go through a screen may depend on how they land on the screen while being shaken. Statistically, results are the same over a large number of tests of the same material.

The nest of sieves is shaken, and each particle settles as far as it can. If no particles remain on the top sieve, then the top sieve size represents one kind of maximum size for the sample. The absolute maximum particle size is somewhere between this top sieve size and the size of the next sieve. If even a few particles remain on the top sieve, it is not known how large the maximum size particles are.

Some particles fall through to the pan, and their size is not known either if the lowest sieve has small enough holes to catch the smallest significant size, the size of particles in the pan is not important. However, excessive quantity of material fine enough to reach the pan is of interest.

In Chinese or American standards, the principal size division is that between fine and coarse aggregates at a particle size of 5 mm. The size of sieves for coarse aggregates may be 4.75 mm, 9.50 mm, 16.0 mm, 19.0 mm, 26.5 mm, 31.5 mm, 37.5 mm, 53.0 mm, 63.0 mm, 75.0 mm, and 90 mm. And the size of sieves for coarse aggregates may be 4.75, 2.36, 1.18, 0.6, 0.3, 0.15, and 0.075 mm. A typical size gradation of sand is given in Fig. 3.8.

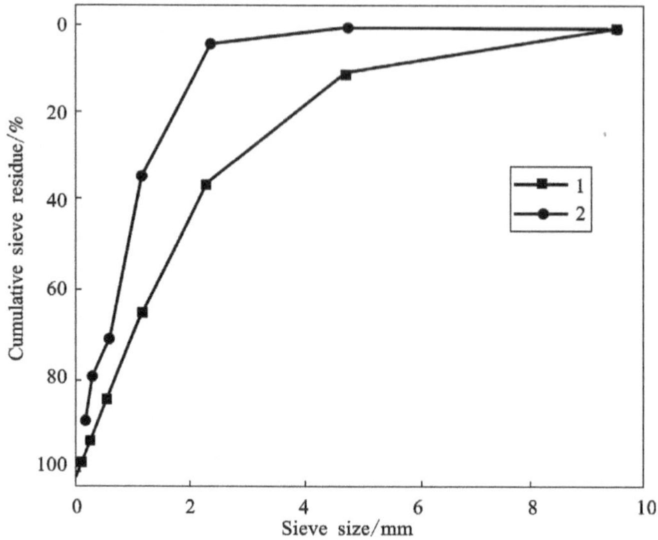

Figure 3.8 Typical sand size gradation curves.

There are many reasons for specifying grading curves; the most important is their influence on the bulk density (or voids between aggregates), and thus, cement and water requirements, workability, pumpability, economy, porosity, shrinkage, and durability of concrete. For example, aggregate with a size of 25 mm will have a higher void than that of aggregates with a size of 25 and 9 mm, as shown in Fig. 3.9. The void may need more cement paste to fill in up. Usually, very coarse sands produce harsh and unworkable concrete mixtures, and very fine sands increase the water requirement (therefore, the cement requirement for a given water-cement ratio) and are uneconomical. Variations in grading can seriously affect the uniformity of concrete from batch to batch. In general, aggregates that do not have a large deficiency or excess of any size and give a smooth grading curve will produce satisfactory results.

In practice, there are three zones for sand, zone I, zone II, and zone III, which correspond to fine sand, medium sand, and coarse sand, as shown in Table 3.4. In most cases, sand falling in the zone II is desirable for batching concrete.

The fineness modulus (FM) of sand is an index of the fineness of an aggregate, which can be calculated by

$$M_x = \frac{(A_2 + A_3 + A_4 + A_5 + A_6) - 5A_1}{100 - A_1} \tag{3.1}$$

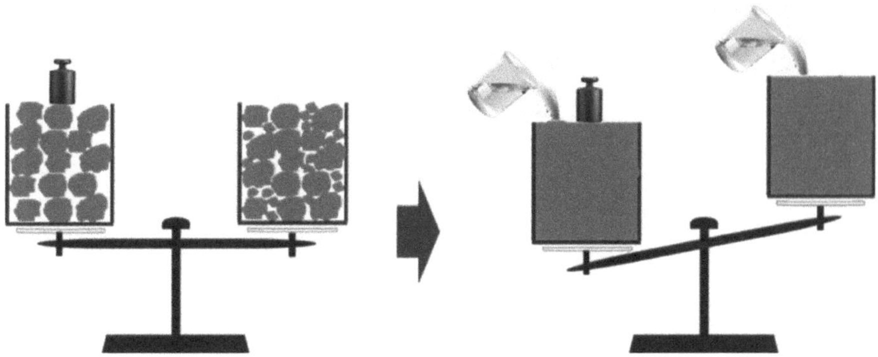

Figure 3.9 Void between aggregates.

Table 3.4 Particle gradation of sand.

Type	Natural sand			Manufactured sand		
Gradation zone	**I zone**	**II zone**	**III zone**	**I zone**	**II zone**	**III zone**
Sieve size/mm	**Cumulative sieve residue/%**					
4.75	10–0					
2.36	35–5	25–0	15–0	35–5	25–0	15–0
1.18	65–35	50–10	25–0	65–35	50–10	25–0
0.60	85–71	70–41	40–16	85–71	70–41	40–16
0.30	95–80	92–70	85–55	95–80	92–70	85–55
0.15	100–90			97–85	94–80	94–75

where M_x is the fineness modulus of sand; A_1, A_2, A_3, A_4, A_5 are the cumulative percentages by mass retained on each of a specified series of sieves, as shown in Table 3.5.

Normally, FM of 3.7–3.1 is coarse sand, FM of 3.0–2.3 is medium sand, and FM of 3.0–1.7 is fine sand. Medium sand is always desirable for engineering applications. If the FM value is outside the required range of 2.3–3.1, the fine aggregate should be rejected unless suitable adjustments are made in proportions of fine and coarse aggregate. If the FM varies by more than 0.2, adjustments may need to be made with regard to coarse and fine aggregate proportions as well as the water requirements for the concrete mixture. The fine aggregate must not have more than 45% retained between any two consecutive standard sieves. The amounts of fine aggregate passing the 300 and 150 μm sieves affect workability, surface

Table 3.5 Relationship between cumulative and each sieve residue.

Sieve size/mm	Each/%	Cumulative/mm
2.36	a_1	$A_1 = a_1$
1.18	a_2	$A_2 = a_1 + a_2$
0.60	a_3	$A_3 = a_1 + a_2 + a_3$
0.30	a_4	$A_4 = a_1 + a_2 + a_3 + a_4$
0.15	a_5	$A_5 = a_1 + a_2 + a_3 + a_4 + a_5$
<0.075	a_6	$A_6 = a_1 + a_2 + a_3 + a_4 + a_5 + a_6$

texture, air content, and bleeding of concrete. Most specifications allow 5%–30% to pass the 300 μm sieve. The lower limit may be sufficient for easy placing conditions or where concrete is mechanically finished, such as in pavements. However, for hand-finished concrete floors, or where smooth surface texture is desired, fine aggregate with at least 15% passing the 300 μm sieve and 3% or more passing the 150 μm sieve should be used.

The most desirable fine-aggregate grading depends on the type of application, the paste content, and the maximum size of coarse aggregate. In leaner mixtures, or when small-size coarse aggregates are used, a grading that approaches the maximum recommended percentage passing each sieve is desirable for workability. In general, if the water-to-cement ratio (w/c) is kept constant, and the ratio of fine-to-coarse aggregate is suitable, a wide range in grading can be used without a measurable effect on strength. However, the best economy will sometimes be achieved by adjusting the concrete mixture proportions to suit the gradation of the local aggregates. It can be observed from Fig. 3.10, the voids between aggregates can be well controlled by adjusting the aggregate gradation.

3.3.3.6 The maximum size of aggregate

The maximum size of aggregate is conventionally designated by the smallest sieve that 100% of a particular aggregate must pass through. The nominal maximum size of an aggregate is the smallest sieve size through which the major portion of the aggregate must pass (typically 85%–95% passing). The nominal maximum-size sieve may retain 5%–15% of the aggregate, depending on the size number.

The maximum size of a coarse aggregate used in concrete has a direct influence on the economy, workability, strength, shrinkage, and permeability of concrete. The larger the maximum aggregate size, the smaller will be the surface area per unit volume, which has to be covered by the cement

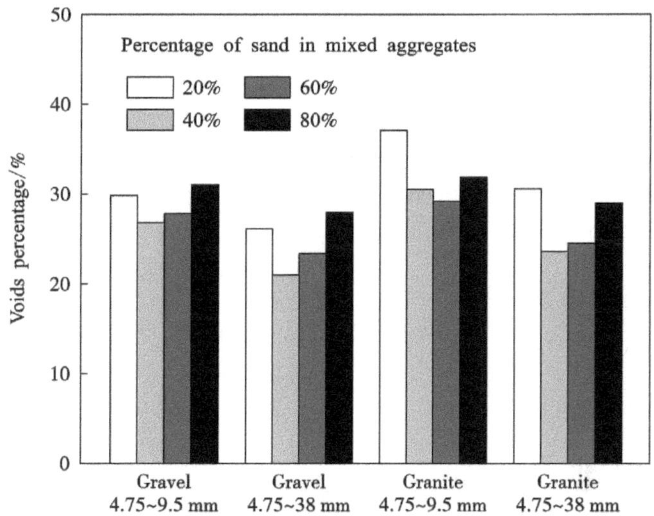

Figure 3.10 Effect of a percentage of sand in mixed aggregates on a percentage of voids.

paste of a given water-to-cement ratio. Generally, mixtures with large maximum size of coarse aggregate tend to produce concrete with better workability, probably because of the decrease in the specific surface. However, there is an optimal maximum size of coarse aggregate that produces the highest strength for a given consistency and cement content. Because the price of cement is much high than aggregate, any action that saves cement without reducing the strength and workability of concrete can result in significant economic benefit.

Many other factors also govern the choice of maximum aggregate size for a concrete mixture. The maximum size of aggregate that can be used generally depends on the size and shape of the concrete member and on the amount and distribution of reinforcing steel. Requirements for limits on the nominal maximum size of aggregate particles are covered by ACI 318. The nominal maximum size of aggregate should not exceed

① One-fifth the narrowest dimension of a vertical concrete member: $D_{max} = 1/5$ B

② Three-quarters the clear spacing between reinforcing bars and between the reinforcing bars and forms: $D_{max} = 3/4$ S, and 3/4 C

③ One-third the depth of slabs: $D_{max} = 1/3$ T.

As large particles tend to produce more microcracks in the interfacial transition zone (ITZ) between the coarse aggregate and cement paste, the maximum aggregate size is generally limited to 19 mm for high-strength concrete mixtures. Similarly, aggregate grading also has a considerable effect on the cement paste requirement of a concrete mixture. In practice, low void contents are achieved by using smoothly graded coarse aggregates with suitable proportions of graded sand.

3.3.3.7 Absorption

Aggregates are also porous materials. They contain some pores which can absorb and hold water. The internal structure of an aggregate particle is made up of solid matter and voids that may or may not contain water. Aggregate porosity may affect durability as freezing of water in pores in aggregate particles can cause surface popouts. It should be noted that absorption can be used as an initial indicator of soundness. Furthermore, aggregates with low absorption tend to reduce shrinkage and creep. The coarse and fine aggregates will generally have absorption levels in the range of 0.2%—4% and 0.2%—2%, respectively. Free-water contents usually range from 0.5% to 2% for coarse aggregate and 2%—6% for fine aggregate.

Before concrete mixing, the aggregate can, therefore, be in one of the four moisture conditions, as shown in Fig. 3.11.

Oven dry—Zero moisture content, fully absorbent;

Air dry—Dry at the particle surface but containing some interior moisture, less than potential absorption;

Figure 3.11 Moisture states of aggregates.

Saturated surface dry (SSD)—Neither absorbing water from nor contributing water to the concrete mixture, equal to potential absorption;

Damp or wet—Containing an excess of moisture on the surface (free water).

In the fresh concrete, aggregate that is in either of conditions of oven-dry or air-dry will absorb some of the mixing water, and aggregate in the condition of surface moisture will contribute water to it. Saturated surface dry aggregate is perhaps most desirable in practice, but is difficult to achieve except in the laboratory.

Absorption of lightweight aggregate is of prime importance to the subsequent concrete properties, since it determines the amount of water available for cement hydration, i.e., the amount that is nonabsorbed or "free"; therefore, to ensure that the required free w/c ratio is obtained, it is necessary to allow for the aggregate moisture condition when calculating the amount of mix water. If the aggregate is drier than saturated surface dry, extra water must be added, and vice versa.

3.3.3.8 Density

Natural aggregates are porous, and its porosity ranges from 2% to 40%. For instance, the porosity of intrusive igneous rocks maybe 2%, and maybe up to 5% for dense sedimentary rocks, and 10% to 40% for very porous sandstones and limestone. The volume occupied by the aggregate particles, including the pores existing within the particles, is important to the concrete mix design. This is called apparent specific gravity, which is relatively constant for many commonly used rocks, ranging between 2.6 and 2.7. For instance, the typical apparent specific density for granite, sandstone, and dense limestone are 2.69, 2.65, and 2.60, respectively.

In addition to apparent specific density, another density called bulk density is needed for particulate materials. It is defined as the mass of the aggregate fragments that would fill a unit volume. Experimentally, the determination of bulk density is by adding aggregate particles into a container with known volume, and the bulk density equals the mass of aggregate divided by the known volume. Apparently, bulk density includes the void content between particles, which affects paste requirements for concrete. Void contents range from about 30% to 45% for coarse aggregates to about 40%—50% for fine aggregate. Angularity increases void content, while larger sizes of well-graded (uniform) aggregate and improved grading decrease void content. The approximate bulk density of aggregates commonly used in normal-weight concrete ranges from 1300 to 1750 kg/m^3.

The relative density of an aggregate is the ratio of its mass to the mass of an equal absolute volume of water. Most natural aggregates have relative densities between 2.40 and 2.90, which means that the aggregates are 2.40–2.90 times as dense as water. Relative density is not generally used as a measure of aggregate quality.

3.3.3.9 Soundness

Soundness is the aggregate resistance to weathering that primarily includes resistance to freezing-thawing cycles (FT), and to a lesser extent, resistance to wetting and drying; and heating and cooling. A sound aggregate has a satisfactory durability factor when used in properly mature concrete with enough air void content. An aggregate is considered to be unsound when the volume changes in aggregate induced by weathering, resulting in the deterioration of concrete. Rocks with a certain characteristic pore structure tend to be unsound. It is also found that concretes containing some cherts, shales, limestones, and sandstones have been found susceptible to damage by frost action or by salt crystallization within the aggregate particle. However, high moisture absorption is not the direct index for unsoundness, because some aggregates such as pumice and expanded clays can absorb large amounts of water but remain sound. It is generally believed that unsoundness of aggregate is related to its pore size distribution rather than to its total porosity. A pore size distribution that allows the aggregate particles to get saturated on wetting (or thawing in the case of frost attack), but prevent easy drainage on drying (or freezing), is capable of causing high hydraulic pressure within the aggregate particles. The soundness of aggregate may be estimated by ASTM C 88 "Soundness of Aggregate by Sodium Sulfate or Magnesium Sulfate," by AASHTO T 103 "Soundness of Aggregates by Freezing and Thawing" or by petrographic examination.

As per ASTM Method C 88, a standard procedure is proposed to directly determine the resistance of aggregate to disintegration on exposure to a saturated sodium or magnesium sulfate solution for five wet-dry cycles.

3.3.3.10 Mechanical properties

The strength of aggregate generally does not influence the strength of conventional concrete as much as the strength of the paste and the paste-aggregate bond. For (ultra) high-strength concrete, aggregate strength does become important. The compressive strength of aggregate can be determined by a cubic or cylindrical specimen; it depends on the characteristics of aggregates and may range from 65 to 270 MPa. In case that there is no cubic or cylindrical specimen, and only aggregate particles are

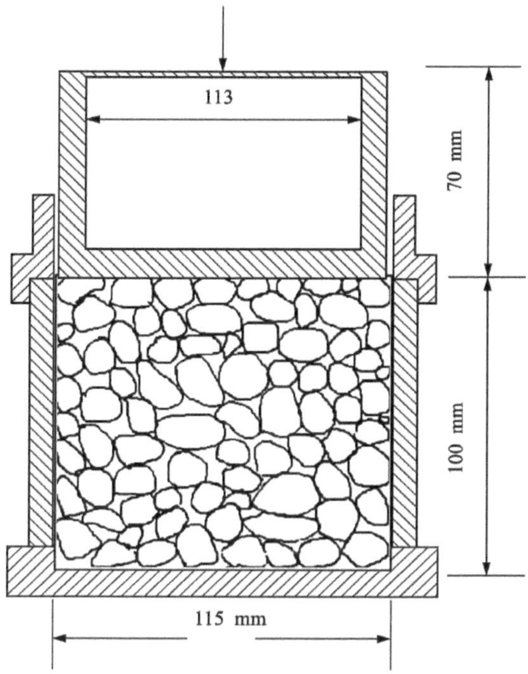

Figure 3.12 Crushing strength test.

available, the strength of aggregates can be evaluated by crushing index, the schematic diagram for this test is given in Fig. 3.12. Crushing index is often used in practice, instead of the strength of cubic or cylindrical specimen.

In this test, a certain amount of aggregate particles is put in a steel-made container. A specified loading of 200 KN is loaded on the particles, and is held for 5 s. Some aggregate particles may be crushed into smaller particles. The crushing index can be obtained by calculating the percentage of the weight of particles greater than 2.36 mm to the initial weight of samples.

Tensile strength of aggregate is seldomly used in practice, its tensile strength ranges from 2 to 15 MPa, and often is determined by Brazil splitting test. Typical values of dynamic elastic modulus for most granite, basalt, trap rock, flint, quartzitic sandstone, and dense limestone aggregates are in the range of 70–90 GPa.

3.3.3.11 Deleterious substances

Deleterious substances are those presenting as minor constituents of either fine or coarse aggregates; the substances are detrimental to the workability, setting and hardening, and durability of concrete. The deleterious

Table 3.6 Harmful materials in aggregates.

Substances	Effect on concrete	Test designation
Organic impurities	Affects setting and hardening, may cause deterioration	ASTM C40 (AASHTO T 21) ASTM C87 (AASHTO T 71)
Materials finer than the 75 μm (No.200) sieve	Affects bond, increase water requirement	ASTM C117 (AASHTO T 11)
Coal, lignite, or other lightweight materials	Affects durability, may cause stains and popouts	ASTM C123 (AASHTO T 113)
Soft particles	Affects durability	ASTM C235
Clay lumps and friable particles	Affects workability and durability, may cause stains and popouts	ASTM C142 (AASHTO T 112)
Chert of less than 2.40 relative density	Affects durability, may cause popouts	ASTM C123 (AASHTO T 113) ASTM C295
Alkali-reactive aggregate	Causes abnormal expansion, map cracking, and popouts	ASTM C227 ASTM C289 ASTM C295 ASTM C342 ASTM C586 ASTM C1260 (AASHTO T 303) ASTM C1293 ASTM C1567

substances may include organic impurities, silt, clay, shale, iron oxide, coal, lignite, mica, and certain lightweight and soft particles (Table 3.6).

Aggregates are potentially harmful if they contain substances known to react chemically with cement paste or interfere with the action of chemical admixture and produce any of the following consequences: ① significant volume changes of the paste, aggregates, or both; ② interference with the normal hydration of cement; and ③ interfere with the action of chemical admixtures, especially SPs.

Organic impurities may delay setting and hardening of concrete, interfere with the interaction of chemical admixtures, reduce strength gain, and probably cause deterioration.

Materials finer than the 0.075 mm, especially silt and clay, may be present as loose dust and may form a coating on the aggregate particles. Even thin coatings of silt or clay on gravel particles can be harmful to the bonding between the cement paste and aggregate. This significantly reduces the strength and durability of concrete. If certain types of silt or clay are present in excessive amounts, water requirements may increase significantly. There is a tendency for some fine aggregates to degrade from the grinding action in a concrete mixer. This effect may alter mixing water, air-entraining admixtures, and slump requirements.

Coal or lignite, or other low-density materials such as wood or fibrous materials, in excessive amounts, will affect the durability of concrete. If these impurities occur at or near the surface, they might disintegrate, pop out, or cause stains. Clay lumps present in concrete may absorb some of the mixing water, cause popouts in hardened concrete, and affect durability and wear resistance. They can also break up during mixing and thereby increase the mixing-water demand. Mica is composed of flat plates that can either act as cleavage planes in coarse aggregate or as severely flat particles in fine aggregate. Either particle size can result in a decrease in concrete strength. The test for mica content uses hazardous materials and is not commonly performed.

3.3.4 Green aggregate

For the sake of environmental protection, there is a growing interest in using waste materials as alternative aggregate materials and significant research is made on the use of many different materials as aggregate substitutes such as coal ash, blast furnace slag, waste plastics, rubber waste, sintered sludge pellets, and others. The consumption of waste materials can be increased manifold if these are used as aggregate into cement mortar and concrete. The use of waste materials is beneficial to both economy and ecology. It can solve problems of lack of aggregate in various construction sites and reduce environmental problems related to aggregate mining and waste disposal. The use of waste aggregates can also reduce the cost of the concrete production.

Depending on their generation, wastes can be separated into two types: wastes directly are obtained from the industry as industrial by-products and some wastes which can be recycled. The first type includes coal ash, slags from metal industries, industrial sludge, waste from industries like pulp and paper mills, mine tailings, food and agriculture, and leather. The second type includes different plastic and rubber wastes. A broad classification of industrial waste aggregate can be made depending on the chemical nature of wastes. Some waste aggregates come from the production and use of organic materials. Plastics, rubber, leather, and some food industries wastes are organic wastes. On the other hand, industrial slags, mining wastes, coal industry wastes, and others are inorganic wastes. Glass reinforced plastics and some industrial sludge may contain both organic and inorganic materials. Another classification of industrial waste aggregate can be done depending on the weight of waste aggregates. Some aggregates are lightweight by nature. Plastics, rubber, most food and agricultural industries wastes, and coal bottom ash are of this kind. On the other hand, most of the industrial slags are heavier than conventional aggregates. Since the aggregates significantly affect the properties of concrete, a thorough evaluation is necessary before using any waste material as aggregate in concrete, not only from the perspective of concrete technology but also from the perspective of environmental protection.

It can be expected that there are many old constructions and will be demolished to make new buildings. Due to these large-scale demolitions, a huge amount of debris is generated all over the world. The use of recycled aggregate in concrete is of great environmental and economic benefits. It becomes more and more popular all around the world.

3.3.5 Supplementary cementing materials

Supplementary cementing materials, also called supplementary cementitious materials (SCMs), have become more popular over recent decades, because of increasingly stringent environmental legislation. In addition, SCMs are used in conjunction with Portland or blended cement, having a positive effect on the properties of concrete through hydraulic or pozzolanic activity or both. The most common SCMs are fly ash, a by-product of coal-fired power plants, ground granulated blast furnace slag, a by-product of steelmaking, and silica fume, a by-product of industrial electric arc furnaces. SCMs are added to the concrete as part of the total cementitious system. The use of SCMs in concrete reduces the amount of Portland cement, as the mineral admixtures act as a partial cement replacement. On the one hand, this reduces the cement production, an energetically expensive and environmentally problematic process; on the other hand, the industrial wastes are well used. SCMs can be either preblended with the cement during its production for sale and used as a blended cement or mixed directly with other components when the concrete is produced.

3.3.5.1 Fly ash

Fly ash, also known as pulverized fuel ash (PFA), is a pozzolanic material. Fly ash is a by-product of the power industry. During combustion, the coal's mineral impurities (such as clay, feldspar, quartz, and shale) fuse in suspension and are carried away from the combustion chamber by the exhaust gases. As the fused material rises with the gas flow, it cools and solidifies into spherical glassy particles called fly ash. The diameter of fly ash particles usually ranges between 1 and 150 μm. Fly ash produced from thermal power plants is a variable material because of several factors: ① type and mineralogical composition of the coal, ② degree of coal pulverization, ③ type of furnace and oxidation conditions, and ④ the manner in which fly ash is collected, handled, and stored before use. Consequently, fly ash is a material with high variation in its compositions and concrete suppliers need to be even more vigilant in assessing the quality of this material before incorporating it into the concrete. Fly ash can react with portlandite in the presence of water to form cementitious compounds. As portlandite is the main hydration product of cement and is present as reactive calcium hydroxide, there is a distinct attraction in adding fly ash to concrete.

There are two types of fly ash, as per the classification in ASTM 618, i.e., Class F and Class C. Class F fly ash has pozzolanic properties, silica (in the presence of SiO_2) being the most important chemical compositions, and alumina (Al_2O_3) and iron oxide (Fe_2O_3) are also active in the presence of reactive calcium hydroxide. Therefore, the sum of these three chemical compositions required are 70% for Class F, and 50% for Class C fly ash, as shown in Table 3.7. Class C ash contains a higher amount of calcium compounds and may have some minor hydraulic cementing value in the absence of cement. Certainly, it is possible to use Class C fly ash in larger proportion than Class F fly ash, due to its higher reactivity. Class C ash may be less effective than Class F ash in providing sulfate resistance and reducing

Table 3.7 Requirements for fly ash as SCM in Portland cement concrete as per ASTM C 618.

	Fly ash classification	
Requirements	**F**	**C**
Chemical requirements $SiO_2 + Al_2O_3 + Fe_2O_3$, min/(%)	70.0	50.0
SO_3, max/(%)	5.0	5.0
Moisture content, max/(%)	3.0	3.0
Loss of ignition, max/(%)	6.0	6.0
Physical requirements. Amount retained when wet sieved on 45 μm sieve, max/(%)	34	34
Pozzolanic activity index, with Portland cement at 28 d, min/(%)of control	75	75
Pozzolanity activity index with lime at 7 d, min/(MPa)	5.5	-
Water requirements, max/(%)of control	105	105
Autoclave expansion or contraction, max/(%)	0.8	0.8
Specific gravity, max variation from average	5	5
Percentage retained on 45 μm sieve, max variation, percentage points from average	5	5

alkali—silica reaction. It is possible to use high replacement levels of Type F fly ash as in the case of roller-compacted concrete and massive elements, since fly ash can significantly reduce the heat liberation of concrete. Depending on the environmental temperature and replacement level of fly ash, it will take a longer time to develop strength.

Carbon is the most important impurity as it can inhibit the action of admixtures, particularly air-entraining admixtures. It is measured by loss on ignition, which should not exceed 6% for both class F and Class C fly ash. However, it is worth to mention that rice husk ash containing up to 23% of carbon was reported to be successfully used in particular circumstances, so possibly higher percentages in fly ash would not necessarily render it useless in all circumstances. In practice, the consistency of carbon content is more important than carbon content, since it has an important influence on the air entrainment. Other impurities are alkalis and magnesium, which need to be limited in cement but are not often a problem.

There are three kinds of effect for the incorporation of fly ash in concrete. They are physical effects on both fresh and hardened concrete, chemical effects on the setting process and hardened concrete, and physical chemistry effects on the setting process.

3.3.5.1.1 Physical effects

The fly ash particles are very similar in size and shape to entrained air bubbles (see Fig. 3.13), and thus has the following effects:

① Water reduction.

② Reduction of bleeding.

③ Improved cohesion and plasticity.

Figure 3.13 Fly ash particles.

④ Improved pumpability.
⑤ Reduced slump loss with time.

3.3.5.1.2 Chemical effect

By reacting with portlandite which is the reaction product of cement, fly ash substantially reduces permeability in the hardened concrete if properly cured. This reaction occurs after cement hydration and generates less heat during hydration. This is generally a valuable property in hot climates and for mass concrete, but maybe a distinct disadvantage in colder climates. The pozzolanic activity of fly ash depends on its ① fineness; ② calcium content; ③ structure; ④ specific surface; ⑤ particle size distribution; and ⑥ loss of ignition (LOI) content.

3.3.5.1.3 Surface chemistry effect

It appears that fly ash can act as a catalyst or a nucleation site for crystal growth in the cement paste.

Advantages of fly ash

① Reduce the heat of hydration in the critical period.
② Allow for more workable fresh concrete, easier to pump, compact, trowel, less bleeding, and segregation usually gives a better off-form surface.
③ Substantially reduced permeability in long-term age.
④ Allows for more durable concrete, which is more resistant to alkali—silica reaction, chloride penetration, and sulfate attack than concrete with Portland cement.
⑤ Allows for higher strengths.
⑥ Cheaper than straight cement in most parts of the world.
⑦ Fly ash concrete reacts extremely well to steam curing with little or no detrimental effect, unlike Portland cement—only concrete.

Disadvantages of fly ash

① Long time for strength development.
② Reduce alkalinity of concrete, and accelerate the rate of carbonation.
③ Coal-fired power plants produce power, not fly ash. The quality can vary.

In summary, the use of a proportion of fly ash is generally desirable. The exceptions are when (a) high early strength is required, (b) heat generation is advantageous, and (c) especially with strength grades below 30 MPa where adequate curing is uncertain and corrosion protection of reinforcement is required. Where fly ash is used, care must be taken to ensure that reported strengths are realistic and not the result of assuming that water-cured cylinders necessarily correctly represent poorly cured in situ concrete.

The circumstances in which it may be worthwhile specifying that fly ash be used would include hot weather concreting, large sections where low heat cement or ice might otherwise be needed, projects in which exceptionally high strength or good pumpability is needed, and projects where high sulfate resistance is needed.

3.3.5.2 Blast-furnace slag

In the production of cast iron, a large number of by-products are yielded, which is called iron blast furnace slag (or slag). It forms by fusion of the gangue, i.e., nonferrous material of the iron ore, mainly silica and alumina compounds, with calcium and magnesium oxides of the thermally decomposed carbonatic flux and combustion residues of the coke. These reactions take place at temperatures between 1300 and 1600°C. The slag floats on top of the liquid iron and is tapped at regular intervals. When leaving the blast furnace, the molten slag has a temperature of around 1450°C and is cooled down, either slowly in contact with air in open pits or rapidly through pelletization or granulation.

When the slag is cooled slowly in the air, the mineral components are usually present as crystalline melilites that do not react with water at ordinary temperature. When ground to very fine particles, the material will be weakly cementitious and pozzolanic. However, if the liquid slag is rapidly quenched from a high temperature by either water or a combination of air and water, most of the lime, magnesia, silica, and alumina are held in a noncrystalline or glassy state. The water-quenched product is called granulated slag due to the sand-size particles; the material quenched by air and a limited amount of water is in the form of pellets and is called pelletized slag. Normally, the former contains more glass. To use as a cementitious material, the granulated slag must be ground as fine or finer than cement. The fineness of grind, together with the chemical composition and extent of glassiness, will determine how rapidly the slag will react in concrete.

Slag is much more reactive than fly ash, due to its chemical composition and production way. Its chemical composition is quite similar to that of cement, as shown in Table 3.8. In general, the reactivity of slag increases with increasing contents of CaO, Na_2O, and Al_2O_3 and with decreasing contents of SiO_2, FeO, TiO_2, MnO. A high CaO/SiO_2 ratio generally indicates a good reactivity. Elevated alumina contents in the slag are favorable for strength development of slag cement in the presence of sulfate, especially at early ages due to enhanced ettringite formation.

Slag cannot be used alone to make concrete but can be used to replace Portland cement in a very high percentage. Slag is either used as a cement constituent replacing Portland clinker and/or as concrete constituent

Table 3.8 Typical chemical composition of slag.

Parameter	Min	Max
SiO_2/%	32	42
CaO/%	35	48
Al_2O_3/%	6	>19
MgO/%	3	14
TiO_2/%	0.2	>2
Na_2O_{eq}/%	0.3	1.2
SO_3/%	1	4
CaO/SiO_2	0.9	1.3
Glass content/%	66	100
Bulk density/kg/L^{-1}	0.6	1.3

replacing cement. Both applications have their advantages and disadvantages. The direct use of SCM in concrete gives the concrete producer maximum flexibility for designing concretes for various applications with a minimum number of silos for cement and SCM. The prerequisite is adequate knowledge about concrete design and performance. A disadvantage of direct SCM addition to concrete is that the performance of the cementitious part cannot be optimized.

The kind of slag utilization differs locally. Typical slag contents in concrete range from 20% to 80% of the total cementitious material. The replacement level of cement by slag depends on the following requirements: ① strength development, ② durability, e.g., resistance to ASR, or attack by sulfates, seawater, or other chemicals or frost and deicing salt, ③ heat of hydration, and ④ setting time.

In Portland blast-furnace cement, the slag may be interground with the cement clinker or added as a separate material. The Portland cement clinker is softer than the slag and therefore will be more finely ground when the materials are interground. Even when sold as a composite "blended cement" (the term is also applied to other blends), the slag cement may have been either interground or postblended.

Concrete using slag cement will tend to develop early strength more slowly than pure Portland cement concrete except when very finely ground. However, if thoroughly cured, it may have as good or better eventual strength. It normally has a greater resistance to chemical attack and is particularly suitable for marine environments. Its normally greater fineness may confer resistance to bleeding in the fresh state and lower permeability when hardened.

Slag may contain small quantities of heavy metals. However, in comparison with Portland cement, slag usually has lower heavy metal content.

3.3.5.3 *Silica fume*

Silica fume (also known as microsilica or condensed silica fume) is a by-product of the manufacture of silicon, ferrosilicon, or the like, from quartz and carbon in electric arc furnaces. Silica fume is produced during the high-temperature reduction of quartz in electric arc furnaces, where the main products are silicon or ferrosilicon alloys. The high-purity quartz is heated to 2000°C in an electric arc furnace with coal, coke, or wood chips added to remove the oxygen. The alloy is collected at the bottom of the furnace. As the quartz is reduced to the alloy, it releases silicon monoxide vapor. In the upper parts of the furnace, this fume oxidizes and condenses into microspheres of amorphous silica (silicon dioxide). The fumes are drawn from the furnace by powerful fans, often through a recollector and cyclone which removes the larger coarse particles of unburned wood or carbon, and are then blown into a series of special filter bags.

It is usually more than 90% pure silicon dioxide, and is a superfine material with a particle size of the order of 0.1 μm and a surface area of over 15,000 m^2/kg (i.e., a 100 times greater than cement or fly ash). Its relative density is similar to that of fly ash at about 2.3. Due to its extreme fineness, it has a very low bulk density of only 200–250 kg/m^3 in its loose form. For this reason, it is usually handled either in a condensed form. In the condensed form, particles are agglomerated by aeration and the bulk density increases to 500–700 kg/m^3.

A very comprehensive guide is available from the American Concrete Institute: ACI 234.06R "Guide to the Use of Silica Fume in Concrete" or Chinese standard GB/T 18736. There is a tendency for silica fume to be regarded as only justified for very high-strength concrete, but this is far from the truth. Its uses are many and varied. It can provide significant reductions in permeability, increased resistivity, and increased durability, and its effects on the properties of fresh concrete are more important for many uses than its effect on hardened properties. These effects include a very substantial increase in cohesion and an almost complete suppression of bleeding or any other form of water movement through concrete (in either the fresh or hardened state). While the suppression of bleeding is desirable in many ways, it does cause exposed flat surfaces of fresh concrete to be very susceptible to evaporation cracking. At low replacement levels (2%–3%), silica fume is a useful pumping aid.

The contribution of silica fume to concrete can be summarized as follows:

① Cementitious effect—silica fume can be used to replace cement up to 25% due to its pozzolanic reactivity. This means significant reductions in Portland cement for a given strength.

② Increased strength can be achieved using silica fume as an addition, and if incorporated into the engineering design, savings in concrete volume can be achieved. Increased strength can also extend a lifetime.
③ Using silica fume to augment multiple blends (OPC with high fly ash or slag volumes) can negate the delay effects usually seen—enabling workable strengths and durability factors at earlier ages.
④ The increased durability factors associated with the incorporation of silica fume into a concrete (water resistance, chloride, sulfate, and ASR resistance) can greatly extend lifetime performance. It is an indispensable component for ultrahigh-performance concrete (UHPC).

3.3.5.4 Metakaolin

Kaolinite, $Al_2Si_2O_5(OH)_4$, is a clay mineral that provides the basis of the plasticity of traditional ceramics in the hydrous state. The name of kaolin is derived from the Chinese term "Kao-ling" meaning high ridge, the name of a hill in the city of Jingdezhen, where this material was mined centuries ago for use in ceramic production. Structurally, kaolinite consists of octahedral alumina sheets and tetrahedral silica sheets stacked alternately, and takes on a platelike morphology.

Metakaolin (MK), which has the approximate chemical formula Al_2O_3 $2SiO_2$ or abbreviated as AS_2, is produced by heating kaolin-containing clays within the temperature range of about 600–900°C. Kaolinite crystals with hexagonal morphology tend to be preserved when heating kaolin at a temperature of 550°C. Heating kaolin above 550°C transforms it into MK by the loss of structural OH groups and a rearrangement in Si and Al atoms leading to a decrease in octahedral Al and the appearance of penta- and tetra-coordinated Al. As the temperature increases (650 and 800°C), the kaolinite flakes become more deformed and locally condensed into disordered material, although the original layered structure of the kaolinite is at least partially retained. Above 900–950°C, MK begins to recrystallize to a siliceous spinel and disordered silica; at higher temperatures, crystalline mullite is generated.

The typical chemical compositions of MK are given in Table 3.9. MK can be considered as a highly pozzolanic material and appears to have excellent potential as supplementary cementing material for high-performance cementitious materials. The optimum level of cement replacement seems to be around 10%–20%, where maximum strength is observed. Incorporation of high reactivity metakaolin as a partial cement replacement between 10% and 15% may also be sufficient to control deleterious expansion due to alkali–silica reaction in concrete, depending on the nature of the aggregate. The addition of metakaolin reduces the ingress of chloride by improving the microstructure and chloride binding behavior.

Table 3.9 The typical chemical compositions of MK.

Component	By weight/%
SiO_2	51.5
Al_2O_3	40.2
Fe_2O_3	1.2
CaO	2.0
MgO	0.1
K_2O	0.5
SO_3	0.0
TiO_2	2.3
Na_2O	0.1
L.O.I	2.0

3.3.5.5 Natural pozzolans

Natural pozzolans can be subdivided into two categories as materials of volcanic origin and materials of sedimentary origin. The first category includes materials formed by the quenching of molten magma when it is projected to the atmosphere upon explosive volcanic eruptions. The explosive eruption has two consequences: ① The gases, originally dissolved in the magma are released by the sudden decrease of pressure. This causes a microporous structure in the resulting material. ② Rapid cooling of the molten magma particles when contacted with the atmosphere results in quenching, which is responsible for the glassy state of the solidified material. Pozzolanic materials of volcanic origin may be found in loose (incoherent) or compacted (coherent) forms in nature. The latter results from the postdepositional processes such as weathering, compaction, cementation, and hardening of the originally loose material. These processes may change the original structure into clayey or zeolitic character. Transformation into clayey structure reduces the pozzolanicity, whereas zeolitization improves it. The second category of natural pozzolans includes clays and diatomaceous earth. Clays have very limited pozzolanic reactivity unless they are thermally treated. Diatomaceous earth, which is a sedimentary rock, consists basically of the fossilized remains of diatoms (a type of algae). It has an amorphous siliceous structure but may contain crystalline phases up to 30%, by mass.

The reactive chemical compositions of natural pozzolans are silica (SiO_2), alumina (Al_2O_3), and iron oxide (Fe_2O_3). The sum of these three oxides is required to a minimum value of 70% by mass for a suitable pozzolan. The chemical compositions of natural pozzolans vary much from location to location. Similar to the variability in chemical compositions, many different minerals may be present in natural pozzolans. However, the fundamental constituent is always a glassy phase of siliceous nature which can be distinguished by XRD techniques. Glass content of natural pozzolans of volcanic origin generally varies from 50% to 97% and the rest are mostly clay minerals, quartz, and feldspars. Glassy phase in diatomaceous earths may be as low as 25%, and some may be almost totally glassy. The remainder is composed of clay minerals, quartz, and feldspars.

The use of natural pozzolans is now accepted by modern cement and concrete industry as supplementary cementitious materials due to the many advantages they offer apart from the obvious environmental benefit of supplementing clinker or cement which require a lot of energy and emit high quantities of CO_2 during their production. They are responsible for higher compressive strength in later ages of cement hydration which continues to increase for a long period of time; they provide self-healing properties to the concrete and significantly improved durability which is the main reason for their use. Particularly, the use of natural pozzolans demonstrates lower chloride migration and improved sulfate resistance, while in most cases, their addition leads to mitigation of ASR. Moreover, they reduce bleeding and segregation considerably, lower the heat of hydration which is beneficial in large concrete structures, and eliminate microcracking. However, their use can also bring some side effects, such as workability loss, and higher water and SP demands, and higher shrinkage.

It is important to note that each natural pozzolan has different effects on cement and concrete properties and can be used in different ratios in a concrete mix. Overall, the benefits of using pozzolans in concrete are well established in most parts of the world, leading to extensive research and application in construction worldwide.

In summary, SCMs have great effects on the fresh and hardened properties of concrete. Table 3.10 summarize general effects that these materials have on the fresh and hardened properties of concrete. It should be noted that these effects vary considerably between and often within classifications of SCMs. Specified performance should be evaluated for specific mixtures.

Table 3.10 The impact of SCM characteristics on the fresh properties of concrete.

	Fly ash				Natural pozzolans		
	Class F	Class C	Slag cement	Silica fume	Calcined shale	Calcined clay	Metakaolin
Water demand	⇩	⇩	⇩	⇧	⇔	⇔	⇧
Workability	⇧	⇧	⇧	⇩	⇧	⇧	⇩
Bleeding and segregation	⇩	⇩	⇳	⇩	⇔	⇔	⇩
Setting time	⇧	⇳	⇧	⇔	⇔	⇔	⇔
Air content	⇩	⇩	⇔	⇩	⇔	⇔	⇩
Heat of hydration	⇩	⇳	⇩	⇔	⇩	⇩	⇔
Early age strength gain	⇩	⇔	⇳	⇧	⇩	⇩	⇧
Long-term strength gain	⇧	⇧	⇧	⇧	⇧	⇧	⇧
Abrasion resistance	⇔	⇔	⇔	⇔	⇔	⇔	⇔
Drying shrinkage and creep	⇔	⇔	⇔	⇔	⇔	⇔	⇔
Permeability and absorption	⇩	⇩	⇩	⇩	⇩	⇩	⇩
Corrosion resistance	⇧	⇧	⇧	⇧	⇧	⇧	⇧
Alkali–silica reactivity	⇩	⇩	⇩	⇩	⇩	⇩	⇩
Sulfate resistance	⇧	⇳	⇧	⇧	⇧	⇧	⇧
Freezing and thawing	⇔	⇔	⇔	⇔	⇔	⇔	⇔
Deicer scaling resistance	⇔ or ⇩	⇔ or ⇩	⇔ or ⇩	⇔ or ⇩	⇔ or ⇩	⇔ or ⇩	⇔ or ⇩

Note: ⇩, Lowers; ⇧, Increases; ⇳, May increase or lower; ⇔, No impact; ⇔ or ⇩, may lower or have no impact.

3.3.6 Chemical admixtures

Several decades ago, the improvement of chemical admixtures in both the fresh and hardened concrete has been long recognized. Tens of types of chemical admixtures are being used in modern concrete. In some countries, chemical admixtures have become a dispensable ingredient for concrete, and almost all concrete produced contains one or more admixtures. Therefore, construction engineers should have some knowledge of the advantages and limitations of commonly used admixtures.

Chemical admixtures are chemicals that are added to the concrete immediately before or during mixing and significantly change its fresh, early age or hardened state to economic or physical advantage. Only small quantities are required, typically lower than 5% by weight of the cement. The effectiveness of a chemical admixture depends up many factors such as the admixture itself (molecular structure, molecular weight, addition rate, time of addition), concrete material (cement type, water content, aggregate shape, gradation, slump, and proportions), and processing factors (mixing way, mixing time, and temperature).

Chemical admixtures vary widely in composition, from surfactants and soluble salts to polymers and insoluble minerals. Generally, they are used in concrete for different purposes, such as improvement of workability, setting time adjusting agent, strength development controlling agent, and improvement of durability (FT action, alkali-aggregate expansion, sulfate attack, and corrosion of the reinforcement). In this section, only four distinct types of chemical admixture are introduced, namely (super)plasticizers, set controlling agents and air-entraining/detraining agents, viscosity enhancing agent, as presented in Table 3.11 Since these chemical admixtures are most often used in practice.

3.3.6.1 Superplasticizers

Plasticizers, also called water-reducing admixtures, are a kind of workability aids, which increase the fluidity or workability of a cement paste or concrete. There are natural and synthetic ones available. The first use of plasticizer may date back to 1930s. Lignosulfonates were first used as plasticizers and water reducers, and ready-mix concrete represents their largest application. Although the performance of natural polymers is limited compared to synthetic ones, it is worth mentioning that they are still widely

Table 3.11 Concrete admixtures by classification.

Type of admixtures		Desired effect	Material
Setting controlling	Accelerators	Accelerate setting and early strength development	Calcium chloride (ASTM D98 and AASHTO M 144) Triethanolamine, sodium thiocyanate, calcium nitrite
	Retarding admixtures	Retard setting time	Lignin, borax, sugars, tartaric acid, and salts
Air controlling agent	Air detrainers	Decrease air content	Tributyl phosphate, dibutyl phthalate, octyl alcohol, water-insoluble esters of carbonic and boric acid, silicones
	Air-entraining admixtures	Improve durability in freeze–thaw, deicer, sulfate, and alkali-reactive environments Improve workability	Salts of wood resins (Vinsol resin), some synthetic detergents, salts of sulfonated lignin, salts of petroleum acids, salts of proteinaceous material. Fatty and resinous acids and their salts, alkylbenzene sulfonates, salts of sulfonated hydrocarbons
	Foaming agents	Produce lightweight, foamed concrete with low density	Cationic and anionic surfactants Hydrolized protein
Viscosity-enhancing agent		Cohesive concrete for underwater placements	Cellulose, acrylic polymer
Natural plastisizer		Increase flowability of concrete Reduce water–cement ratio	Lignosulfonates Casein
Synthetic superplasticizers		Significantly increase flowability of concrete Significantly reduce water–cement ratio	Sulfonated melamine formaldehyde condensates Sulfonated naphthalene formaldehyde condensates Polycarboxylates

used in the concrete industry mainly because of their low cost of production. Synthetic ones include some of the most commonly used SPs, such as polynaphthalene sulfonates (PNS), polymelamine sulfonates (PMS), and vinyl copolymers. The higher dispersing ability of these compounds qualifies them as true SP or high-range water reducers. Thanks to the formulation and large-scale use of self-compacting concrete, many more effective SPs have been developed. The development of comb-shaped SPs in the 1980s represented a breakthrough in concrete technology, as they allow the use of a very low water/cement ratio (*w/c* of 0.20 or less) while keeping good workability.

Lignosulfonates are obtained as by-products of bisulfite pulping of wood, which is used to separate pure cellulose fibers by the dissolution of hemicellulose and lignin. Lignin is a natural and renewable biopolymer present in wood and is, after cellulose, the second most abundant organic molecule on Earth. Lignosulfonates contain numerous functional groups, such as carboxylic acid, phenolic hydroxyl, cathechol, methoxyl, sulfonic acid, and various combinations of these. The molecular weight of commercial Lignosulfonates ranges from a few thousand to 150,000 Da, showing a much greater polydispersity than synthetic polymers. Lignosulfonates as dispersants in concrete show a limited water reduction capability (8%—10%), with an averaged dosage of about 0.1%—0.3% by weight of cement.

Comb-shaped copolymers are the last-generation SPs, and this category includes many types of SPs exhibiting the same common comblike structure. The structure of comb-shaped SPs generally consists of the main chain, the so-called backbone, bearing carboxylic groups, to which nonionic side chains made of polyethers are attached. These SPs are also called polycarboxylate ethers, polycarboxylate esters, or polycarboxylates (PCEs). The key to the success of PCE SPs lies in the fact that they offer a broad range of possible molecular structures. Since the molecular structure greatly affects the performance of PCEs, tailoring their molecular structures enables the production of SPs with quite different properties that can be used in a broader range of applications. The main factors determining the performance of polycarboxylates are

① length of the backbone
② chemical nature of the backbone (acrylic, methacrylic, maleic, etc.)
③ length of the side chains

④ chemical nature of the side chains (PEG, polypropylene oxide, etc.)
⑤ distribution of the side chains along the backbone (random, gradient)
⑥ anionic charge density
⑦ linkage between backbone functionalities and side chain (ester, ether, amide, etc.).

Generally speaking, the mechanisms of SP can be classified into two types, i.e., electrostatic and steric repulsions. In case of electrostatic repulsion, SP molecules adsorb onto the cement particles; the surfactant imparts a strong negative charge, which helps to lower the surface tension of the surrounding water considerably and greatly enhances the fluidity of the system. Another mechanism is steric repulsion. Comb-shaped copolymers act through steric repulsion. Instead of electrostatic repulsion as the dominant cement dispersion mechanism, inhibition of reactive sites through dispersion is the dominating mechanism by which these new generations of SPs work. In steric repulsion, short-range physical barriers are created between the cement particles. One side of the polymer chain gets adsorbed on the surface of the cement grain, while the long unadsorbed side creates the steric repulsion. The grafted side chains of comb SPs protrude and extend from the adsorbed—cement—surface site to hinder neighboring cement particles to reach the range within which van der Waal's force of attraction would be effective. Fig. 3.14 illustrates these two mechanisms. As the effect of steric repulsion is more efficient than electrostatic repulsion, the former has a greater influence over the slump retention and larger water-reducing ratio, even with dosages that are considerably lower than the naphthalene or melamine-sulfonate type SPs.

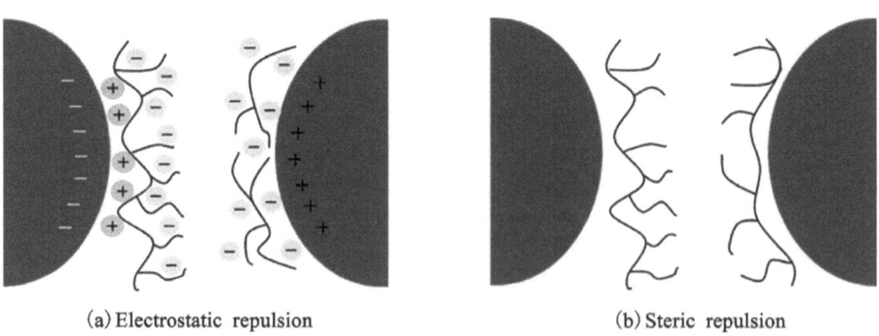

Figure 3.14 (A) Electrostatic and (B) steric repulsion of superplasticizers.

High-range water reducers or SPs are generally more effective than regular water-reducing admixtures in producing workable concrete. The effect of certain plasticizers in increasing workability or making flowing concrete is short-lived. This period is followed by a rapid loss in workability or slump loss. High temperatures can further aggravate slump loss. Due to their propensity for slump loss, these admixtures are sometimes added to the concrete mixer at the job site. They are available in liquid and powder forms. Extended-slump-life plasticizers added at the batch plant help reduce slump-loss problems. An increase in strength is generally obtained with water-reducing admixtures as the water-to-cement ratio is reduced. For concretes of equal cement content, air content, and slump, the 28 d strength of a water-reduced concrete containing a water reducer can be 10%—25% greater than concrete without the admixture. Using a water reducer to reduce the cement and water content of a concrete mixture, while maintaining a constant water—cement ratio, can result in equal or reduced compressive strength, and can increase slump loss by a factor of two or more.

Despite a reduction in water content, SPs might cause increases in drying shrinkage. Luckily, the effect of SPs on drying shrinkage is small when compared to other major factors that affect shrinkage cracking in concrete. High-slump, low-water content, plasticized concrete tends to develop less drying shrinkage than a high-slump, high-water-content conventional concrete. However, high slump plasticized concrete has similar or higher drying shrinkage than conventional low-slump and low-water-content concrete.

Water reducers can be modified to give varying degrees of retardation, while others do not significantly affect the setting time. Some water-reducing admixtures may also entrain some air in concrete. Concretes with SPs can have larger entrained air voids and higher void-spacing factors than normal air-entrained concrete. Air loss can also be significant when compared to concretes without high range water reducers held at constant water-to-cement ratios.

3.3.6.2 Set controlling agents

In practice, there is a need to adjust the hydration rate of cement particles, either accelerate or retard it. A set accelerating admixture is used to shorten the setting time and accelerate the rate of hydration and strength development of concrete at an early age. Set retarding admixtures are used to delay the rate of setting of cement-based materials. Retarders are sometimes used for different purposes. First, it can offset the accelerating effect of hot weather on the setting of concrete. Secondly, it can delay the initial set of concrete or grout when difficult or unusual conditions of placement occur. Thirdly, it can delay the set for special finishing techniques, such as an exposed aggregate surface, or anticipating long transport time or delays between batching and placement.

Obviously, cement hydration is a chemical process, and the chemical reaction rate can be controlled by changing the chemical balance. It is generally accepted that the early reactions of Portland cement compounds with water occur through the solution. In other words, the cement minerals first dissolve into water, then ionize and then the hydration products can be formed in the solution. Since the solubility of hydration products are very limited, the hydration products crystallize out, and the stiffening, setting, and hardening phenomena with Portland cement pastes are directly related to different stages of the progressive crystallization process. It is, therefore, reasonable to assume that, by adding certain soluble chemicals to the Portland cement and water system, either the rate of ionization of cement compounds or the rate of crystallization of the hydration products can be altered. As a consequence, this will affect the setting and hardening characteristics of the cement paste. To understand the mechanism of acceleration or retardation, consider a hydrating Portland cement paste as being composed of calcium cations and silicate and aluminate anions, the solubility of each being dependent on the type and concentration of the acid and base ions present in the solution. Because most chemical admixtures readily ionize in water, it is possible to change the type and concentration of the ionic constituents in the solution phase by adding these admixtures to the cement and water system, thus influencing the dissolution of the cement minerals as per the following guidelines:

① An accelerating agent should promote the dissolution of cations or anions from the cement. The accelerator should be able to promote the dissolving of the silicate which has the lowest dissolving rate during the early hydration period.

② A retarding agent should impede the dissolution of the cement cations or anions, preferably that the aluminate which has the highest dissolving rate during the early hydration period.

③ The presence of monovalent cations in solution (i.e., K^+ or Na^+) reduces the solubility of Ca^{2+} ions but tends to promote the solubility of silicate and aluminate ions. The former effect is dominant in the case of small concentrations, and the latter effect becomes dominant in case of large concentrations.

④ The presence of certain monovalent anions in solution (i.e., Cl^-, NO_3^-) or SO_4^{2-} reduces the solubility of silicates and aluminates but tends to promote the solubility of calcium ions. The former effect is dominant in case of small concentrations, and the latter effect becomes dominant in case of large concentrations.

In general, some reduction in strength at early ages accompanies the use of retarders. However, increased long-term strength may result from retarding the initial rate of hydration. Excessive addition rates of a retarding admixture may permanently inhibit the hydration of the cement. The effects of these materials on the other properties of concrete, such as shrinkage, permeability, and durability, may be unpredictable. The incorporation of retarders can affect some of the other properties of concrete, including slump, bleeding, and early-age strength development. Therefore, acceptance tests of retarders should be made with actual job materials under anticipated job conditions. It should be noted that set retardation is not the same thing as workability retention. Mixes containing water-reducing retarders may lose slump more rapidly than plain concrete in some circumstances. Delayed addition may be very important because a greater effect is obtained by a delay of the order of 5 min after the water has been in contact with the cement.

Set accelerators cause a reduction in the time to both the initial and final sets, the effect generally increasing with admixture dosage. Most accelerators tend to increase shrinkage. The trend is for high early strength to be achieved by the use of nonretarding SP in many applications rather than chemical accelerators to avoid any detrimental effect. There will also typically be an increase in the early-age strength development. Calcium chloride is by far the most economical and effective accelerator. The widespread use of calcium chloride as an accelerating admixture has provided much data and experience on the effect of this chemical on the properties of concrete. Besides accelerating strength gain, calcium chloride causes an increase in potential reinforcement corrosion and may lead to discoloration (a darkening of concrete). The incorporation of calcium chloride can also affect other properties of concrete, including shrinkage, long-term strength, and resistance to FT, sulfates, and ASR. An overdose of calcium chloride can result in placement problems and can be detrimental to concrete. It may cause rapid stiffening, a large increase in drying shrinkage, corrosion of reinforcement, and loss of strength at later ages.

3.3.6.3 Air-entraining agents

Since the accidental discovery of air entrainment in the concrete in the mid-1930s, the practice of AE in concrete has been regulated in standards in cold climate conditions. Discovered by chance when grinding aid for improving cement grinding ended up chemically air entrainment in concrete that has been used to enhance its durability in a cold environment. Up to now, various commercial air-entraining admixtures have had years of success that facilitate the air void system (AVS) in concrete for improving FT resistance of concrete.

The entrained bubble diameters are generally in the range 0.02−0.1 mm, with an average spacing of about 0.25 mm. They are sufficiently stable to remain in fresh concrete during the placing, compaction, and setting. Entrained air is different from entrapped air in terms of size and shape. Entrained air is also good for the workability of concrete mixtures. It is worth to note that because air-entraining surfactants render the cement particles hydrophobic, any overdose of the admixture would cause an excessive delay in cement hydration. Also, air-entrained mixtures depending on the amount of entrained air suffer a corresponding strength loss. Approximately, increase in 1% of air content may decrease compressive strength by 5%. Apart from FT cycling resistance, the benefits of entrained air include the following:

① Increase the volume of the paste.
② Fluidification of the cement paste.
③ Improve concrete rheology.
④ Decrease the absorptivity and permeability of the hardened concrete.
⑤ The dissipation of the energy concentrated at the tip of fissures when, for any reason, hardened concrete starts to crack.
⑥ The presence of a free volume, which can receive deposits of expansive components (ettringite, silica gel) that can be formed in a concrete.
⑦ Increase the fire resistance of concrete by reducing pore pressure and thermal stress at elevated temperature.

Air-entraining agent is a kind of surfactant. Most air-entraining admixtures consist of one or more of the following materials: wood resin (Vinsol resin), sulfonated hydrocarbons, fatty and resinous acids, and synthetic materials. Its chemical formula consists of a nonpolar hydrocarbon chain with an anionic polar group. Different from SP, air-entraining admixtures concentrate at the air−water interface and reduce the surface tension encouraging the formation of microscopic bubbles during the mixing process. In addition, the air-entraining admixture stabilizes those bubbles, make the bubble film stronger, and impede bubble coalescence.

The air-entraining admixture acts at the air—water interface. Air-entraining admixtures typically have a negatively charged head which is hydrophilic and attracts water, and a hydrophobic tail which repels water. The polar end, which is hydrophilic, orients itself toward the water. The stirring and kneading action of mechanical mixing disperses the air bubbles, and strong shearing makes bubbles become finer. The charge around each bubble leads to repulsive forces that prevent the coalescence of bubbles. The surface charge causes the air bubble to adhere to the charged surfaces of cement and aggregate particles. The fine aggregate particles also act as a three-dimensional grid to help hold the bubbles in the mixture. This improves the cohesion of the mixture and further stabilizes the air bubbles.

Concrete should contain some entrained air in order to improve its rheology, decrease the risks of bleeding and segregation, as well as improve the visual aspect of the surface of concrete elements. Only when concrete is submitted to FT cycles, the spacing factor of the bubbles network should be a concern.

Of course, air entrainment could decrease the compressive strength of concrete. It is worth to mention that it is not the compressive strength of concrete that conditions its durability, but rather it's *w/c*. Therefore, for given workability, air-entrained concrete has a lower *w/c* than a corresponding non—air-entrained concrete, which means that it will be more durable.

3.3.6.4 Viscosity-modifying agents

Viscosity-modifying admixtures (VMAs) are essential to control the stability and cohesion of concrete with very specific rheological requirements, such as self-compacting concrete, underwater concrete, or shotcrete. VMAs increase the stability of cementitious systems due to a combination of different physicochemical phenomena that depend on the nature of the VMA and its concentration.

Viscosity-modifying agents are a family of admixtures designed for specific applications. They are used to

① Reduce segregation in highly flowable/self-compacting concrete.
② Reduce washout in underwater concrete.
③ Reduce friction and pressure in pumped concrete.
④ Compensate for poor aggregate grading, especially a lack of fines in the sand.
⑤ Increase cohesion of shotcrete.
⑥ Reduce bleeding in concrete.
⑦ Improve green strength in semidry concrete.

Most VMAs are based on high-molecular-weight polymers with a high affinity to water, and they mainly act through the following mechanisms:

① Bridging flocculation involves the adsorption of a chain of a high-molecular-weight polymer onto two or more cement particles, physically holding them together.
② Polymer-polymer association. Associative polymers contain segments distributed along the chain that has a tendency to interact with each other. As a consequence, intramolecular and intermolecular associations between the polymer chains can develop, producing a three-dimensional network and an increase of the viscosity of the interstitial solution.
③ Entanglement. At high concentrations, the chains of VMA polymers can entangle and increase the apparent viscosity of the interstitial solution and cement suspension.
④ Depletion flocculation occurs because nonadsorbed polymers are depleted from a "volume exclusion shell" around larger particles. The difference of polymer concentration in bulk solution with respect to the depleted zone leads to an increase of the osmotic pressure in the system, which causes its flocculation. This mechanism does not modify plastic viscosity of suspensions, but it leads to an increase of the yield stress.

Some VMAs are based on inorganic materials such as colloidal silica which is amorphous with small insoluble, nondiffusible particles, larger than molecules but small enough to remain suspended in water without settling. By ionic interaction of the silica and calcium from the cement, a three-dimensional gel is formed, which increases the viscosity and/or yield point of the paste.

This three-dimensional structure/gel contributes to the control of the rheology of the mix, improving the uniform distribution and suspension of the aggregate particles and so reducing any tendency to bleeding, segregation, and settlement.

Most VMAs are supplied as a powder blend or are dispersed in a liquid to make dosing easier and improve dosing accuracy. The dosage depends on the application but typically ranges from 0.1% to 1.5% by weight of cement but can be varied for specific applications. Most VMAs have little effect on other concrete properties in either the fresh or hardened state but some if used at high dosage can affect setting time and/or the content and stability of entrained air.

3.4 Concrete at fresh state

3.4.1 Batching, mixing, and transporting

Batching is the process of measuring quantities of concrete mixture ingredients by either mass or volume and introducing them into the mixer. To produce concrete of uniform quality, the ingredients must be measured accurately for each batch. Most specifications require that the batching of concrete ingredients be carried out by mass rather than by volume. This is because although water and liquid admixtures can be batched accurately either by volume or weight, powder, sand, and coarse aggregate can be only accurately measured by mass. Nowadays, most concrete is batched and mixed by ready mixed concrete plants, where the batching is generally automatic or semiautomatic rather than manual, as shown in Fig. 3.15. Ready-mixed concrete is defined as concrete that is manufactured for delivery to a purchaser in a plastic and unhardened state. During the last several decades, the ready-mixed concrete industry has experienced tremendous growth worldwide, and most countries have accepted the way of batching concrete. Batching concrete in this way can guarantee the large-scale production of concrete.

In most specifications, the accuracy of weighing materials is required to be cementitious material $\pm 1\%$, aggregates $\pm 1\%$, water $\pm 1\%$ of the total mixing water, and admixtures $\pm 0.5\%$.

All concrete should be mixed thoroughly until its ingredients are uniformly distributed, and a homogeneous mixture is obtained. Mixers should not be loaded above their rated mixing capacities and should be operated at

Figure 3.15 Concrete mixing plant.

the mixing speed and for the period, either based on revolutions or time, recommended by the manufacturer. The rated mixing capacity of revolving drum truck mixers is limited to 63% of the gross volume of the mixer. For stationary plant mixers, the mixing capacity varies depending on the design. The increased output should be obtained by using a larger mixer or additional mixers, rather than by speeding up or overloading the equipment on hand. If the blades of a mixer become worn or coated with hardened concrete, mixing action will be less efficient, and the mixing blades should be changed, or the coated hardened concrete should be removed. If the concrete has been adequately mixed, samples taken from different portions of a batch should have essentially the same strength, density, air content, slump, and coarse aggregate content, with some allowance for testing variability.

Concrete is often transported to the job site by truck mixer, as shown in Fig. 3.16. From the batching plant to the job site, concrete should be mixed all the time to maintain its flowability and homogeneousness. Before transporting concrete, a good plan should be made to avoid a long time delay. This will seriously affect the quality of the finished work.

3.4.2 Placing, finishing, and curing

After arrival at the job site, the ready-mixed concrete should be placed as near as possible to its final position. Belt conveyors, truck-mounted chutes, and mobile-boom pumps are the most commonly used today for concrete placement. Pumping concrete is more and more popular for the construction industry Fig. (3.17). Since high pressure is required to pump

Figure 3.16 Concrete truck mixer.

Figure 3.17 Concrete pumping.

concrete to the casting site, depending on the length or height, pumping concrete should be well designed to minimize segregation and bleeding. This increases the risk of cracking, and thus decreases the durability of concrete. In addition, pumping is believed to modify the characteristics of the air-bubble system and the related FT and scaling durability. Especially for air-entrained high-performance concrete, when pumping is applied to concrete, a significant drop in air content is often observed.

Generally, for a massive structure, concrete cannot cast in one time and should cast in different layers. The concrete mixture is often deposited in horizontal layers of uniform thickness, prior to placing the next layer, each layer is thoroughly vibrated and compacted. The placement rate should be well designed based on the volume of the structure, and one layer should be cast rapid enough so that the layer immediately below is still plastic when a new layer is deposited. If the structure is too large and the layer cannot be deposited completely within the setting time of concrete, retarding agent should be used. This prevents cold joints, flow lines, and planes of weakness that occur when fresh concrete is placed on hardened concrete.

Consolidation is of several purposes: ① compacting fresh concrete to eliminate stone pockets, honeycombs, and entrapped air; ② to mold it within the forms and around embedded items and reinforcement. There are many consolidation ways for concrete. The method chosen depends on the consistency of the mixture and the placing conditions, such as the complexity of the formwork and amount and spacing of reinforcement. Generally, mechanical methods using either internal or external vibration are preferred. Highly workable concrete can be consolidated by hand rodding; inserting a tamping rod or other suitable tool repeatedly into the concrete. The tamping rod should be long enough to reach the bottom of the form or lift and thin enough to easily pass between the reinforcing steel and the forms. The consolidation points should be well distributed on the formwork, avoid missing consolidation. Spading can be used to improve the appearance of formed surfaces. A flat, spade-like tool should be repeatedly inserted adjacent to the form. This forces the larger coarse aggregates away from form faces, and assists entrapped air voids in their upward movement toward the top surface where they can escape. Concrete should also not be overconsolidated; in this case, segregation can happen to fresh concrete.

Vibration is the most widely used technique for consolidating concrete. When concrete is vibrated, the viscosity of concrete is decreased due to the action of shearing and the concrete behaves like a liquid. As a consequence, fresh concrete can flow into every corner of the formwork, and the large entrapped air voids can easily escape from the concrete to the surface. Since concrete is a thixotropic material, the structure builds up progressively after stopping vibration. Vibrators, whether internal or external, are usually characterized by their frequency of vibration and the amplitude of vibration. The frequency of vibration can be measured using a vibrating reed tachometer. When vibration is used to consolidate the concrete, a standby vibrator should be on hand at all times in the event of a mechanical breakdown.

Internal or immersion-type vibrators, often called spud or poker vibrators, are commonly used to consolidate concrete in walls, columns, beams, slabs, and other heavily reinforced structure. Flexible-shaft vibrators consist of a vibrating head connected to a driving motor by a flexible shaft. Inside the head, an eccentric weight connected to the shaft rotates at high speed, causing the head to revolve in a circular orbit. The motor can be powered by electricity, gasoline, or air. The vibrating head is usually cylindrical with a diameter ranging from 19 to 175 mm. Some vibrators have an electric motor built directly into the head, which is generally at least 50 mm in diameter. The dimensions of the vibrator head as well as its frequency and amplitude in conjunction with the workability of the mixture affect the performance of a vibrator. Vibrators should not be used to move concrete horizontally since this causes segregation. Whenever possible, the vibrator should be lowered vertically into the concrete at regularly spaced intervals and allowed to descend by gravity. It should penetrate to the bottom of the layer being placed and at least 150 mm into any previously placed layer.

External vibrators can be form vibrators or surface vibrators such as vibratory screeds, plate vibrators, vibratory roller screeds, or vibratory hand floats or trowels. Form vibrators are designed to be securely attached to the outside of the formworks, which are very useful for consolidating concrete in members that are very thin or congested with reinforcement, stiff mixtures where internal vibrators cannot be used. Attaching a form vibrator directly to the concrete form generally is unsatisfactory. Rather, the vibrator should be attached to a steel plate that in turn is attached to steel I-beams or channels passing through the form stiffeners. Loose attachments can result in significant vibration energy losses and inadequate consolidation. Form vibrators are often electrically operated. They should be spaced to distribute the intensity of vibration uniformly over the form.

Depending on the intended service use, the surface of concrete may be required for different appearance; concrete should be finished in different ways. Some surfaces may require only strike off and screed to proper contour and elevation, while other surfaces may require a broomed, floated, or troweled finish may be specified. Screeding or strike off is the process of cutting off excess concrete to bring the top surface of a slab to proper grade. The manually used template is called a straightedge, although the lower edge may be straight or slightly curved, depending on the surface specified. It should be moved across the concrete with a sawing motion while advancing forward a short distance with each movement. Straightedges are sometimes equipped with vibrators that consolidate the concrete and assist in reducing the strike off work. This combination of straightedge and vibrator is called a vibratory screed.

In order to eliminate high and low spots and to embed large aggregate particles, a bull float should be used immediately after strike off. Edging is required along with all edge forms and isolation and construction joints in floors and outdoor slabs such as walks, drives, and patios. Edging densifies and compacts concrete next to the form. Afterward, the surface should be floated with a hand float or with a finishing machine using float blades. Where a smooth, hard, dense surface is desired, floating should be followed by steel troweling. Troweling should not be done on a surface that has not been floated. Troweling after only bull floating is not considered adequate finishing.

All newly placed and finished concrete structures should be well cured. On the one hand, water evaporation should be prevented from plastic shrinkage. On the other hand, the temperature of the concrete structure should be well controlled. In case of high temperature, special measures should be taken to decrease the temperature of the ingredient of concrete. In case of temperature lower than 0°C, measures should be taken to prevent freezing damage, for instance, heating, covering, insulating, or enclosing the concrete. Before concrete gets hardened, cautions should be taken to prevent mechanical damage.

There are many curing methods for concrete in practice, as follows: ① ponding or immersion concrete in water, spraying or fogging on the surface of the concrete. These measures can provide additional moisture to concrete and prevent the fast moisture evaporation and maintain the mixing water in the concrete during the early hardening period; ② steam-cured concrete: in this method, concrete is put in a chamber filled with steam moisture with high temperature. The strength gain of concrete is accelerated. ③ Cover the concrete with plastic sheets or wet clothes to prevent evaporation. ④ Spaying curing compound onto the surface of concrete, which can form a thin film layer which is impervious and mixing water in concrete can be retained.

3.4.3 Workability

Mixing, handling, and placing fresh concrete can give concrete workers a subjective understanding of whether the concrete is "friendly" or not. From a subjective point of view, the works can describe the concrete harsh or cohesive, lean or rich. However, this is a quantitative description, and has different meanings to different workers. Whether a concrete is friendly can be judged by the following guidelines: (1) concrete should be able to flow and fill in the form easily with the assistance of whatever equipment is available. (2) Compactability: nearly all of the air entrapped in fresh concrete during mixing and handling should be capable of being removed by the compacting system being used, such as poker vibrators. (3) Segregation should not happen to concrete, and remain as a homogeneous uniform mass.

It has been universally accepted that workability is used to describe the fresh properties of concrete. Although workability has different definitions, the meaning and testing method are the same all around the world. American Concrete Institute defines workability of concrete as "that property of freshly mixed concrete or mortar which determines the ease and homogeneity with which it can be mixed, placed, consolidated and finished." American Society for Testing and Materials defines workability of concrete as "that property determining the effort required to manipulate a freshly mixed quantity of concrete with minimum loss of homogeneity." Neither of these two definitions makes any reference to a quantitative, measurable property. The most commonly used, simplest, and crudest testing method, is the slump test (see Fig. 3.18). The equipment for the slump test is very simple and cheap. It consists of a tamping rod and a truncated cone which is 300 mm height and 100 mm diameter at the top, and 200 mm diameter at the bottom. In this test, concrete is first filled in the truncated cone by 1/3 full by volume, and then rod it with 25 strokes using a steel rod. Uniformly distribute strokes over the cross section of each layer. Secondly, fill the cone 2/3 full by volume (half the height) and again rod 25 times with rod just penetrating into, but not through, the first layer. Distribute strokes evenly as the first layer. Thirdly, fill the cone to

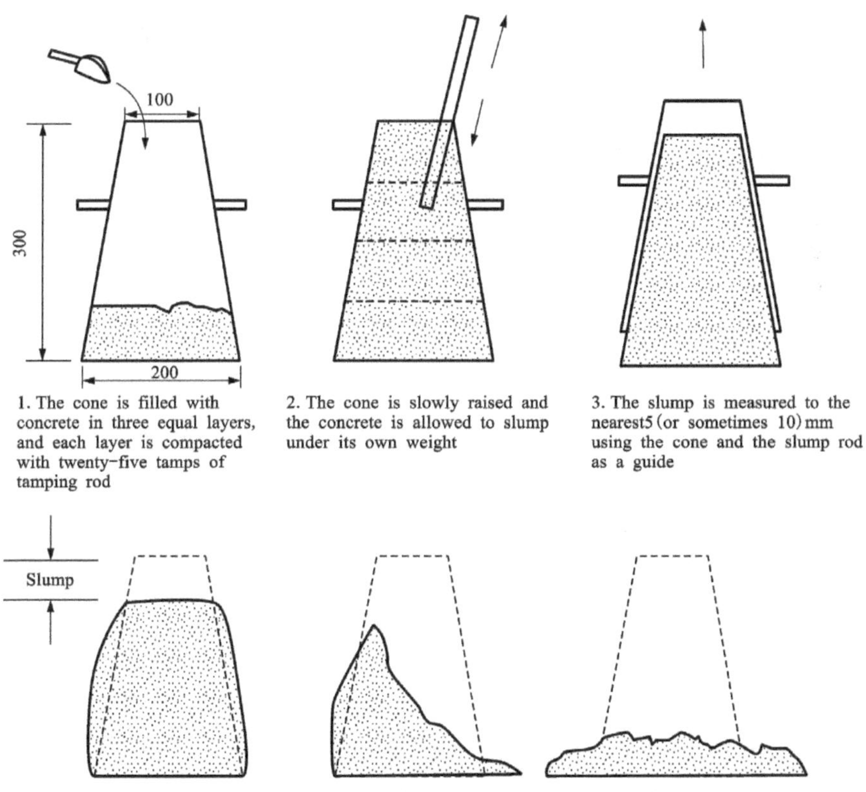

Figure 3.18 Slump test.

overflowing and again rod 25 times with rod just penetrating into, but not through, the second layer. Uniformly distribute strokes again. Fourthly, strikes off an excess concrete from the top of the cone with the steel rod so that the cone is exactly level full. Clean the overflow away from the base of the cone mold. Fifthly, steadily lift up the cone within 5 ± 2 s. Note that there is no lateral or torsional motion being imparted to the concrete. The entire operation from the start of the filling through the removal of the mold shall be carried out without interruption and shall be completed within an elapsed time of 2−1/2 min. Finally, place the steel rod horizontally across the inverted mold so that the rod extends over the slumped concrete. Measure the distance from the bottom of the steel rod to the displaced original center of the specimen. This distance is the slump of the concrete, as shown in Fig. 3.18. The slump value indicates the flowability of concrete. The higher the slump value is, the more flowable concrete will be.

In addition, the slump can be used to roughly evaluate two other properties of fresh concrete, i.e., bleeding and cohesiveness. The concrete retains the overall shape of the cone after lifting the cone and does not collapse. Slightly tap the slumped concrete, and watch the collapse of concrete. If the slumped concrete collapses suddenly, the cohesiveness of concrete is bad. If the slumped concrete collapses slowly, the cohesiveness of concrete is good. The water ring at the bottom of the slumped concrete indicates the water-holding capacity. If too much water can be found at the bottom, the water-holding capacity of concrete is bad, and vice versa.

The slump test is not suitable for measuring the consistency of a highly flowable or highly stiff concrete mixture. For highly flowable concrete, the final diameter or 'flow' of the concrete is often measured. This is called the slump flow. For highly stiff concrete, Vebe test is often used.

Generally speaking, slump is classified into four classes:

Slump1: 0−10 mm.

Slump2: 15−30 mm.

Slump3: 35−75 mm.

Slump 4: 80−155 mm.

Slump 5: 160−220 mm.

Slump 6: >220 mm.

Different slump values are used for different engineering purposes. For instance, a slump of 0 mm can be used for roller-compacted concrete. Concrete with a slump of 160−220 mm can be used for pumping concrete, and a slump greater than 220 mm is for self-compacting concrete.

Theoretically, the workability of concrete is directly related to its rheological properties. Rheological measurement is quite complicated. The term of rheology was first proposed by Professor E.C. Bingham in 1920. In the following years, rheology was born as a new subject with the establishment of the Rheology Society and Journal of Rheology. Rheology is an important tool for scientists and engineers in many industries, including plastics, paint, printing inks, detergents, oils, etc. The rheology of cement and concrete mainly involves the evolution of viscosity, plasticity, and elasticity of cement-based materials under shear stress. The rheological properties have significant influence on constructing, forming, or casting process.

The fundamental physical parameters for describing rheological properties mainly include yield stress and plastic viscosity. The yield stress is the minimum stress when concrete initiates to flow. It is caused by the force required to break down the network structure which is formed by the colloidal interaction and rigid links between cement particles, as well as the adhesion and friction between aggregate particles. The magnitude of yield stress depends on the solid volume fraction, packing fractions, size and surface roughness of solid particles, interactions between particles, and specifically on the effect of SP. The yield stress can be used to characterize the filling ability and stability of fresh concrete. The plastic viscosity is defined as the proportional coefficient between shear stress and shear rate under a state of steady shear, and it is greatly affected by colloidal particle interaction forces, Brownian forces, hydrodynamic forces, and viscous forces between particles. The value of plastic viscosity depends on the ratio of volume fraction and packing density of particles in the mixture, and it is a good indicator for the compactibility, machinability, and segregation resistance. The fundamental physical quantities can be obtained using rheometers. However, the rheometers are usually not available at most concrete laboratories. Therefore, it is necessary to investigate the relationship between the fundamental parameters and the traditional parameters, such as slump, slump flow, and flow time. For fresh concrete, the yield stress and plastic viscosity are correlated to the slump or slump flow and the flow time, respectively. An excellent relationship between the yield stress and the slump has been obtained.

The concentric cylindrical shearing system is often used to measure the rheological properties of concrete, as shown in Fig. 3.19. Due to the presence of coarse aggregate, segregation may happen to concrete under the action of shearing, and the gap between the two cylinders should be wide to avoid direct contact of aggregates during shearing. The measurement of the relationship between rotation speed and shear torque is

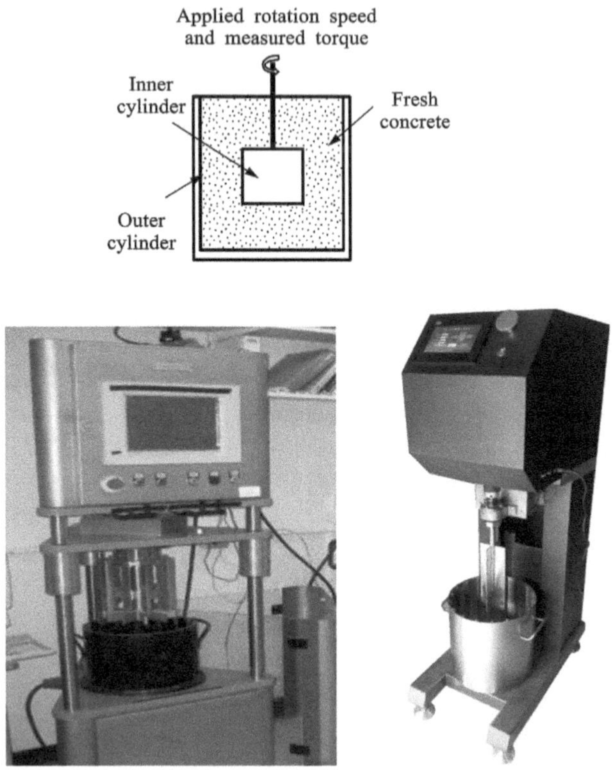

(a) Contec rheometer (b) TCR rheometer

Figure 3.19 Rheological measurement and rheometer for concrete.

normally called the flow curve. There are also many complicated mathematical processings for the transformation of torque-rotation speed to the rheological parameters (i.e., yield stress and viscosity).

Due to the gap width for viscometer of concrete, some general rheometers are not applicable for concrete. Some viscometers specialized for concrete were developed for concrete which has a coarse aggregate of up to 30 mm, as shown in Fig. 3.19. Unlike rheological properties for general flow, the rheological properties of concrete cannot be obtained by the up–down flow curve. Instead, rheological parameters for concrete should be obtained by the stepwise procedure shown in Fig. 3.20, and the dataset of shear rate versus torque can be transformed into shear rate versus shear stress, and then different rheological models are used to fit the dataset. There is general agreement that the behavior of fresh paste, mortar, and concrete all approximate reasonably closely to the Bingham model

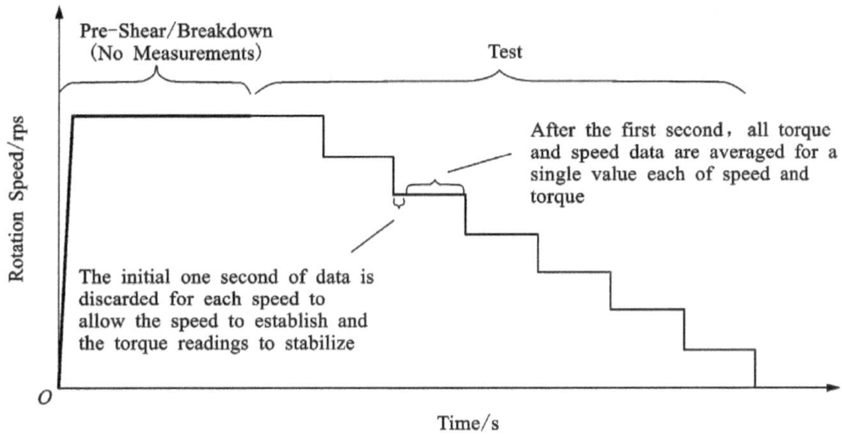

Figure 3.20 Stepwise procedure for determining the rheological parameters.

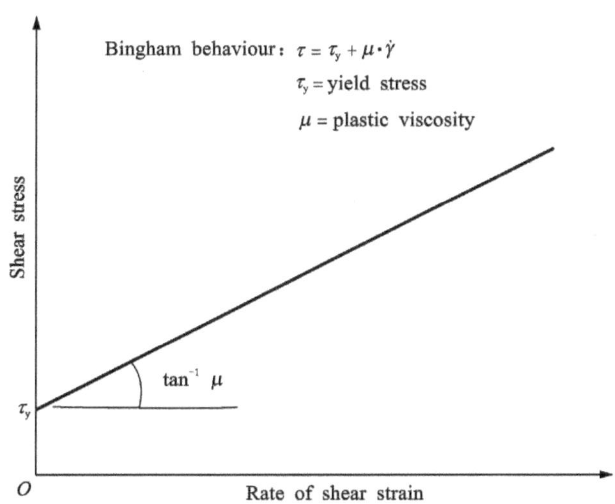

Figure 3.21 Bingham flow.

illustrated in Fig. 3.21. Flow only starts when the applied shear stress reaches yield stress sufficient to overcome the interparticle interference effects, and at higher stresses, the shear rate varies approximately linearly with shear stress, the slope defining the plastic viscosity. Thus, the two Bingham constants, yield stress and viscosity, are required to define the behavior, unlike Newtonian fluid which does not have yield stress, and which therefore requires only a single constant, viscosity. Nonlinear flow curves of concrete are also found in literature, and concrete may behave as a shear-thinning or shear-thickening behavior, depending on water-to-binder

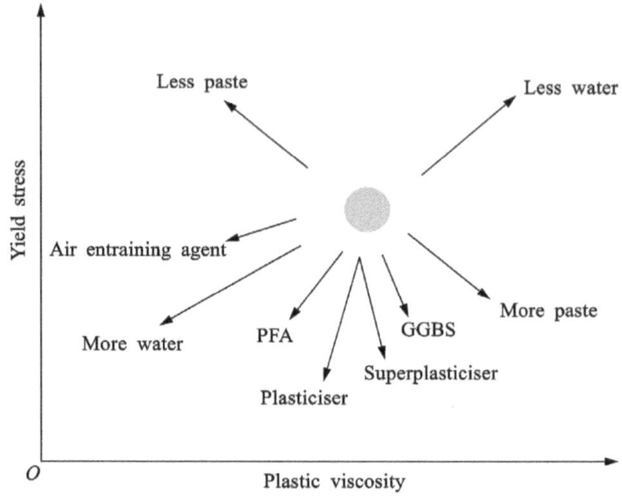

Figure 3.22 Summary of the effect of varying the properties of concrete constituents on the Bingham constants.

ratio (w/b ratio) and the dispersion state of cement particles. For the sake of simplicity, the Bingham model is the most used model for concrete. The constituent factors, such as water, cement paste, aggregate, affect the rheological properties of concrete, as shown in Fig. 3.22.

The workability loss is very common in practice. This is due to

① Mixed water reduction caused by either adsorption by the aggregate or evaporation to the air.

② Mixed water reduction caused by cement hydration reactions.

③ Interactions between admixtures (particularly SPs) and the cementitious constituents of the mix.

Slump loss is defined as the loss of slump in fresh concrete with elapsed time. This is a normal phenomenon with all concrete mixtures because it results from the gradual stiffening and setting of a hydrating Portland cement paste, which is associated with the formation of hydration products, such as calcium hydroxide, ettringite, and calcium silicate hydrates.

Evaporation of mix water can be reduced by keeping the concrete covered during transport and handling as far as possible. Most available data relate to loss of slump, which increases with higher temperatures, higher initial slump, higher cement content, and higher alkali and lower sulfate content of the cement. The rate of loss of workability can be reduced by continued agitation of the concrete, e.g., in a ready-mix truck, or modified by admixtures. In principle, retempering, i.e., adding water to compensate for slump loss, should not have a significant effect on strength if only that water which has been lost by evaporation is replaced. It has been shown that water can be added during retempering to slightly increase the water/cement ratio by 5% without any loss in 28d strength. However, retempering might not be well controlled in site; thus, this should be avoided.

3.4.4 Properties at early age

3.4.4.1 Bleeding and segregation

Segregation is defined as the separation of the light components from the heavy component in a fresh concrete mixture so that they are no longer uniformly distributed. There are two kinds of segregation. The first one is characterized by the separation of mortar or paste from the coarse aggregate in concrete. The second one is characterized by the separation of water from the body of concrete. This is called bleeding, and is also a form of segregation. Water is the lightest component, and aggregate is the heaviest component. Actually, cement particle has a higher density than that of aggregate, the interaction between particles form flocs and the flocs have a lower density than that of aggregate. Due to the density difference in the constituents of fresh concrete, segregation and bleeding often happen simultaneously. The paste is not viscous enough to hold water and aggregate suspending in the paste, and water tends to move upward, and aggregate tends to move downward.

With the cement hydration going on, more and more cement hydration products are formed, and the paste becomes more and more viscous. Thus, water and aggregate tend to be stable in concrete with hydration continuing. In practice, it is necessary to reduce the tendency for segregation in a concrete mixture because full compaction, which is essential for achieving the maximum strength potential, is not possible in a segregated concrete mixture. Furthermore, bleeding and segregation of concrete may bring about many other issues. On the one hand, when bleeding water moves upward, it may be blocked by aggregate, and numerous bleed-water pockets can be formed. If it is reinforced concrete and there are some horizontal reinforcing bars in the concrete, the water moving upward may also be blocked. The bonding between paste and aggregate and steel bar will be reduced. As a consequence, the upper half of a reinforced concrete beam or column may be weaker than the lower half. On the other hand, aggregate moving downward may cause honeycomb, as shown in Fig. 3.23.

Both characteristics of raw materials, mixture design and processing, may cause bleeding and segregation, such as

① Too much high *w/c* ratio.
② Low viscosity of the paste.

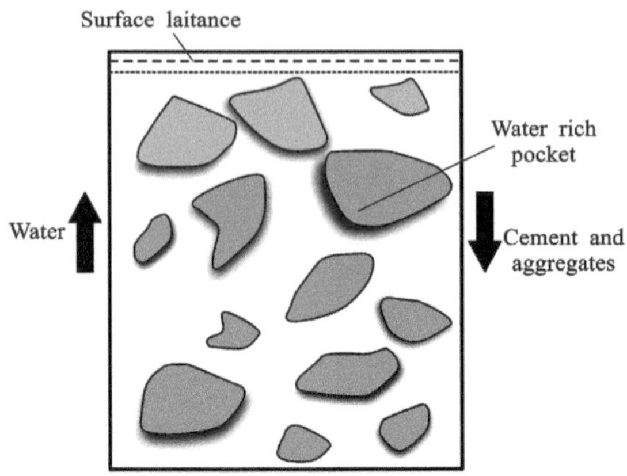

(a) schematic diagram of bleeding and segregation

(b) Homeycomb of concrete

Figure 3.23 Segregation and bleeding in placed concrete.

③ Low sand ratio.

④ Inappropriate placing and compacting methods.

Therefore, bleeding and segregation can be reduced or eliminated by paying attention to the selection of materials, mixture design, and concrete handling and compaction methods. And the effective ways of reducing bleeding and segregation can be given as follows:

① Reduce the water content, water-to-cementitious material ratio, and slump.

② Use finer cementitious materials.

③ Increase sand ratio or the amount of fines in the sand.
④ Use chemical admixtures that permit reduced water-to-cementitious materials ratios or provide other means capable of reducing the bleeding of concrete.
⑤ Air-entrainment.

3.4.4.2 Plastic shrinkage and cracking

During the fresh state, concrete undergoes volumetric changes that strongly influence the properties of the hardened material. The volumetric changes include plastic settlement and plastic shrinkage, which in combination with other factors may lead to cracking of concrete in this fresh state where the tensile strength of concrete is still very low.

Plastic settlement and plastic shrinkage are associated with plastic settlement cracking and plastic shrinkage cracking, respectively. The plastic settlement involves a vertical volumetric reduction in the gravitational direction, while plastic shrinkage is generally known to be a three-dimensional volume reduction dictated by the orientation and size of the micropore structure of the plastic concrete. Typically, plastic settlement cracking precedes plastic shrinkage cracking, and both physical phenomena occur during the plastic stage of freshly cast concrete.

There are five basic volumetric altering mechanisms in fresh concrete.

① Hydration. The beginning of hydration means loss of free water on which the plastic nature of concrete is based. This water loss is usually via evaporation, absorption by unsaturated aggregates, the formation of hydration products, etc. The progression of hydration gives an indication of the physical state of concrete. The start of the exothermic processes of hydration is preceded by a dormant stage which offers numerous advantages with the handling and placing of concrete. Hydration in plastic concrete is usually associated with arbitrary terms such as initial and final set. These terms give an indication of the initial setting and final setting time, which are influenced by a number of factors and may greatly vary between laboratory and field tests. Since hydration is characterized by loss of free water, the rate at which this happens directly influences the progression of other mechanisms of plastic shrinkage cracking, such as free settlement, bleeding, and capillary pressure.
② Free settlement. The gravitation drives free settlement of solid particles in fresh concrete, and it works in such a way to cause denser packing of the solid particles. Settlement of solid particles displaces water upward,

imparting an upward force due to the viscous drag caused by the flowing water. The free settlement remains active until the point where hydration of cement starts, and can be ceased mechanically before the end of the dormant period if the solid particles physically come in contact with one another hindering further settlement. The outcome of this phenomenon is a vertical volume contraction.

③ Bleeding. This has been described above.

④ Evaporation. The main factors that influence the evaporation rate are wind speed, relative humidity, air temperature, concrete temperature, the difference between air and concrete temperature, as well as solar radiation.

⑤ Capillary pressure. Once the concrete surface bleed water starts evaporating up to a point exposing the solid surface, the solid particles at the surface then form menisci which create a negative pressure in the capillary water. The negative capillary pressures act to draw more water to the surface of concrete, and in the process, act to bring the solid particles as close as possible.

Plastic shrinkage cracking of concrete occurs when the stresses arising in the concrete due to a volume deformation exceeds its tensile strength. Obviously, the volume shrinkage and tensile strength of concrete develop simultaneously. Cracks often develop near the surface of concrete, and typical plastic shrinkage cracks (Fig. 3.24) are parallel to one another and are 0.3−1 m apart and 25−50 mm deep.

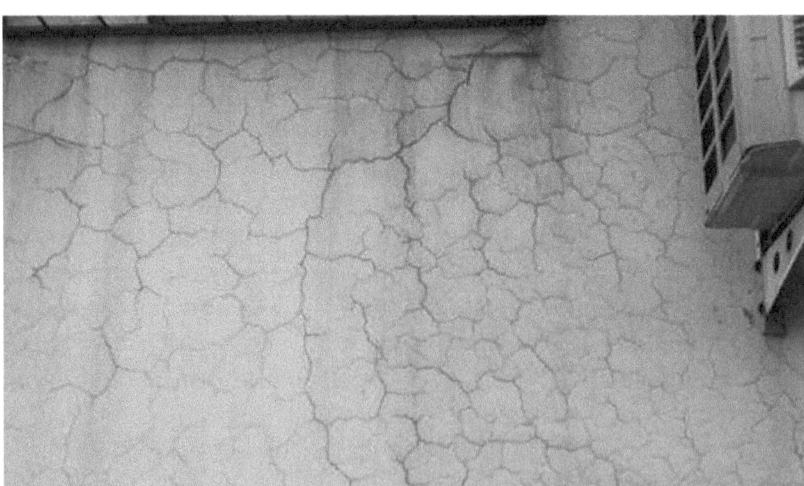

Figure 3.24 Plastic cracking.

As stated above, a variety of causes contribute to plastic cracks in concrete. In order to avoid cracking caused by volume change in the fresh state, some measures, singly or collectively, can be used, as follows:

① Moisten the subgrade and forms.
② Moisten aggregates that are dry and absorptive.
③ Erect temporary windbreaks to reduce wind velocity over the concrete surface.
④ Erect temporary sunshades to reduce concrete surface temperature.
⑤ Keep the fresh concrete temperature low by cooling the aggregate and mixing water.
⑥ Protect concrete with temporary coverings such as polyethylene sheeting, wet burlap, fog spray, or a curing compound.
⑦ Reduce the time between placing and start of curing by eliminating delays during construction.
⑧ Settlement cracks may be eliminated by revibration of concrete when it is still in the plastic state. Revibration also improves the bond between concrete and reinforcing steel, and enhances the concrete strength by relieving the plastic shrinkage stresses around the coarse aggregate particles.

3.5 Mechanical properties

3.5.1 Compressive strength

The mechanical properties of concrete are the properties most valued by designers and quality control engineers, including compressive strength, tensile strength, elastic modulus, Poisson's ratio, etc. In concrete design and quality control, strength is the property generally specified. This is because, compared to most other properties, testing of compressive strength is relatively easy. In addition, many properties of concrete, such as elastic modulus, water tightness or impermeability, and resistance to weathering agents including aggressive water, are believed to be dependent on the strength and may therefore be deduced from the compressive strength. The strength of a material is defined as the ability to resist stress without failure. Failure is identified with the appearance of cracks or complete fracture. However, it is shown that concrete contains many initial microcracks during the formation of its structure. Therefore, the strength of concrete is related to the stress required to cause failure, and it is defined as the maximum stress the concrete sample can withstand. In tension testing, the fracture of the test piece usually signifies failure.

The compressive strength test can be performed either on a cubic or cylindric concrete specimens. The concrete specimens must be large enough to ensure that an individual aggregate particle does not have a dominant influence on the result. Generally, 100 mm is recommended for maximum aggregate sizes of 20 mm or less, 150 mm for maximum sizes up to 40 mm. The cubes are usually cast in lubricated steel or plastic molds which should have accurate dimension. In particular, the opposite faces of the molds are smooth and parallel to each other. The concrete is fully compacted by external vibration or hand tamping, and the top surface should be troweled smooth. Since curing conditions have a great influence on the strength development of concrete, concrete specimens are often cured in a standard condition which is specified in standards or specifications. Normally, the standard curing condition is $20 \pm 2°C$, and relative humidity >95%. At the specified age, which is often 28 or 56 d, the compressive strength can be measured. The cube-testing machine has two heavy platens through which the load is applied to the concrete. The bottom one is fixed, and the upper one has a ball seating, which allows rotation to match the top face of the cube at the start of loading. This then locks in this position during the test. It is worth to mention that the load is applied to a pair of faces which were cast against the mold, i.e., with the troweled face to one side. This ensures that there are no local stress concentrations which would result in a falsely low average failure stress. A specified load is used for testing, which depends on the strength of concrete and ranges from 0.3 to 1.0 MPa/s. A fast rate of loading gives overestimated strengths; if the loading rate is too low, a low strength may be obtained. It is important that the specimen should be properly made and cured. Then, the test will give a true indication of the properties of the concrete, unaffected by such factors as poor compaction, drying shrinkage cracking, etc.

The failure modes within the cube (Fig. 3.25A) produce a double pyramid shape after failure. This is because of the difference in the elastic modulus between the steel plate and the concrete, and the friction between the two, resulting in lateral restraint forces in the concrete near the plate. In this case, concrete is, therefore, in a triaxial stress state, instead of pure compression state, with consequent higher failure stress than the true, unrestrained strength. This is the major objection to the cube test. The test is, however, relatively simple, and relatively high accuracy, which can be used as a reference method comparing different concretes.

The use of cylindrical specimens can partly overcome the end-restraint problem. Cylinders with a height/diameter ratio of 2, most commonly

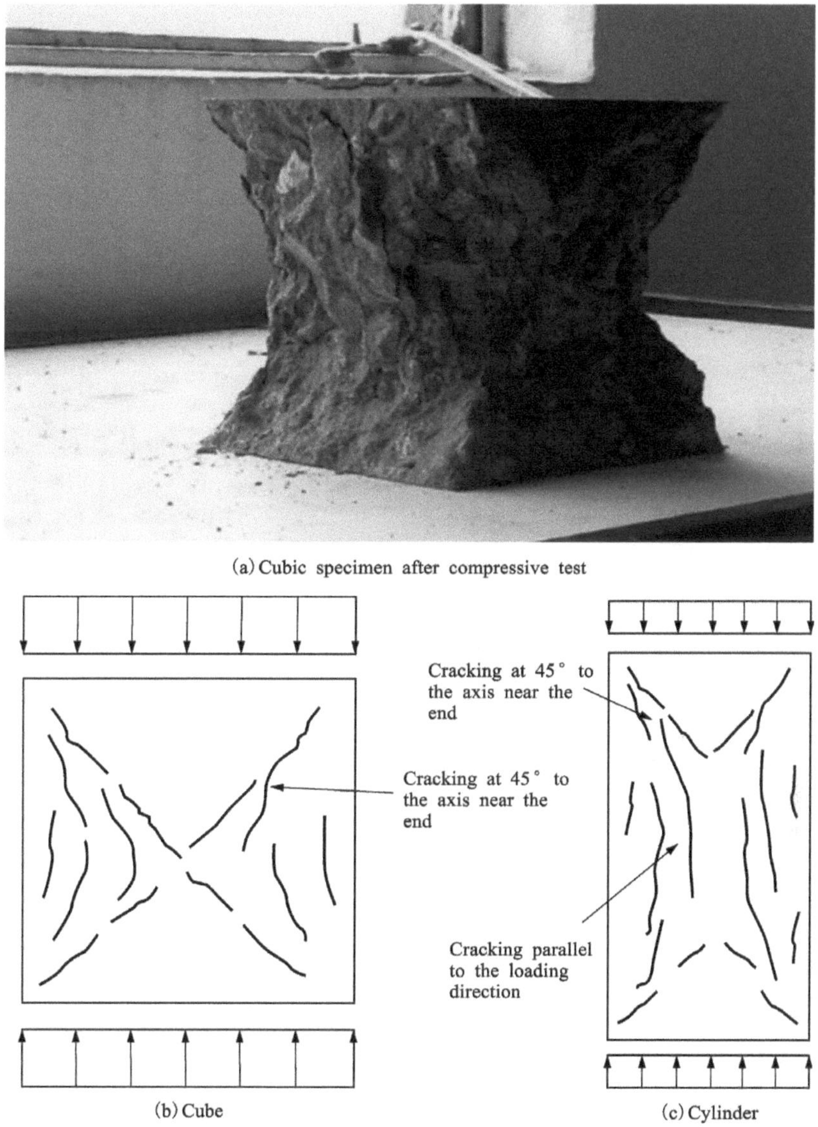

Figure 3.25 Cracking patterns during testing of concrete specimens in compression.

300 mm high and 150 mm diameter, are tested vertically, as specified in American standards. The effects of end restraint are much reduced over the central section of the cylinder, which fails with near uniaxial cracking (Fig. 3.25B, C), indicating that the failure stress is much closer to the pure compressive strength. As a rule of thumb, it is often assumed that the strength

obtained from a cubic specimen is about 25% higher than the strength of a cylindrical specimen, but this ratio has been found to depend on several factors.

Although cylindrical specimens have some advantages, the testing of cylinders has one major disadvantage. That is, the top surface of cylinders are finished by a trowel and is not plane and smooth enough for full contact with the plate. Thus, further preparation for the top surface is necessary. Some measures can be used. Firstly, the surface can be ground, but this is very time-consuming. Secondly, the cylinder and the normal procedure is to cap it with a thin (2−3 mm) layer of high-strength gypsum plaster, molten sulfur, or high early strength cement paste, applied a day or two in advance of the test. Alternatively, the end of the cylinder can be set in a steel cap with a bearing pad of an elastomeric material or fine dry sand between the cap and the concrete surface. Apart from the inconvenience of having to carry this out, the failure load is sensitive to the capping method, particularly in high strength concrete.

It is worth to mention that the compressive strength of concrete is a variable, and it obeys normal distribution. The concept of strength grade is used in practice, which is defined statistically. Usually, 95% of confidence is required. In other words, to test a series of specimen's strength following a specific standard, the compressive strength of 95% of the specimens is higher than the value. This value is the standard value of compressive strength, noted as fcu,k. The strength grade is expressed by starting with letter C. such as C20, C30, C40, C50...C120. It can be noted that strength grades are often integral tens digit.

3.5.2 Tensile strength

Concrete is a brittle material, and relatively weak in tension. Thus, reinforced concrete uses the advantages of both concrete and steel; the former is strong in compression, and the latter is strong in tension. Although the tensile strength of concrete is not utilized in the design of reinforcement, cracking in concrete will occur when the stress exceeds the tensile strength, and therefore this property is specified in different applications. For a couple of reasons, it is difficult to directly determine the tensile strength of concrete. Firstly, a direct tensile test is difficult to conduct without causing eccentric stresses, which results in high variability and unrealistically low values. Secondly, the clapping of the end of a specimen is not easy work. Thus, the indirect method, i.e., "Brazil splitting test," is the most common testing method for the tensile strength of concrete, as shown in Fig. 3.26. It is worth mentioning that the frequency of tensile strength

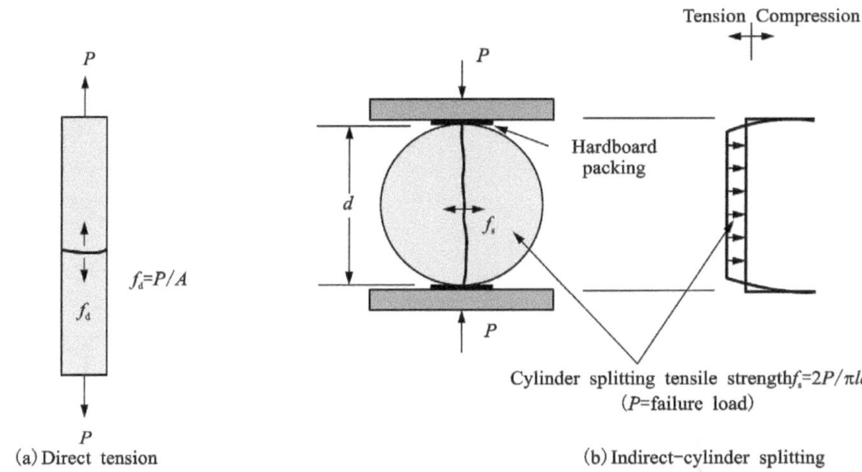

Figure 3.26 Tensile strength test.

testing is much lower than that of the compressive strength test in practice, and it is seldom to use Brazil splitting test as a quality control test.

In the Brazil splitting tension test, a concrete cylinder is subjected to compression loads along two axial lines, which are diametrically opposite. The load is applied continuously at a constant rate until the specimen fails. The compressive stress produces transverse tensile stress, which is uniform along the vertical diameter. Thus, the tensile strength can be calculated from the determined splitting load, as follows:

$$T = \frac{2P}{\pi l d} \tag{3.2}$$

where T is tensile strength; P is failure load; l is length, d is diameter of the specimen.

3.5.3 Elastic modulus

Deformation behavior of concrete under stress is often obtained from the stress—strain curve. Concrete is not a pure elastic or plastic material. The typical stress—strain curve of concrete is given in Fig. 3.27. When concrete is subjected to compression, concrete may undergo four stages. The first stage is below 30% of the ultimate load. At this stage, concrete is almost an elastic material. This means there is no crack in concrete; therefore, the stress—strain curve remains linear. When a load increases above 30% of the ultimate load, the stress—strain curve starts to deviate from a straight line. On the micro-structural level, cracks start to form, mainly form at the interface transition

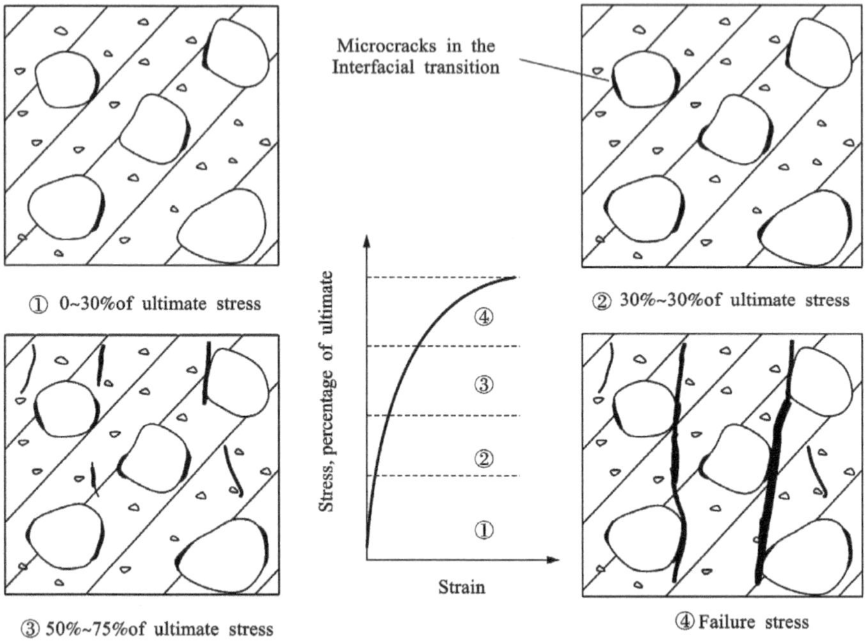

Figure 3.27 Diagrammatic representation of the stress–strain behavior of concrete under uniaxial compression.

zone. This is stage 2. Until about 50% of the ultimate stress, the microcracks at the interface transition zone appears to be stable. When a load is over 50%, cracks begin to form in the matrix, not only at the interface transition zone. This is stage 3. With further increase in stress level up to about 75% of the ultimate load, all the crack system becomes unstable and starts to propagate, causing the stress–strain curve to bend considerably toward the horizontal. Above 75% of the ultimate load, the increasing stress causes the rapid propagation of the crack system. This is stage 4.

According to the strain–stress curve, there are several ways to define the elastic modulus of concrete, i.e., initial tangent elastic modulus, tangent elastic modulus, secant elastic modulus, as shown in Fig. 3.28.

The original tangent modulus is also called Young's elastic modulus, which is calculated from the initial tangent. Young's elastic modulus is the most used parameter in practice. In reinforcement concrete structural design, Young's elastic modulus is a necessary input parameter, which is used to calculate the deformation of the structure. In particular, high-rise buildings and long-span structures are designed based on stiffness, and therefore a high elastic modulus may be specified. Low elastic modulus is

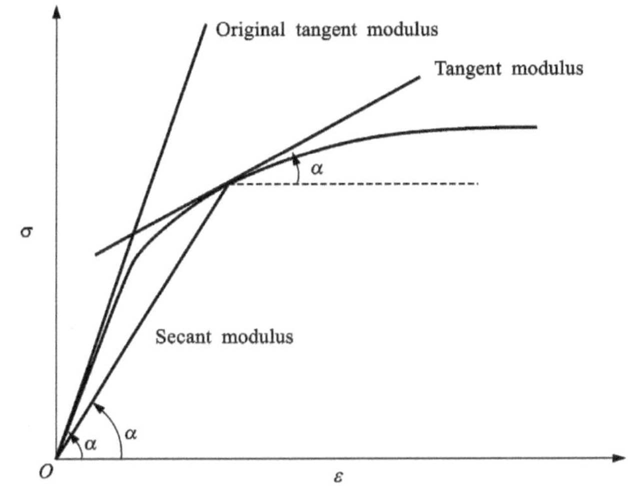

Figure 3.28 Types of elastic modulus.

desired for some cases, such as roads or dams, where strains caused by the settlement can be accommodated without cracking.

In comparison with compressive strength, the elastic modulus is more difficult to measure. Two parameters of loading and deformation should be monitored simultaneously to obtain the loading–deformation curve. A special gauge for the deformation of concrete is needed, as shown in Fig. 3.29.

When the concrete is under uniaxial compression, a lateral strain will develop simultaneously with vertical strain. The ratio of lateral to vertical strain is called Poisson's ratio. A common value for concrete is 0.20–0.21, but the value may vary from 0.15 to 0.25 depending upon the aggregate, moisture content, concrete age, and compressive strength. Poisson's ratio is generally of no concern to the structural designer. It is sometimes used in advanced structural analysis.

3.5.4 Factors affecting mechanical properties

It is well known that concrete is a porous inhomogeneous material. Generally speaking, the mechanical properties of concrete is closely related to its porosity. For an ideal homogeneous porous material, it can be described by the expression

$$S = S_0 f(p) \tag{3.3}$$

where S is mechanical property of the material which has a given porosity p; S_0 is intrinsic mechanical property at zero porosity.

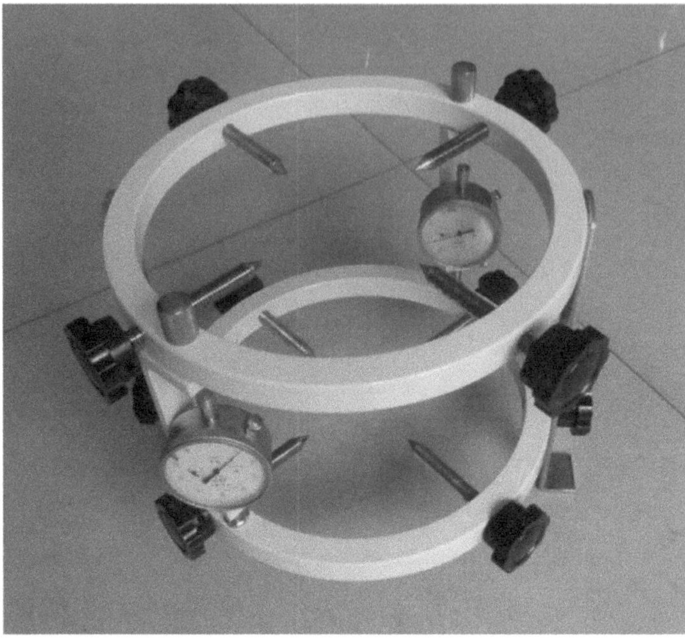

Figure 3.29 Gauge for the deformation of concrete under compression.

Compressive strength is the easiest mechanical property to measure. In contrast, tensile strength and elastic modulus are not easy to determine. Although there is no direct relationship between compressive and other mechanical properties, the other mechanical properties can be estimated from compressive strength. For instance, the ratio of tensile to compressive strength depends on strength grade. The higher strength is, the lower ratio is. For low-strength concrete: 0.10–0.11, moderate-strength: 0.08–0.09, and high-strength: 0.07. There are also many empirical equations established between compressive strength and elastic modulus. However, the relationship also depends on concrete types.

Compressive strength can be affected by many factors, such as water-to-binder ratio, curing time, aggregate. Although the actual response of concrete to applied stress is a result of complex interactions between various factors, to facilitate a clear understanding of these factors, they can be classified into three categories: ① characteristics of raw materials and mix proportions, ② curing conditions and loading conditions, and ③ testing parameters. Although all the factors affect the mechanical properties of concrete, the effects of these factors on different mechanical properties may be different. In general, the detailed factors affecting the mechanical properties of concrete are given in Fig. 3.30.

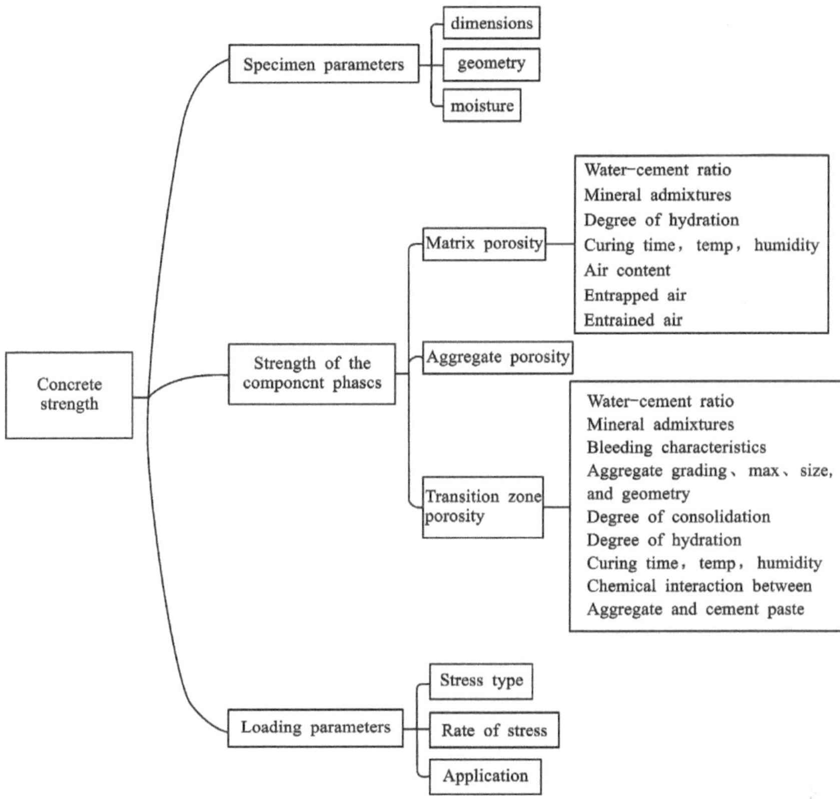

Figure 3.30 Factors influencing the mechanical properties of concrete.

It is worth mentioning that the curing condition has a significant influence on the strength development of concrete, as shown in Fig. 3.31. It can be seen that the curing at an early age is very important for the early strength. The later curing conditions also have a great influence on strength development at a later age.

3.6 Deformation

Concrete is prone to cracking when suffering from the surface water evaporation, the severe temperature change, the volume deformation and the steel bar corrosion, etc. Macro and micro cracks induced by restrained deformation of concrete greatly degrade the durability. Among all the factors affecting the cracking behaviors of concrete, the volume shrinkage is responsible for most scenarios of concrete cracking. Deformation of concrete may be caused by different factors, such as environmental effects

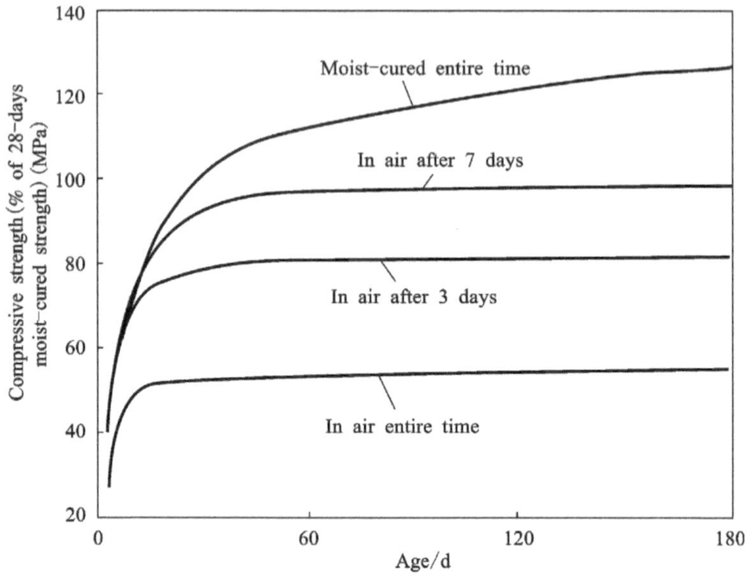

Figure 3.31 Effect of curing conditions on the concrete strength development.

(moisture gain or loss, heat expansion), and externally applied stress in both the short term and long term. A general view of the nature of the deformation behavior under the real environment is given in Fig. 3.32. This figure demonstrates the deformation of concrete under uniaxial compressive stress and in a drying environment. Before t1, which is the loading time, there is a drying or autogenous shrinkage strain. Once the loading is applied to the material, elastic deformation happens simultaneously, and the elastic modulus can be defined from the ratio of stress and deformation. The deformation strain increases with the loading time, which is called a creep strain. The creep strain develops very fast at an early age, and slowly develops at a later age. Although reducing in rate with time, the creep does not tend to a limiting value. It is worth mentioning that during the time between t1 and t2, drying shrinkage occurs all the time. Once the load is removed, at time t2, there is an immediate (elastic) strain recovery, which is often less than the initial strain on loading. This is followed by a time-dependent creep recovery, which is less than the preceding creep, i.e., there is a permanent deformation, but, unlike creep, this reaches completion in due course.

3.6.1 Drying shrinkage

Fresh and young concrete is submitted to delayed deformations developing without external loading, and inducing a volume reduction named

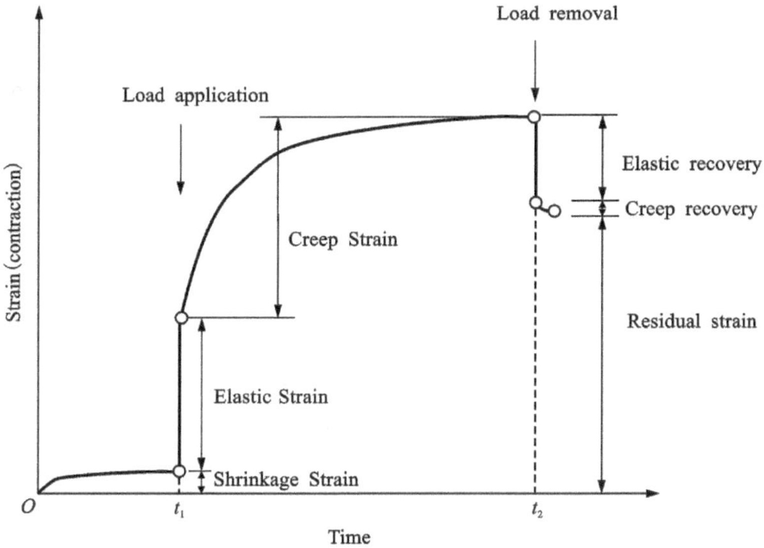

Figure 3.32 The response of concrete to compressive stress applied in a drying environment.

shrinkage. This reduction is caused by the loss of a part of the mixing water by evaporation, resulting in a reduction of cement paste volume. It depends on several parameters related to the composition of the concrete, the quality of its components, the sample size, and the external conditions of conservation.

The shrinkage of concrete is normally measured in the laboratory or on structural elements by determination of length change, instead of volumetric change, and it is therefore expressed as a linear strain. A considerable part of this is irreversible, i.e., is not recovered on subsequent rewetting. Further drying and wetting cycles result in more or less completely reversible shrinkage; hence, there is an important distinction between reversible and irreversible shrinkage. The water content first increases to make up for the self-desiccation during hydration, and to keep the paste saturated. Second, additional water is drawn into the C—S—H structure to cause the net increase in volume. Four principal mechanisms have been proposed for shrinkage and swelling in cement pastes, which are summarized below.

3.6.1.1 Capillary effect

The capillary effect is defined as the spontaneous flow of a liquid into a narrow tube or porous material. This movement often acts in opposition to gravity. The capillary effect is sometimes called capillary motion, or wicking. The capillary effect is caused by the combination of surface tension of the liquid and the adhesive forces between the liquid and tube material. Surface tension and adhesion are two types of intermolecular forces. These forces pull the liquid into the tube. In order for wicking to occur, a tube needs to be sufficiently small in diameter.

Cement-based materials are porous materials with various capillary pores. When water starts to evaporate due to a lowering of the ambient vapor pressure, the free surface becomes more concave, and the surface tension increases. The relationship between the radius of curvature, r, of the meniscus and the corresponding vapor pressure, p, is given by Kelvin's equation:

$$\ln\left(\frac{p}{p_0}\right) = \frac{2T}{R\theta\rho r} \tag{3.4}$$

where p_0 is the vapor pressure; T is the surface tension of the liquid; R is the gas constant; T is the absolute temperature; ρ is the density of the liquid. The tension within the water near the meniscus can be shown to be $2T/r$, and this tensile stress must be balanced by compressive stresses in the surrounding solid. Hence, the evaporation, which causes an increase in the tensile stress, will result in increased compressive stress, which is in the surrounding solid, and lead to a decrease in volume, i.e., shrinkage. Therefore, when cement-based materials are subjected to a drying condition, the water in the pore is continuously evaporated, and the corresponding vapor pressure is continuously decreased, and thus the pressure increases steadily. The capillary pores gradually empty from bigger size to smaller size. Higher water/cement ratio cement-based materials with higher porosities have higher shrinkage. It is worth to mention that the capillary effect is the result of the meniscus of liquid. When the water in pores is completely evaporated, the imposed stresses on the surrounding solid will also disappear. Thus, the capillary effect only takes effect in an environment with relative humidity higher than 50%.

3.6.1.2 Disjoining pressure

The sheet-like model for C—S—H gel is generally accepted. The two sheets of C—S—H are prevented from moving apart by an interparticle van der

Waals type bond force. The adsorbed water forms a layer about five molecules on the solid surface at saturation, which is under pressure from the surface-attractive forces. In regions narrower than twice this thickness, the interlayer water will be in an area of hindered adsorption. This results in the development of a swelling or dis-joining pressure, which is balanced by a tension in the interparticle bond. On drying, the thickness of the adsorbed water layer reduces, as does the area of hindered adsorption, hence reducing the disjoining pressure. This results in an overall shrinkage.

3.6.1.3 Movement of interlayer water

The third type of evaporable water, the interlayer water, may also have a role in drying shrinkage. Since the strong interaction between interlayer water and solid surface, the loss of interlayer water is only in very harsh conditions. A steep hygrometric energy gradient is needed to move it, and such movement is likely to result in significantly higher shrinkage than the movement of an equal amount of free or adsorbed water. The above discussions apply to the reversible shrinkage only, but the reversibility depends on the assumption that there is no change in a structure during the humidity cycle.

3.6.2 Creep

When concrete is loaded, there are two parts of the deformation caused by the load: ① the elastic deformation that occurs immediately with loading; ② the time-dependent deformation that begins immediately but continues at a decreasing rate as long as the concrete is loaded, as shown in Fig. 3.32. The second deformation is called creep. Creep of concrete is a manifestation that the relationship between stress and strain is a function of time and, since moisture movement readily occurs in concrete under loading, there are many influencing factors affecting creep in compression: type and content of aggregate, w/c ratio, stress/strength ratio, type of cement, age at loading, size and shape of member, storage environment, type of load, time under load, chemical and mineral admixtures, temperature, etc.

Within normal stress ranges, creep is proportional to stress. In relatively young concrete, the change in volume or length due to creep is largely unrecoverable; in older or drier concrete, it is largely recoverable. The magnitude of the creep strains can be several times higher than the elastic strains on loading, and they, therefore, often have a highly significant influence on structural behavior. Particularly, creep is crucial to prestressed concrete. Creep of prestressed concrete causes the loss of prestress. As a consequence, the structural performance of the prestressed element might be affected. In addition, the creep of concrete does not show a limit, even at the duration of 30 years. The creep is substantially increased when the concrete is simultaneously drying, i.e., creep and shrinkage are interdependent. In practice, creep always happens together with drying shrinkage. In order to clearly understand the shrinkage behavior of concrete, free shrinkage, basic creep, and total creep are defined, as shown in Fig. 3.33. Free shrinkage (ε_{sh}) is defined as the shrinkage of the unloaded concrete in the drying condition, and basic creep (ε_{bc}) as the creep of a similar specimen under load, but not drying, i.e., sealed so that there is no moisture movement to or from the surrounding environment. The total strain (ε_{tot}) is the strain measured on the concrete under simultaneously shrinking and creeping. It is found that

$$\varepsilon_{tot} > \varepsilon_{sh} + \varepsilon_{bc} \tag{3.5}$$

The difference, i.e., $\varepsilon_{tot}-(\varepsilon_{sh}+\varepsilon_{bc})$, is called the drying creep (ε_{dc}).

Since the creep process is occurring within the cement paste, and the moisture content and movement have a significant effect on its magnitude.

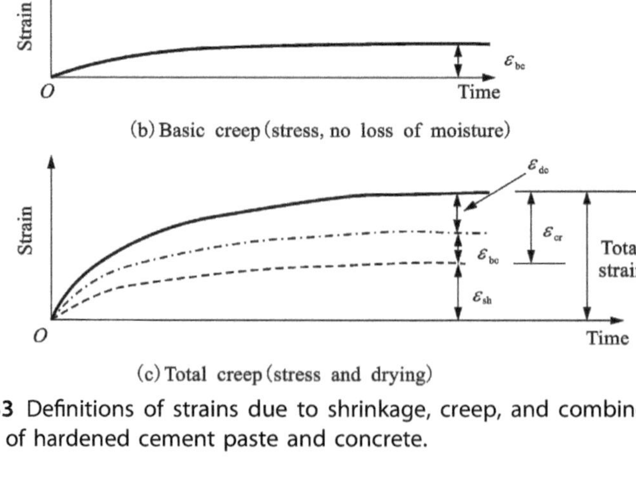

Figure 3.33 Definitions of strains due to shrinkage, creep, and combined shrinkage and creep of hardened cement paste and concrete.

Thus, shrinkage and creep have some similarities. The mechanisms for creep are described as follows:

3.6.2.1 Moisture movement

When concrete is under compression, which causes changes in the internal stresses and strain energy, change in the thermodynamic equilibrium occurs. As a consequence, moisture then moves from regions with high energy to the regions with low energy due to the free energy gradient. This moisture movement can occur at several levels:

① Capillary level. The water in the capillary pore can move rapidly after the action of stress, and this is normally reversible.

② Adsorbed water. The adsorbed water to the cement paste move gradually from regions of hindered adsorption, and this movement should be reversible;

③ C—S—H lever. The interlayer water in C—S—H gel pores diffuses very slowly out of the layer under the action of stress. Some extra bonding may then develop between the solid layers, and so this process may not be completely recoverable. This is the main mechanism for basic creep.

3.6.2.2 Structural adjustment or microcracking

There are many defects and microcracks in hardened cement concrete. When relatively high stress is applied to the concrete, propagation of these cracks and the formation of new cracks will contribute to the creep strains. And the structure of the solid may be adjusted because of both microcracking and moisture movement. The mechanisms are essentially irreversible.

3.6.2.3 Delayed elastic strain

Water in the capillary or gel pores is responsible for the creep of hardened cement and concrete. When the water is redistributed due to the action of stress, stress redistribution will occur, and the stress will be transferred to other solids, such as unhydrated cement particles, calcium hydroxide, and ettringite crystals, even the aggregates. As a result, elastic deformation happens. The process acts in reverse on the removal of the load, so that the material finally returns to its unstressed state; thus, the delayed elastic strain would be fully recoverable.

3.6.3 Chemical shrinkage

Chemical shrinkage refers to the volume change during the early ages of hydration resulted by the formation of hydration products with lower volume in comparison with the volume of the initial reactants (water and cement) during the hardening process. Since chemical shrinkage is based merely on the volumes of initial and final products, it is possible to calculate it based on molecular weights. The basic reactions of cement clinker can be defined by the following symbolic equations of the four clinker phases:

$$
\begin{gathered}
2C_3S + 6H \rightarrow C_3S_2H_3 + 3CH \\
2C_2S + 4H \rightarrow C_3S_2H_3 + CH \\
C_3A + 6H \rightarrow C_3AH_6 \\
C_4AF + 2CH + 10H \rightarrow C_3AH_6 + C_3FH_6 \qquad (3.6)
\end{gathered}
$$

As seen from the above equations, each of the four cement clinker phases requires water for reacting. These processes are exothermic and result in a decreased volume of the reaction products. This volume reduction, or chemical shrinkage, begins immediately after mixing of water and cement, and the rate is greatest during the first hours and days. The magnitude of chemical shrinkage can be determined using the molecular weight and densities of the compounds as they change from the basic to reaction products. Based on the published data, the chemical shrinkage of each clinker phase can be calculated, and are shown as follows:

① C_3S: 0.0532 cm^3/g
② C_2S: 0.04 cm^3/g
③ C_3A: 0.1113 cm^3/g
④ C_4AF: 0.1785 cm^3/g

It can be seen that the cement chemistry will affect the shrinkage due to the varying chemical shrinkage at the very early age. For instance, if cement has a high C_3A or C_4AF content, it is expected that there will be greater shrinkage than a comparable cement with lower C_3A or C_4AF. The chemical shrinkage of cement paste is not affected by the *w/c* ratio. The *w/c* ratio and cement fineness will only affect the rate of the chemical shrinkage. The final magnitude of the shrinkage as the degree of hydration approaches 100% will only be influenced by the chemical composition of the cement.

It is worth to mention that the chemical reactions of the clinker are far more complicated than the equations mentioned above, and the density of the reaction products is also more complicated. Thus, the chemical shrinkage values of each clinker phase are just a rough estimation.

3.6.4 Autogenous shrinkage

In general, the part of shrinkage which does not include any volume change due to loss or ingress of substances, temperature variation, and application of an external force and restraint can be considered as autogenous shrinkage. Therefore, it is also referred to as self-desiccation shrinkage. A technical committee on autogenous shrinkage at the Japan Concrete Institute defined autogenous shrinkage as the macroscopic volume reduction of cementitious materials when cement hydrates after the initial setting. To some extent, the autogenous shrinkage can be referred as the measured deformation of cement paste in a closed system. The significant influence of autogenous shrinkage only limit to high- or ultrahigh-performance concrete in which a high amount of cementitious materials and low water-to-binder ratio are applied, and the self-desiccation is highly triggered in the paste. In addition, autogenous shrinkage may be the only form of shrinkage occurring in the center of a large mass of concrete, but its magnitude is normally at least an order of magnitude less than that of drying shrinkage. It can, however, be larger and more significant in concrete with very low water-to-cement ratios.

The development of drying shrinkage of concrete is lengthy relative to the autogenous shrinkage. Accurate measurement of drying shrinkage is a challenge as the autogenous shrinkage deformation of the sealed specimens should be deducted from the total measured deformation of the concrete due to their distinct physical definition. The drying shrinkage value measured by the traditional method contains part of autogenous shrinkage; however, it is not a simple superposition, since the drying condition has a serious effect on the hydration of cement.

3.6.5 Thermal expansion

In common with most other materials, cement paste and concrete expand on heating. Knowledge of the coefficient of thermal expansion is needed in two main situations. Firstly, it can be used to calculate stresses due to thermal gradients arising from heat of hydration effects or continuously varying diurnal temperatures. Secondly, it can be used to calculate overall dimensional changes in structures such as bridge decks due to ambient temperature variations. The measurement of thermal expansions on laboratory specimens is relatively straightforward, provided sufficient time is allowed for thermal equilibrium to be reached (at most a few hours). However, the in situ behavior is complicated by the observation that, unlike most other materials, the thermal movement is time-dependent, and, as with shrinkage, it is difficult to estimate movement in structural elements from those on laboratory specimens.

The thermal expansion causes problems when the rate of temperature change is too severe, and when gradients exist over the concrete's cross section. The gradient resulting from uneven temperatures will cause strains and may result in cracking. During the early ages, differential temperatures within a concrete specimen cause thermal strains as the exterior surface will have a different temperature (due to environmental exchanges) than the interior. The gradient develops when temperature equilibrium cannot be reached, thus generating stresses and risking cracking. This becomes even more of a risk with massive concrete structures (i.e., the thickness of structure is greater than 1 m) since it takes much longer to obtain a temperature equilibrium.

Each concrete has a thermal expansion coefficient which is dependent on the individual material properties (such as aggregate and *w/c*). During early ages, the thermal dilation coefficient is changing very rapidly as the concrete strength develops, depending mainly on the moisture content. The thermal expansion coefficient reaches a constant level of approximately 12 $\mu\varepsilon/^{\circ}C$ ($12 \times 10^{-6}/^{\circ}C$) after 24 h. For comparison, the thermal expansion coefficient of water at 23°C is 237 $\mu\varepsilon/^{\circ}C$.

The thermal expansion coefficients of the most common rock types used for concrete aggregates vary between about $(6-10) \times 10^{-6}$ per °C, i.e., lower than the values for pure cement paste. The thermal expansion coefficient for the concrete is, therefore, lower than that for cement paste. Furthermore, since the aggregate occupies 60% to 80% of the total concrete volume, there is a considerable reduction of the effects of humidity that are observed in the paste alone, to the extent that a constant coefficient of thermal expansion over all humidities is a reasonable approximation. The value depends on the concrete mix proportions, chiefly the cement paste content, and the aggregate type; for normal mixes, the latter tends to dominate.

3.7 Durability

The durability of Portland cement concrete is defined as its ability to resist weathering action, chemical attack, abrasion, or any other process of deterioration. In other words, a durable concrete will retain its original form, quality, and serviceability when exposed to its intended service environment. No material is inherently durable. As a result of environmental actions, the microstructure and the properties consequently change with time. A material is assumed to reach the end of service life when its properties, under given conditions of use, have deteriorated to the extent that its continued use is ruled either unsafe or uneconomical.

Concrete has long been regarded as very durable material. During the last decades, more and more durability problems of concrete occurred all around the world. Premature failure of concrete infrastructure has caused huge economic loss, due to the durability problem. Therefore, concrete structure design based on durability has become more and more popular. In fact, concrete is a quite vulnerable material to many aggressive substances, such as sulfate ions, CO_2, FT cycles. In the meantime, concrete may also suffer from the durability problem caused by inappropriate raw materials, such as alkali-aggregate reaction (AAR). In addition, concrete is often used in combination with a steel bar; the corrosion of steel bar in concrete is a major durability problem for reinforced concrete. Thus, it is also necessary to distinguish between degradation of the concrete itself and loss of protection and subsequent corrosion of the reinforcing or prestressing steel contained within it.

In most cases, the rate of the degradation processes is controlled by the rate at which moisture, air, or other aggressive agents can penetrate into the concrete. This penetrability is a unifying theme when considering durability, and for this reason, we shall first consider the various transport mechanisms through concrete. However, different durability issues should also be discussed separately, since the deterioration mechanisms are different.

It is worth to mention that deterioration of concrete is seldom due to a single cause in practice. Usually, at an advanced stage of a material's degradation, more than one deleterious phenomenon is at work. In general, the physical and chemical causes of deterioration are so closely intertwined and mutually reinforcing that separation of the causes from their effects often becomes impossible. Therefore, a classification of concrete deterioration processes into neat categories should be treated with some caution. The purpose of such a classification is to explain systematically and individually the various phenomena. However, one must not overlook the interactions that occur when several phenomena are present simultaneously.

3.7.1 Permeability

As an important physical property of various porous materials, such as soils, rocks, and concretes, permeability attracts great attention in both research and engineering communities. Permeability also plays an important role in quantitatively analyzing the mass transport process and, thus, the durability performance of concrete structures under various environmental actions. Starting from the hydration of cement, part of the water is reacted with cement, and part of the water remains in the pore of the cement matrix. When the evaporable water in the concrete is lost, leaving the pores empty or unsaturated. The empty or partially saturated pores are connected, and makes concrete a percolated structure. In addition, many microcracks develop during the process of hydration. These cracks and pores are the transporting channels of the mass. A satisfactory fluid transport property factor is difficult to determine because of the effect of unpredictable changes in the pore structure upon penetration of fluid from outside. Note that the fluid transport property of the material is changing continuously because of cycles of narrowing and widening of the pores and microcracks due to ongoing physical—chemical interactions between the penetrating fluid and the minerals of the cement paste. Permeability is defined as the property that governs the rate of flow of a fluid into a porous solid. For steady-state flow, the coefficient of permeability (K) can be determined from Darcy's expression:

$$\frac{dq}{dt} = K\frac{\Delta HA}{L\mu} \tag{3.7}$$

where d_q/d_t is rate of fluid flow; m is viscosity of the fluid; ΔH is pressure gradient; A is surface area; L is thickness of the solid.

In Chinese standard, the water permeability was measured by the testing setup shown in Fig. 3.34.

In this test, six concrete specimens were mounted on the testing setup, and the water pressure is exerted on the concrete specimen from the bottom gradually. The maximum water pressure that the specimen can stand for 8 h without water penetration is the water resistance grade of concrete.

The permeability coefficient of concrete to gases and water vapor is much lower than the coefficient for liquid water; therefore, tests for measurement of permeability are generally carried out using water that has no dissolved air. Due to their interaction with cement paste, the permeability values for solutions containing ions would be different from the water permeability.

Theoretically, the intrinsic permeability of solid porous material is independent of the inert percolating liquids and gases if the compressibility and slippage effect of gas flow are taken into account. Actually, the permeability to water and gas for some rocks are observed to be similar or at

Figure 3.34 The testing setup for water permeability of concrete.

least at the same order of magnitude. However, it has been broadly observed that the water permeability of modern cement-based material is always anomalously lower than that permeability coefficients to many other fluids like ethanol and nitrogen gas by as large as two to three orders of magnitudes.

Numerous efforts have been paid to discover the physical principles about the anomalous low water permeability, which are mostly attributed to the ambiguous interactions between water and cementitious materials. It is said that the microcracks induced by sample preparation may dramatically increase the following measured gas permeability, which is obviously larger than water permeability. Since newly developed microcracks will, of course, obviously increase the permeability, it sounds plausible but cannot reasonably give the reason why the permeability to ethanol is rather close to the measured intrinsic gas permeability. In another aspect, although the continued hydration, autogenous crack healing, and self-sealing effect, which is triggered by the drying action, all make the water permeability decrease with time during experimental testing, the water permeability of mature intact undried mortar material is still at least an order of magnitude lower than its permeability to isopropanol. It was pointed out that the anomalous low water permeability is physically ascribed to the swelling of C—S—H gel specific to water, which makes the pore structure finer at a water-saturated state. In contrast, the water removal treatment through isopropanol exchange or drying will remarkably coarsen the pore structure of cement-based material due to the contraction of C—S—H gel. Isopropanol exchange can extract interlayer water of cement-based material without replacement and make the interlayer pores collapse, whereas water can penetrate into the amorphous C—S—H gel and make it swell. This sensitivity of pore structure to water is closely related to the disjoining pressure of an aqueous solution.

Usually, aggregates have quite low porosity, and thus low permeability. Theoretically, the introduction of low-permeability aggregate particles into a high-permeability cement paste is expected to reduce the permeability of the system because the aggregate particles should intercept the channels of flow within the cement paste matrix. Concrete with the same water—cement ratio and degree of maturity should have a lower coefficient of permeability than that of a neat cement paste. However, the addition of aggregate to a cement paste or a mortar increased the permeability considerably. The larger the aggregate size, the greater the coefficient of permeability.

ITZs between aggregate and the cement paste are responsible for this phenomenon. ITZs are characterized by high porosity, microcracks. Consequently, ITZs are very porous and highly permeable. The aggregate size and grading affect the bleeding characteristic of a concrete mixture that, in turn, influences the ITZ.

Generally speaking, the durability issues can be classified into two categories.

① Permeability related: sulfate attack, chloride ingression, CO_2 ingression, etc.

② Permeability unrelated: abrasion, FT cycle, AARs, fire resistance, etc. They are addressed as discussed below.

3.7.2 Sulfate attack

Sulfate ions can come from seawater, contaminated aggregates, groundwater containing sulfates from clay soils, fertilizers and industrial effluent, etc. Thus, concrete infrastructures are often in contact with sulfate ions. Chemical or/and physical reactions may occur between concrete and sulfate ions. Thus, sulfate attack can be classified into both physical sulfate attack and chemical sulfate attack.

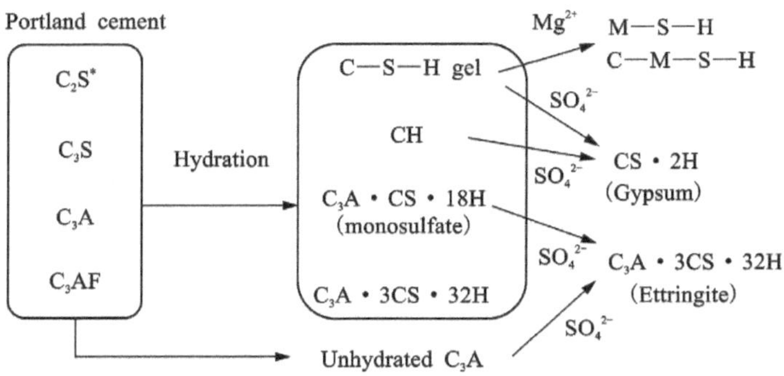

Figure 3.35 Reactions of sulfates with the portlandite and tricalcium aluminate in hydrated concrete.

A prevalent form of physical attack is the reversible change of anhydrous sodium sulfate into decahydrate. If crystallization takes place in the pores at or near the surface of concrete, large pressure may develop, with consequent deleterious action. It is reported that, tensile hoop stress of 10—20 MPa develops by the crystallization of mirabilite (decahydrate) from a saturated solution of thenardite (anhydrite) at 20°C. The tensile stress of such magnitude would inevitably disrupt concrete. The development of crystallization pressure requires supersaturation in the liquid film between the crystals and the pore wall.

In the case of chemical sulfate attack, sulfate ions can react with the portlandite and tricalcium aluminate in hydrated concrete to produce ettringite or gypsum, which causes disruptive expansion, as shown in Fig. 3.35. In addition, if magnesium ions present, C—S—H may transform into M—S—H, which has no binding capacity.

Despite numerous studies and expertise, some degradation mechanisms remain unclear and controversial. The formation of gypsum and its effects are still far from clear. Even some researches argue that the formation of gypsum does not cause expansion. In the meantime, the formation of ettringite is not crystal clear, and the expansion consequence is also questioned by some researches. It is worth to mention that sulfate attack is seldom the sole phenomenon responsible for the concrete deterioration in practice, and the threat of structural failures due to sulfate attack seems to be even less of a threat than that caused by alkali—silica reaction.

$$Ca(OH)_2 + NaSO_4 + 2H_2O \rightarrow CaSO_4 \cdot 2H_2O + NaOH \quad (3.8)$$

$$Ca(OH)_2 + MgSO_4 + 2H_2O \rightarrow CaSO_4 \cdot 2H_2O + Mg(OH)_2 \quad (3.9)$$

$$CaSO_4 \cdot 2H_2O + 3CaO \cdot Al_2O_3 + 26H_2O \rightarrow 3CaO \cdot Al_2O_3 \cdot CaSO_4 \cdot 32H_2O \quad (3.10)$$

The Bureau of Reclamation of America has played a preeminent early role in studying sulfate attack on concrete and in developing sulfate-resisting concrete. An extensive and independent long-term investigation of the behavior of concrete exposed to sodium sulfate was started by the Bureau of Reclamation in the 1950s. The specimens were continuously exposed to sulfate over a period of more than 40 years. In addition, the Bureau conducted accelerated tests.

Many valuable information are obtained from the study by the Bureau of Reclamation. The time of failure of these samples is influenced by their water-to-cement (*w/c*) ratio, cement composition, and percent replacement of cement with fly ash. The analysis indicates that there is a "safe zone" for concrete made with *w/c* ratio lower than 0.45 and cement with unhydrated tricalcium aluminate (C_3A) content lower than 8% where failure did not occur within the 40-year exposure period. As expected, concrete samples cast with a high amount of C_3A fail after a relatively short time of sulfate exposure. Expansion tests indicate that types of cement containing high amounts of C_3S might lead to premature failure of concrete, even when moderate *w/c* ratios are used. Samples prepared with 25% and 45% replacement of cement with fly ash show significantly less expansion than comparable mixtures containing no pozzolans.

Based on their severity on concrete, America Concrete Institute (ACI) classifies the exposure conditions based on the sulfate concentrations in soil and water, as shown in Table 3.12.

In order to make sulfate-resistant concrete, the following guidelines should be followed:

① Lower C_3A content of sulfate-resisting Portland cement.

② Lower water-to-cement ratio of the concrete; higher quality concrete is less vulnerable due to its lower permeability. This is a more significant factor than the C_3A content.

③ Incorporate supplementary cementitious materials, which can decrease the permeability, reduce the amount of free lime in the paste, and effectively "dilute" the C_3A.

Table 3.12 ACI 201.2R Guide to durable concrete.

	Concentration of sulfates as SO_4^{2-}	
Class of exposure	**In groundwater ($\times 10^{-6}$)**	**Water- soluble in soil (%)**
Mild	<150	<0.10
Moderate	150–1500	0.10–0.20
Severe	1500–10,000	0.20–2.00
Very severe	>10,000	>2.00

3.7.3 Acid attack

The hardened cement paste contains more than 20% calcium hydroxide; therefore, concrete is alkaline, and the pH value of the pore solution in concrete normally reaches 13 or even higher. Of course, concrete with supplementary cementitious materials may have lower pH value, due to pozzolanic reaction. Concrete is vulnerable to acid, due to the neutralization reaction. Examples of acids that commonly come into contact with concrete are dilute solutions of CO_2 and SO_2 in rainwater in industrial regions, and $CO_{3}{}_{2-}$ and H_2S-bearing groundwater. The acids attack the calcium hydroxide within the cement paste, converting it, in the case of CO_2, into calcium carbonate and bicarbonate. The latter is relatively soluble, and leaches out of the concrete, destabilizing it. The process is thus diffusion-controlled, and progresses at a rate approximately proportional to the square root of time. The C—S—H may also be attacked, as can calcareous aggregates such as limestone. The rate of attack increases with reducing pH. As mentioned above, the attack process is diffusion rate controlled. In order to improve the acid resistance of concrete, it is possible to produce concrete which is adequately durable for many common circumstances by giving attention to low permeability and good curing. In addition, concretes containing cementing supplementary materials also have greater resistance due to the lower calcium hydroxide content as a result of the pozzolanic reaction. In cases where some extra acid resistance is required, such as in floors of chemical factories, the surface can be treated with diluted water glass (sodium silicate), which reacts with the calcium hydroxide forming calcium silicates, blocking the pores. As a consequence, the permeability of concrete is greatly decreased. In more aggressive conditions, the only option is to separate the acid and the concrete by, for example, applying a coating of epoxy resins or other suitable paint systems to the concrete.

3.7.4 Freezing—thawing cycle

In cold areas, damage may occur to concrete infrastructures, such as pavements, retaining walls, bridge decks, and railings, due to the action of FT cycles. FT damage is a contributing factor in eventual rupture and erosion in concrete structures; continuous FT cause cracks to spread and weakens the concrete to the point of failure. FT damage is one of the major problems requiring heavy expenditures for the repair and replacement of structures.

The most common deterioration forms of concrete due to FT damage are cracking and spalling. Concrete slabs exposed to FT cycles in the presence of moisture and deicing salts are susceptible to scaling. Also, some coarse aggregates in concrete slabs are known to cause cracking, usually parallel to joints and edges, which eventually acquires a pattern resembling a large capital letter D (cracks curving around two of the four corners of the slab). This type of cracking is described by the term D-cracking. Frost damage causes the progressive development of cracks; consequently, the mechanical properties decay, such as dynamic elastic modulus, compressive strength. Especially, the dynamic elastic modulus is often used as an indicator to evaluate the damage extent of concrete due to FT cycles.

No agreement exists among researchers regarding the mechanism responsible for FT damage in cement paste. Researchers attribute the damage to hydraulic pressure buildup that forced water away from the freezing point, osmotic pressure gradients that compel water to move toward freezing points, vapor pressure potentials, and a combination of all these. Two classical theories are responsible for damage in concrete proposed by Powers. According to hydraulic pressure theory, liquid in air voids can freeze with no damage to a concrete structure. The water in capillary pores freezes first, thus, force the liquid to move to the rest portion of the concrete structure. In contrast, according to theoretical considerations and test results, the water moves in the opposite direction during the freezing process. That is to say, the water in pores moves toward the frozen part of the matrix, as shown in Fig. 3.36. Powers deduced, in osmotic pressure

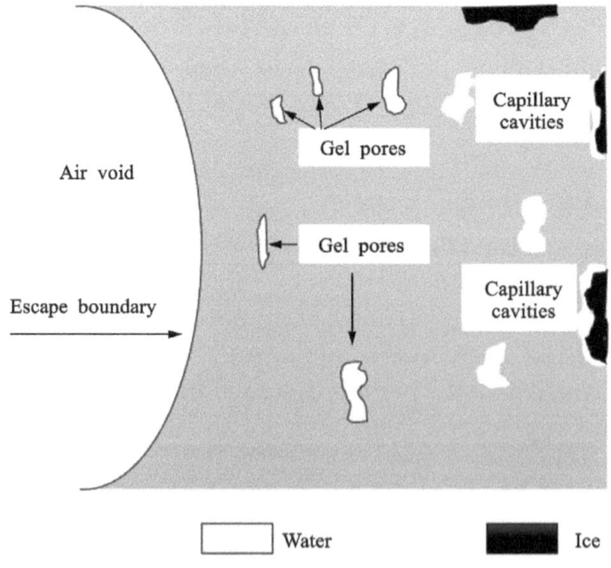

Figure 3.36 Schematic diagram of hydrostatic pressure hypothesis.

theory, that unfrozen water moved toward the frozen part due to the concentration of dissolved substances (mainly Ca^{2+}, Na^{+}, and K^{+}). It is also believed that water in smaller pores or adsorbed on the surface cannot freeze because of water and surface interaction. Due to variation in the vapor pressure of ice in cement paste and unfrozen and supercooled liquid in smaller pores, the supercooled liquid will move toward the location where it can freeze, for example, outer surface or larger pores. The accumulation of ice in cracks and crevices and partial desiccation of the paste are the results of this process. Concrete structure fails in case of poor redistribution of water either because of high *w/c* ratio, rapid cooling, or the lack of air entrainment (AE) in concrete. Such freezing makes a semiamorphous solid, which induced the greater internal stress. Through AE, cement paste can be made resistant to FT damage in case of sufficient strength and maturity (unless filling of air void due to unusual exposure conditions). However, only AE in cement paste does not make concrete resistant to freezing; therefore, freezing in aggregates should be taken into account.

Parameters for the air void system (AVS) in concrete include total air void volume, air void size and distribution, specific surface, and spacing factor. The shortest length between any point in cement paste and the nearest boundary of the air void is called a spacing factor. According to Powers, the FT resistance of concrete highly depends on the air void size and distribution. Many researchers believe that the FT resistance of concrete, besides mechanical strength, mainly depends on the spacing factor and volume of air content. The importance of spacing factor for improving FT resistance of concrete and stress the importance of the total surface area of air voids for enhancing FT resistance of concrete are also questioned.

Although AE in concrete has been extensively studied, ensuring adequate and stable AVS in concrete is still challenging since many factors affect it in both fresh and hardened states. Firstly, concrete ingredients have a great influence on the AE, such as content, fineness, and chemical composition of Portland cement; supplementary cementitious materials and chemical admixtures, for example, high-range water reducer; amount, grading, and maximum size of aggregates; water-binder ratio and slump of concrete; chemical characteristics of mixing water. Secondly, the production procedures are also important affecting factors, such as mixing capacity, mixing time and speed, and sequence of materials addition. Thirdly, construction practices and field conditions have a great influence on the AE as well, such as transporting, delivery, retempering, placement, consolidation, finishing, and ambient temperature. Some of the published results are contradictory to each other.

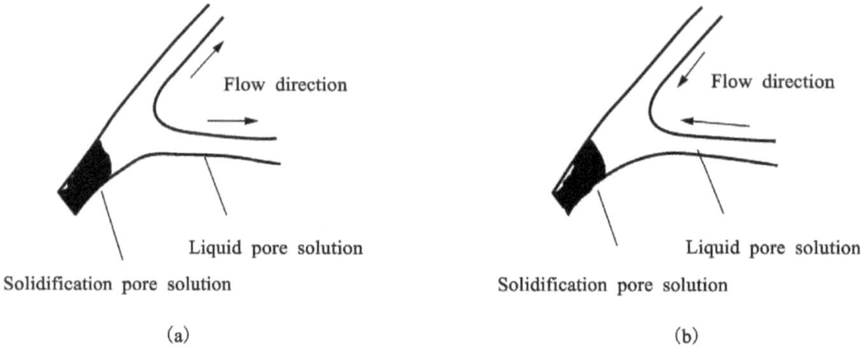

Figure 3.37 The difference between hydrostatic pressure hypothesis and osmotic pressure hypothesis.

The hydraulic pressure theory has long been regarded as a classic explanation for the freezing behavior of concrete. The hydraulic pressure theory is now recognized to be incorrect, as it is established that the freezing water tends to move to capillary pores, opposing the water flowing from capillary pores assumed in the hydraulic pressure theory. The complete osmotic pressure theory explains that with decreasing temperature, the water first begins to freeze in capillary pores, raising the concentration of dissolved chemicals, forcing water in smaller pores to travel to larger pores to reestablish the equilibrium, as shown in Fig. 3.37. If the specimen is saturated, the mentioned process builds internal pressure, which may damage the specimen, provided passed a certain level.

Another theory argued that the water in capillary pores did not freeze in situ; when the temperature drops below 0°C, water becomes supercooled and tends to travel to a surface to freeze, which results in desiccation of the specimen. The damage is the result of the desorption process, which happens when the concentration of water is much higher than it should be according to equilibrium. The air voids reduce the traveling distance to a freezing surface, thereby facilitating the process of desorption, permitting more water to leave the pores and protecting the specimen.

In order to combat the frost damage on concrete, many strategies have been proposed by researchers. They can be categorized into four types.

3.7.4.1 Providing extra space for ice expansion using air bubbles

The development of air-entrained concrete is one of the greatest advancements in concrete technology. Air entrainment involves the incorporation of small air bubbles into the concrete mix by using air-entraining agents so that the concrete contains tiny air bubbles uniformly distributed throughout the cement paste. The AVS can be characterized by three parameters: the bubble size, the spacing factor, and the pore size distribution. Spacing factor is believed to be the most important factor influencing the FT resistance. It has been generally accepted that utilization of air-entraining agents is an effective method combating FT cycles. The side effect is the reduction in compressive strength. Moreover, air entrainment can improve the workability of concrete.

3.7.4.2 Reducing porosity and refining pores using pozzolans and fillers

The second method developed by researchers for minimizing the FT effect on the concrete material is refining the pores and reducing the porosity of concrete in order to decrease water penetration into the concrete. The expansion of pores originates from increasing hydraulic pressure and osmotic phenomenon, which both are responsible for generating microcracks. To achieve better frost resistance, pore refinement and impermeability are needed, so consolidation agents, including micro- and nanosized pozzolans and fillers, can be used to control susceptible points.

3.7.4.3 Containing cracks using fibers, tubes, and sheets

The third mechanism to control the FT damage in concrete is to prevent crack propagation by the incorporation of microfibers, nanotubes, and nanosheets. These additives can bridge cracks in concrete and consume the energy from water expansion due to frost.

3.7.4.4 Reducing water absorption through hydrophobic concrete

Integral waterproofing by using hydrophobic or superhydrophobic admixtures in the batching process can result in better frost resistance through a reduction in water absorption.

3.7.5 Fire resistance

Concrete is incombustible and does not emit any toxic fumes when exposed to elevated temperatures. It is thus a favored material, both in its own right and as protection for steelwork, when fire resistance of structure safety is being considered. However, both physical and chemical changes occur under the elevated temperatures. Although it can retain some strength for a reasonable time at high temperatures, it will eventually degrade. The changes of concrete, paste, and aggregate at elevated temperatures are summarized as follows:

① Aggregate. Free water starts to evaporate at 100°C. At 570°C, quartz starts to expand. Limestone undergoes decarbonation between 600 and 800°C and gets completely disintegrated between 800 and 1200°C.

② Cement paste. Free water starts to evaporate at 100°C, and interlayer and capillary water loss between 100 and 300°C. Ettringite starts to decompose above 70°C. In the range of 300–700°C, calcium silicate hydrate gel completely loses water and decomposes. Similarly, the calcium hydroxide decomposes into lime and water in the temperature range of 350–550°C.

③ At 300°C, microcracks start to appear in the concrete matrix due to the expulsion of water vapors. Above 400°C, these cracks grow further, and their intensity rises with the successive rise in temperature. Furthermore, a rise in porosity is witnessed due to the microstructural deterioration upon exposure to high temperatures. The differential expansion between the hardened cement paste and aggregate, that is initiated in the transition zone, results in thermal stresses and cracking.

Exposure to fire may lead to the occurrence of spalling in concrete, which adversely affects the integrity of concrete, compromising its mechanical properties, especially the compressive strength. Spalling of concrete may even become the source of the collapse of critical structural units. The nonuniform distribution of thermal stresses and accumulation of vapor pressure in pores are the two main causes for explosive spalling.

Exposure to high temperatures causes the vaporization of the entrapped water. Water from the inner portion of concrete tends to move toward the outer portion. This vaporization results in the development of thermal gradient, as the inner layers of concrete become cooler due to vapors condensation and formation of a saturated layer. The formed saturated layer further resists the movement of vapor from the inner portion of concrete, which in turn increases the pore pressure in the inner layers. This pore pressure can only reduce if the saturated layer comprising of condensed vapors somehow dissipates or finds escape to the surrounding air. If the heating rate is high and the pore structure of the concrete is sufficiently dense, the escape of the vapor layer is not fast enough. This causes the development of immense pore pressure and results in explosive spalling and, subsequently, the dismemberment of structural units if the tensile capacity of concrete is not adequate. Another reason is the development of a thermal gradient in concrete at elevated temperatures. When the surface temperature of the concrete rises, some stresses that are compressive in nature develop parallel to the surface of the heated surface, while some perpendicular tensile stresses are developed as well. When these differential stresses exceed the tensile capacity of concrete, it spalls off.

The incorporation of thermally unstable fibers, i.e., poly-propylene fibers, is an effective way to improve the fire resistance of high strength concrete. In the case of the addition of polypropylene fibers, the ability to reduce cracking is due to the fact that concrete permeability increases suddenly between 80 and 130°C, and polypropylene, once it has reached the melting point, flows through the cracks and produces channels allowing the water vapor and gases to be evacuated releasing the pore pressure. Fire resistance can also be increased by air entrainment in concrete. Air bubbles can provide desire porosity, thus reduce pore pressure and thermal stresses, and enhance concrete resistance to heat-induced spalling.

3.7.6 Alkali-aggregate reaction

Aggregates account from 60% to 80% vol. in concrete and are considered as one of its major components, which justify the crucial importance of their properties. In addition to their physical properties, aggregates should be chemically neutral without any harmful components. Metastable reactive silica in aggregates could be very destructive because of their possible reaction with the cement alkalis and production of a harmful gel called alkali-aggregate reaction (AAR). AAR may lead to expansion and cracking of the concrete, leading to loss of strength and elastic modulus. In fact, it is considered as one of the major durability problems of concrete all over the world, which can lead to the decreased service life of the infrastructures and the increased maintenance and reconstruction costs. It is called cancer of concrete, since it is almost no cure for concrete suffered from AAR damage.

In regard to alkali-reactive aggregates, depending on the time, temperature, and particle size, all silicate or silica minerals, as well as silica in hydrous (opal) or amorphous form (obsidian, silica glass), can react with alkaline solutions, although a large number of minerals react only to an insignificant degree. Feldspars, pyroxenes, amphiboles, micas, and quartz, which are the constituent minerals of granite, gneiss, schist, sandstone, and basalt, are classified as innocuous minerals. Opal, obsidian, cristobalite, tridymite, chalcedony, chert, andesite, rhyolites, and strained or metamorphic quartz have been found to be alkali-reactive in the decreasing order of reactivity.

A reaction occurs between the hydroxyl ions (OH^-) in the pore solution and the reactive silica in the aggregate, i.e., silica is not directly attacked by the alkalis (Na_2O and K_2O). Then, hydroxyl ions attack the nonbridge oxygen atoms in silica structure, leading to the decomposition of the silica structure and penetration of the hydroxyl ions into the interior phases of the poorly crystallized silica. On the other hand, at the well-crystallized silica the reaction takes place only at the surface of the structure so slowly that it cannot decompose the silica structure, causing the AAR to stop. The penetration of hydroxyl ions ruptures the Si—O—Si bonds and loosens the network. The silanol groups (≡Si—OH) on the surface of the reactive particles generate a negative charge density, which increases by increasing the pH and ionic strength of the ambient solution, which is related to the surface adsorption and penetration of OH^- within the particle, and could be balanced by diffusing alkali cations (K^+, Na^+). When the alkali cations are diffused into the silica structure, a reaction known as ion exchange happens where alkalis are replaced with hydrogen. Produced osmotic pressure by hydrophilic groups creates an imbibition pressure, which leads to absorb water. This adsorption causes the gel to expand to manifold volume, which produces cracks and propagates them to the vicinities. The continued availability of water to the concrete causes enlargement and extension of the microcracks, which eventually reach the outer surface of the concrete. The crack pattern is irregular and is referred to as map cracking.

The following measures can be taken to control the AAR:

① The alkali content in the raw materials. Alkali may come from different sources, such as cement, chemical admixtures, supplementary cementing materials, salt contaminated aggregates, and penetration of seawater or deicing salt solution. Therefore, the total alkali content in concrete is often strictly controlled. In practice, the alkali content is the sum of Na_2O and K_2O.

② Alkali-reactive aggregate. The amount, size, and reactivity of the alkali-reactive constituent present in the aggregate.

③ The availability of moisture to the concrete structure.

3.7.7 Corrosion of steel bar

Corrosion of steel bars is the primary factor affecting the durability of reinforced concrete structures. Premature deterioration of structures caused by reinforcement corrosion has resulted in enormous economic losses and resource wastes in human society. In the alkaline environment provided by concrete, steel bars are normally covered by a thin iron oxide film that becomes impermeable and strongly adherent to the steel surface. This passivates the steel and prevents corrosion. However, in the presence of chloride ions or carbonation, the passive film may be destroyed, and the corrosion of steel may be initiated. When carbonation reaches down to the steel, the alkalinity of the pore solution is neutralized, and hydration products are dissolved. This destroys the protective layer on the steel bars, exposing the steel to the atmosphere, which, in the presence of sufficient moisture and oxygen, leads to a significant electrical potential difference. The resulting carbonation creates two regions—anode and cathode: Anode where the steel is dissolved, and a cathode where hydroxide ions are formed. Thus, carbonation leads to the initiation of corrosion uniformly over the steel surface. The ingress of chloride ions leads to the breakage of the passive film, and results in localized corrosion initiation (pitting corrosion).

Corrosion of steel in concrete is an electrochemical process (Fig. 3.38). An electrical circuit needs to be formed, i.e., anode, cathode, and electrolyte. The electrochemical potentials to form the corrosion cells may be generated in two ways:

① Composition cells may be formed when there are two different points which have different electrical potential because of differences in chemical composition.

② In the vicinity of reinforcing steel concentration, cells may be formed due to differences in the concentration of dissolved ions, such as alkalis and chlorides.

The transformation of metallic iron into rust is accompanied by an increase in volume, which depends on the state of oxidation and may reach up to 600% of the original metal (Fig. 3.39). This volume increase is believed to be the principal cause of concrete cover expansion and cracking of reinforced concrete elements. Also, like the swelling of poorly crystalline ettringite, the poorly crystalline iron hydroxides may have a tendency to absorb water and expand. It is worth to mention that the anodic reaction involving the ionization of metallic iron will not progress far unless the electron flow to the cathode is maintained by the consumption of electrons.

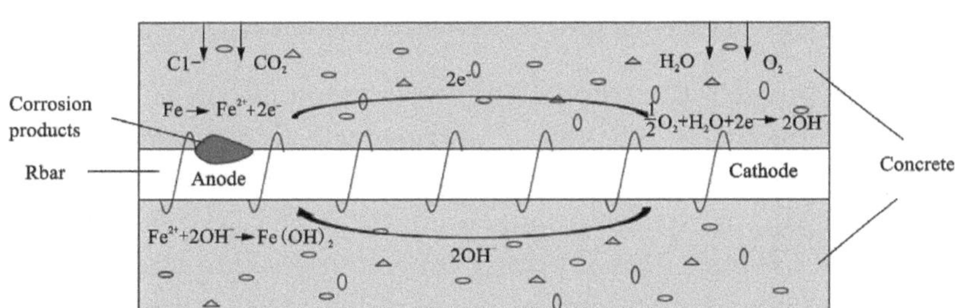

Figure 3.38 Electrochemical process of steel bar corrosion.

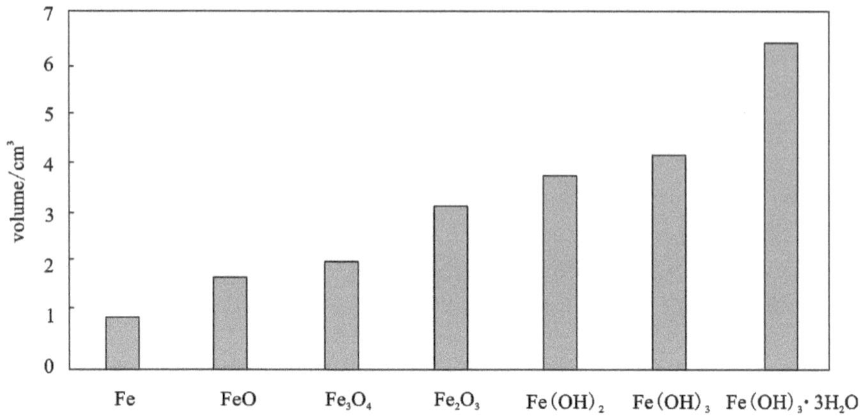

Figure 3.39 Solid volume increase due to oxidation of steel.

For the cathode process, therefore, the presence of both air and water at the surface of the cathode is absolutely necessary. This is the reason why reinforced concrete immersed in seawater has a very low risk of steel corrosion.

In order to prevent the steel corrosion in concrete, there are many measures or techniques that have been developed or proposed.

① Densify concrete cover. Because water, oxygen, and chloride ions play important roles in the corrosion of embedded steel and cracking of the concrete, it is obvious that the permeability of concrete is the key to control the various processes involved in the phenomena.

② Specify minimum concrete cover. To increase the concrete cover, the time taking for the deleterious substances to reach steel bar will be prolonged.

③ Reinforcing bar coatings. Protective coatings for reinforcing steel are of two types: anodic coatings (e.g., zinc-coated steel) and barrier coatings (e.g., epoxy-coated steel). They are often more expensive than producing a low-permeability concrete through quality, design, and construction controls.

④ Cathodic protection method. This involves suppression of current flow in the corrosion cell, either by supplying externally a current flow in the opposite direction or by using sacrificial anodes. Due to its complexity and high cost, the system is finding limited applications.

⑤ Corrosion inhibitor. When mixing concrete, corrosion inhibitor can be added into concrete and lower the corrosion risk of steel bars, such as calcium nitrate.

3.8 Mix design

The objective of mix design of concrete is to select the appropriate proportions of cement, water, fine and coarse aggregates to produce an economic concrete mix with the required fresh and hardened properties. For modern concrete, supplementary cementing materials and various chemical admixtures are also indispensable ingredients. The process of achieving the right combination of cement, aggregates, water, and admixtures is called mix proportioning or mix design. In this book, the term "design" is used, since it is the intention to imply that each element of the problem is approached with a deliberate purpose in view which is guided by a rational method of accomplishment.

Generally speaking, there are economic and technical purposes of mix design. One purpose of mix design is to obtain a product that will meet certain predetermined technical requirements, i.e., the workability of fresh concrete, the strength of hardened concrete at a specified age, and the durability of concrete. Another purpose of mix design is to obtain a concrete mixture satisfying the performance requirements at the lowest possible cost. This involves decisions regarding the selection of ingredients that are not only suitable but also available at reasonable prices and locally available. The underlying goal of concrete mix design is to strike a reasonable balance between the workability, strength, durability, and cost of concrete.

Because of the complexity and even uncertainties, the mix design of concrete is regarded as more like an art than science. For example, the addition of water to a stiff concrete mixture with a given cement content will improve the flowability of fresh concrete; however, the segregation of concrete may be affected in the opposite manner. Appropriate air content may improve the FT resistance; however, the strength may be decreased. The rough surface of aggregate is good for the mechanical properties of concrete; however, it affects the flowability in the opposite manner. The process of mixture proportioning boils down to the art of balancing various conflicting requirements.

Prior to proportioning concrete, the characteristics and chemical compositions of locally available materials should be collected. Only with proper selection of materials and correct understanding of mixture characteristics can the good mix design be made. A trial batch should be produced in the laboratory, and should also be produced by the mixer that will be used for the project. All the properties of concrete, such as workability, density, air content, and strength, should be measured to ensure that the mixture performs as desired. Appropriate adjustments should be made if the mixture does not meet the requirement, and subsequent trial batches produced. This iterative process continues until a satisfactory combination of the constituent materials is identified. The entire process should establish the most economical and practical combination of readily available constituent materials that will meet the requirements of the mixture design. These proportions are designed to meet the requirements for both fresh and hardened concrete. It is worth to mention that a properly designed concrete needs to be properly batched, mixed, transported, placed, consolidated, finished, textured, and cured. Any process can affect the final quality of cast concrete structures.

There is no standard method for concrete mix design, and numerous procedures for computing the concrete mixture proportions are available in most countries of the world. In this chapter, the mix design method of normal concrete in China is introduced. This method is mainly developed on the following principles:

① Bolomy equation. It describes the relationship between compressive stress and *w/c* ratio, coarse aggregate, and binder strength.

② Constant water law. Water content depends on the workability, type of aggregate, and maximum aggregate size.

There are two methods that can be used to calculate the mix proportion, i.e., the weight method and absolute volume method. The former one is considered less precise but does not require the information on the specific gravity of the concrete-making materials. The later one is considered more precise. Both procedures involve a sequence of steps given below, the first six steps being common. To the extent possible, the following background data should be gathered before starting the calculations:

① Sieve analysis of fine and coarse aggregates.

② The fineness modulus of sand.

③ Dry-rodded unit weight of coarse aggregate.

④ The bulk specific gravity of materials.

⑤ Absorption capacity or free moisture in the aggregate.

⑥ Variations in the approximate mixing water requirement with the slump, air content, and grading of the available aggregates.

⑦ Relationship between strength and w/c ratio for available combinations of cement and aggregate.

⑧ Job specifications, such as maximum w/c ratio, minimum air content, minimum slump, maximum size of aggregate, and designed strength.

Regardless of whether the concrete characteristics are prescribed by specifications or left to the mixture designer, the batch weights can be computed using the following sequence of steps:

Step 1: Calculation of w/b ratio

① Calculate the strength of trial mix. The strength of concrete is a random variable, and its probability density distribution is normal. In order to make sure that 95% of the concrete strength can be higher than the designed strength. The statistical parameters of concrete strength should be obtained in advance. The compressive strength of the trial mix can be calculated by

$$f_{cu,0} = f_{cu,k} + t\sigma \tag{3.11}$$

where $f_{cu,o}$ is the compressive strength of trial mix; $f_{cu,k}$ is the design strength of concrete; δ is the standard deviation of strength, 3.0 for strength grade is lower than 20 MPa, 4.0 for strength grade is between 25 and 45 MPa, 5.0 for strength grade is between 50 and 55 MPa; t is the coefficient. When the probability is 95%, t is 1.645.

② Calculation of w/b ratio. The w/b ratio can be calculated by

$$w/b = \frac{\alpha_a f_b}{f_{cu,0} + \alpha_a \cdot \alpha_b \cdot f_b} \tag{3.12}$$

where w/b is the water-to-binder ratio; f_b is the compressive strength of binder at 28 d; α_a and α_b are the empirical parameters, and they are 0.53 and 0.20 for crushed stone, and 0.49 and 0.13 for scree.

The compressive strength of binder at 28 d can be obtained from

$$f_b = \gamma_f \gamma_s f_{ce} \tag{3.13}$$

where γ_f, and γ_s are affecting factors of fly ash and slag, respectively, which can be taken from Table 3.13.

Table 3.13 Affecting factors of fly ash (γ_f) and slag (γ_s).

Affecting factors Content/%	γ_f	γ_s
0	1.00	1.00
10	0.85–0.95	1.00
20	0.75–0.85	0.95–1.00
30	0.65–0.75	0.90–1.00
40	0.55–0.65	0.80–0.90
50	—	0.70–0.85

The strength of the binder can be estimated by

$$f_{ce} = \gamma_c f_{ce,g} \tag{3.14}$$

where $f_{ce,g}$ is the strength grade of cement,; γ_c is the surplus coefficient. 1.12, 1.16, and 1.10 for cements of 32.5, 42.5, and 52.5, respectively.

In order to meet the durability requirement, the calculated w/b ratio should be compared with the data in Table 3.14, and take the lower w/b ratio for the following calculation.

Step 2: Choice of water.

The water content can be determined according to the specified slump, maximum aggregate, and aggregate type. If SP is used, the water content can be calculated (Table 3.15).

$$m_{w0} = m'_{w0}(1 - \beta) \tag{3.15}$$

where β is the water reducing rate of SP.

The addition level of SP is

$$m_{a0} = m_{b0} \cdot \beta_a \tag{3.16}$$

where m_{a0} is the content of SP; m_{b0} is the content of the binder; β_a is the addition level of SP.

Step 3: Determination of binder.

According to the w/b ratio and water content determined in steps 1 and 2, the binder content can be calculated.

$$m_{b0} = m_{w0}/w/b \tag{3.17}$$

The binder content is the sum of mineral admixture and cement content.

$$m_{b0} = m_{f0} + m_{c0} = m_{b0}\beta_f + m_{c0} \tag{3.18}$$

where β_f is the replacement level of mineral admixtures, and the limit for the replacement level of mineral admixtures is given in Table 3.16.

The calculated binder content should be compared with the minimum binder content due to the durability requirements, as shown in Table 3.14.

Step 4: Choice of sand ratio.

The sand ratio depends on the gradation of sand and aggregate, and should be determined by experiments, based on the workability of concrete. Alternatively, the sand ratio also can be selected according to Table 3.17.

Table 3.14 Maximum water-to-binder ratio and minimum binder content in concrete.

<table>
<tr><th rowspan="2">Environment level</th><th rowspan="2">Environment type</th><th rowspan="2">Max, w/b</th><th colspan="3">Minimum cement content/kg</th></tr>
<tr><th>Plain concrete</th><th>Reinforced concrete</th><th>Prestressed concrete</th></tr>
<tr><td>1</td><td>Dry environment
No aggressive water</td><td>0.60</td><td>250</td><td>280</td><td>300</td></tr>
<tr><td>2a</td><td>Humid environment; No freezing disaster and no aggressive water or soil
No aggressive water or soil above the frost line in cold and severe cold areas</td><td>0.55</td><td>285</td><td colspan="2">300</td></tr>
<tr><td>2b</td><td>Wet-dry alternate environment; frequent fluctuation of the water table
Open-air environment in cold and severe cold areas
No aggressive water or soil above the frost line in cold and severe cold areas</td><td>0.50</td><td colspan="3">320</td></tr>
<tr><td>3a</td><td>Environment affected by deicing salt and sea breeze
Fluctuation zone of the water table in cold and severe cold areas</td><td>0.45</td><td colspan="3">330</td></tr>
<tr><td>3b</td><td>Environment with salinized soil, deicing salt or near the seacoast</td><td>0.40</td><td colspan="3">330</td></tr>
</table>

Table 3.15 Unit water demand for harsh and plastic concrete (kg/m^3).

Consistency of mixture		Max., particle size of scree/kg·m^3				Max., particle size of gravel/kg·m^3			
Items	Indexes	10.0	20.0	31.5	40	16.0	20.0	31.5	40.0
Vebe consistency/s	16–20	175	160	–	145	180	170	–	155
	11–15	180	165	–	150	185	175	–	160
	5–10	185	170	–	155	190	180	–	165
Slump/mm	10–30	190	170	160	150	200	185	175	165
	35–50	200	180	170	160	210	195	185	175
	55–70	210	190	180	170	220	205	195	185
	75–90	215	195	185	175	230	215	205	195

Table 3.16 Maximum content of mineral admixtures in reinforced concrete and prestressed concrete.

		Reinforced concrete		Prestressed concrete	
Types of mineral admixtures	**w/b**	**Portland cement**	**Ordinary Portland cement**	**Portland cement**	**Ordinary Portland cement**
Fly ash	≤0.40	45	35	35	30
	>0.40	40	30	25	20
GGBS	≤0.40	65	55	55	45
	>0.40	55	45	45	35
Silica fume	—	10	10	10	10
Composite	≤0.40	65	55	55	45
admixtures	>0.40	55	45	45	35

Table 3.17 Sand ratio in concrete.

	Max., particle size of scree/mm			Max., particle size of gravel/mm		
w/b	**10.0**	**20.0**	**40.0**	**16.0**	**20.0**	**40.0**
0.40	26–32	25–31	24–30	30–35	29–34	27–32
0.50	30–35	29–34	28–33	33–38	32–37	30–35
0.60	33–38	32–37	31–36	36–41	35–40	33–38
0.70	36–41	35–40	34–39	39–44	38–43	35–41

Step 5: Calculation of fine and coarse aggregates.

After completing step 4, all the parameters have been estimated, except the content of fine and coarse aggregates. Their contents can be determined either by the "weight" method or by the "absolute volume."

$$m_{s0} + m_{g0} + m_{c0} + m_{b0} + m_{w0} = m_{cp} \tag{3.19}$$

$$\beta_s = \frac{m_{s0}}{m_{s0} + m_{g0}} \tag{3.20}$$

where β_s is the sand ratio, m_{cp} is the assumed bulk density of concrete 2350–2450 kg/m^3; m_{s0} and m_{g0} are the contents of sand and aggregate, respectively.

$$\frac{m_{s0}}{\rho_s'} + \frac{m_{g0}}{\rho_g'} + \frac{m_{c0}}{\rho_c} + \frac{m_{f0}}{\rho_f} + \frac{m_{w0}}{\rho_c} + 0.01\alpha = 1 \tag{3.21}$$

where ρ_c, ρ_w, ρ_g, ρ_s, and ρ_f are the density of cement (normally 3100 kg/m^3), water (1000 kg/m^3), fine aggregate, coarse aggregate, and mineral admixtures.

Step 6: Adjustments for the aggregate moisture.

Generally, the stock aggregates are moist; without moisture correction, the actual w/c ratio of the trial mix will be higher than the calculated ones, and the saturated-surface dry (SSD) weights of aggregates will be lower than estimated ones. The mixture proportions determined are assumed to be on an SSD basis. For the trial batch, depending on the amount of free moisture in the aggregates, the mixing water is reduced, and the amounts of aggregates correspondingly increased, as shown later by sample computations.

Step 7: Trial batch adjustments.

Because of many assumptions underlying the foregoing theoretical calculations, the mix proportions for the actual materials to be used must be checked and adjusted by means of laboratory trials consisting of small batches (e.g., 15 L of concrete). Fresh concrete should be tested for the slump, workability (freedom from segregation), unit weight, and air content; specimens of hardened concrete cured under standard conditions should be tested for strength at the specified age. After several trials, when a mixture satisfying the desired criteria of workability and strength is obtained, the mixture proportions of the laboratory-size trial batch are scaled up for producing full-size field batches.

Sample Computations

Step 1:

Slump is 35–55 mm; maximum size aggregate is 20 mm. Refer to Table 3.15, it can be found that the water content for 1 m^3 concrete is 195 kg.

Step 2:

$$f_{cu,0} = f_{cu,k} + 1.645\sigma = 30 + 1.645 \times 5 = 38.2 \text{ MPa}$$

For crushed stone, w/c ratio:

$$w/c = \frac{\alpha_A f_{ce}}{f_{cu,0} + \alpha_A \cdot \alpha_B \cdot f_{ce}} = \frac{0.46 \times 46}{38.2 + 0.46 \times 0.07 \times 46} 0.53$$

Refer to Table 3.14, the maximum w/c ratio is 0.6, thus, 0.53 of w/c ratio is selected.

Step 3:

$$m_c = \frac{m_w}{w/c} = \frac{195}{0.53} = 368 \text{ kg/m}^3$$

Refer to Table 3.14, the minimum cement content is 280 kg, thus, 368 kg/m^3 is selected.

Step 4:
$w/c = 0.53$, maximum size aggregate is 20 mm;
Refer to Table 3.17, sand ratio is 35%–40%;
Take sand ratio as 36%.
Step 5:

$$\begin{cases} 368 + 196 + m_s + m_g = 2400 \\ 0.36 = \dfrac{m_s}{m_s + m_g} \end{cases}$$

Finally, the preliminary mix proportion can be obtained $m_s = 661$ kg, $m_g = 1176$ kg, $m_w = 195$ kg, $m_c = 385$ kg, and trial batch is need to perform to check the properties of concrete.

3.9 Self-compacting concrete and its application in high-speed rail

3.9.1 Introduction

Self-compacting concrete or self-consolidating concrete (SCC) is an innovative material that does not require vibration for compaction. It can flow under its own weight without segregation, completely filling formwork and achieving full compaction with no need of vibration. It even performs much better in case of congested reinforcement than traditional vibrated concrete. The hardened concrete is dense, homogeneous, and has the same mechanical properties and durability as traditional vibrated concrete.

In 1980s, the concept of SCC was first proposed by Professor Hajime Okamura in the University of Tokyo. The project team, led by Okamura, attributed insufficient compaction to the most common cause of deterioration of concrete structures, which were fast built after the Second World War. Full-scale trials and demonstrations were carried out, and in the early 1990s, the "self-compacting" concrete was first used in a gas tank in Osaka, Japan. Due to its significant economic and technical benefits, a very wide range of applications of SCC with tremendous success on numerous projects throughout the world have been reported since its first uses in the early 1990s. SCC has been successfully used in precast, cast-in-place, structural, architectural, vertical, horizontal, large, and small projects. During the past several decades, many guidelines or specifications on SCC have been issued across the world, as follows:

① "AIJ Recommended Practice for High Fluidity Concrete for Building Construction," Architectural Institute of Japan, 1997
② "Recommendation for Construction of Self-Compacting Concrete," Japan Society of Civil Engineers, 1998
③ "The European Guidelines for Self-Compacting Concrete: Specification, Production and Use," Self-Compacting Concrete European Project Group, 2005
④ "ACI 237R-07, Self-Consolidating Concrete," American Concrete Institute, 2007.
⑤ "JGT/T 2835-2012, technical specification for application of self-compacting concrete," Ministry of housing and urban-rural development of China. 2012.

In this chapter, the normal application of SCC will not be discussed, which can be found in many publications or books. A special application of SCC in high-speed rail (HSR) in China will be introduced. The ballastless track has been widely implemented in HSR around the world. Despite high initial investment and noise problems, it has obvious advantages over ballast track system, such as ① increase capacity, ② increase speed, ③ reduce maintenance and life cycle costs, and ④ reduce the number of track maintenance operations and thereby increase safety. In order to implement ballastless track wider and better, many countries are developing a new ballastless track form. For the last decade, China has established the China Rail Track System (CRTS), which includes five types of ballastless tracks, including CRTS I and II double-block ballastless tracks, CRTS I, II, and III slab ballastless tracks. The first four types of tracks are developed by Chinese railway companies based on the transferred technology from Germany and Japan, and CRTS III slab ballastless track is independently developed by Chinese railway companies. It is believed that CRTS III Slab ballastless track combines the advantages of CRTS I and II types slab tracks.

The structure section and layout of CRTS III slab ballastless track are shown in Fig. 3.40. It can be easily observed that the track structure consists of four layers, which are, from top to bottom, prefabricated prestressed slab, filling layer, isolated geotextile layer, and base plate. Concrete, instead of the cement emulsified asphalt mortar applied in CRTS I and CRTS II track, is used to construct the filling layer of CRTS III track system. The concrete filling layer is cast-in-place, and is required to have a strong bonding with the above prefabricated prestressed concrete slab; as a consequence, the two layers function as a composite plate. The loadings of the high-speed train are transferred to the roadbed by the composite plate. Therefore, the interface

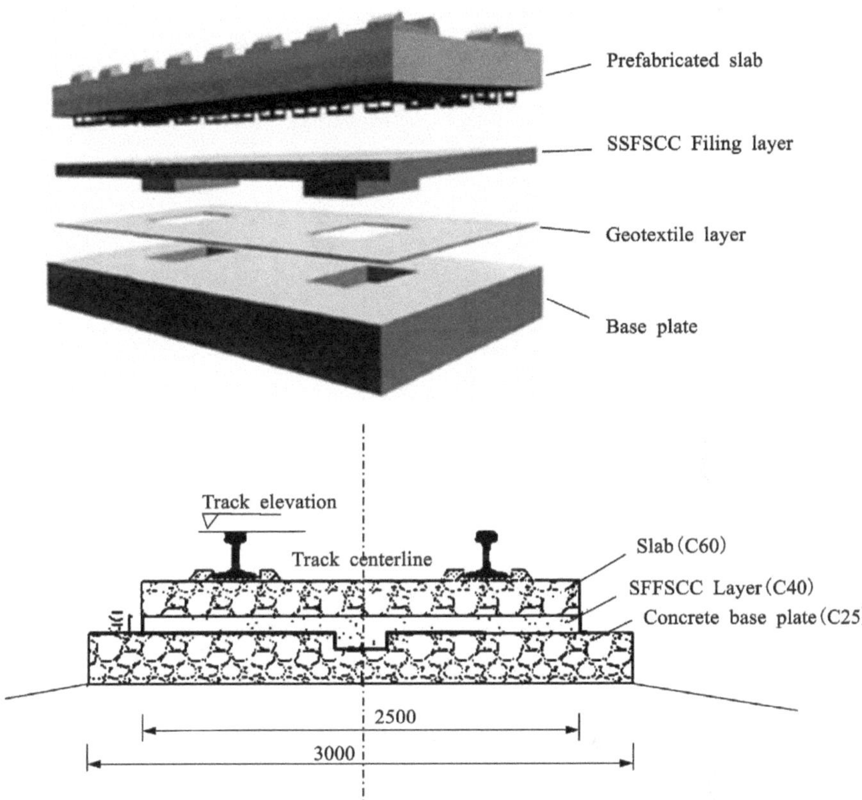

Figure 3.40 The structure section and layout of the CRTS III slab ballastless track.

bonding between prefabricated slab and filling layer should be strong and durable enough to secure the serviceability of the slab track. The thickness of the filling layer, which is cast-in-place, can be adjusted within a range according to the presetting elevation position of above prefabricated slab position. By doing so, the high smoothness of the track line can be attained. The track form implements a concept of "decreasing stiffness from top to bottom."

The CRTS III ballastless slab track is constructed as the steps shown in Fig. 3.41. It can be seen that the construction procedures mainly include four steps. The first step is to place the bottom parts in the sequence of the base plate, geotextile layer, and reinforced mesh. Secondly, the prefabricated slab is put on the top of base plate, and adjusted to its design position with a tolerance of 0.5 mm. Then, the space between the base plate and slab is sealed, and the position of the slab is secured by clamps. Finally, SCC is cast into the sealed space from the pouring hole. SCC applied in the slab track is required to fill up the sealed space between slab

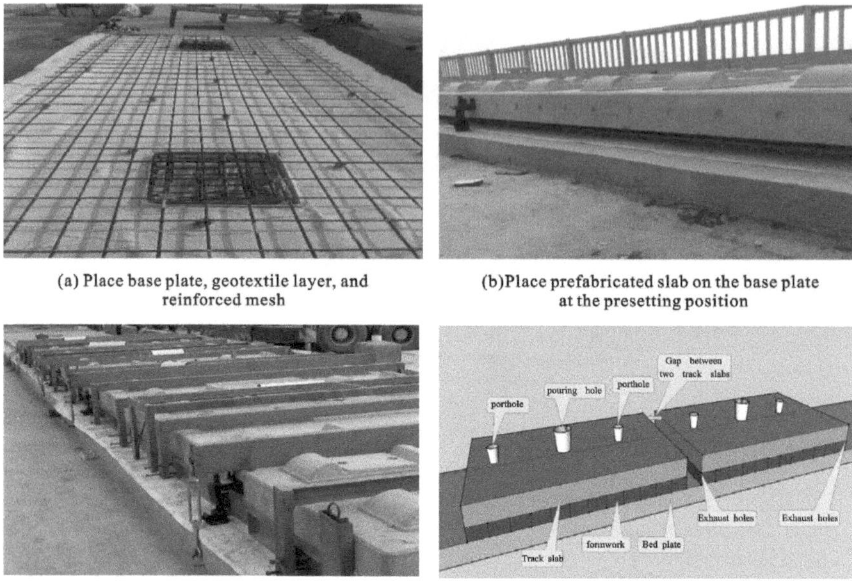

(a) Place base plate, geotextile layer, and reinforced mesh

(b) Place prefabricated slab on the base plate at the presetting position

(c) Install formwork and clamps

(d) Cast concrete from the top pouring hole

Figure 3.41 The main construction steps of the CRTS III slab ballastless track.

and base plate under its own weight. It is a special type of SCC, but different from normal SCC because of the following reasons: ① Normal SCC is often cast into an open formwork. The top surface of SCC can be smoothed by the process of finishing, so slight bleeding and rising of bubbles are acceptable. However, SCC applied in the slab track is grouted into a flat, narrow, and sealed space with the dimension of 90 mm × 2500 mm × 5600 mm. Strong bonding between the top surface of the SCC layer and prefabricated slab is required. Thus, bleeding and rising of bubbles should be strictly controlled. ② Normal SCC often flows on a rigid surface. However, SCC applied in the slab track system flows on a flexible geotextile. It increases the flow resistance. In addition, geotextile may absorb the water from SCC, and affect the flow of SCC. ③ An HSR line is often hundreds of kilometers, and it is usually one-time completion. Construction sites along HSR line may extend hundreds of kilometers, and various raw materials will be used; they may present large variations in properties and compositions. It's quite challenging for the quality control of concrete because of the variation in local raw materials.

Due to the above-mentioned characteristics, SCC for CRTS III slab ballastless track is different from normal SCC and puts forward stricter requirements for SCC. Since the first characteristic is the most important and unique one, hence, we name this type of SCC as sealed-space-filling self-compacting concrete (abbreviated as SSFSCC).

CRTS III slab ballastless track has been implemented more than 3000 km in the HSR network, including Cheng–Guan line, Wuhan metropolitan area intercity line, Panjin–Yingkou line, and Zhengzhou–Xuzhou line. Approximately, more than 1,400,000 m^3 SSFSCC has been employed in the construction of high-speed rail in China up to now. It is estimated that other millions of m^3 of SSFSCC will be consumed in China in the near future, since CRTS III slab track is the main track system employed in the construction of high-speed rail in China.

In the following sections, the property requirements in fresh and hardened states, mix proportioning, and construction technology of SSFSCC are introduced.

3.9.2 The property requirements of SSFSCC

The hardened concrete fulfills the structural functions in the system, depending on its mechanical properties, durability, and volume stability. Fresh properties of concrete significantly affect the quality of hardened concrete. The fresh properties of SSFSCC are particularly important since the intricate slab track system put forward some special requirements, as stated above.

3.9.2.1 Properties in a hardened state

The designed service life of the CRTS III slab track is 60 years. During service, the slab track requires maintaining its original properties without major repair. In order to fulfill its function, the hardened properties of SFSCC have been proposed based on the structural design requirements, as shown in Table 3.18. These properties include mechanical properties, durability, and volume stability. More importantly, the strong interface bonding between the prefabricated slab and SSFSCC layer needs to be ensured during the whole service life.

Table 3.18 Properties requirement of hardened SSFSCC for CRTS III slab ballastless track (56 d).

Properties	Value
Compressive strength/MPa	$\geq$40.0
Flexural strength,/MPa	$\geq$6.0
Elastic modulus/GPa	30–38
6-h charge passed/c	$\leq$1000
Spalling under salt solution freezing and thawing test/g/m^{-2}	$\leq$1000
Shrinkage/$\times 10^{-6}$	$\leq$400

3.9.2.2 *Properties in a fresh state*

Properties of fresh SSFSCC determine the performance of a hardened matrix. Hence, it is very important to choose some existing testing methods, or even develop new testing methods, to evaluate the fresh properties of SSFSCC. Of course, appropriate testing results should be specified as the acceptance criteria for the purpose of practice. As stated above, bleeding, segregation, surface settlement, and instability of air bubble of SSFSCC have to be minimized. On one hand, very flowable SSFSCC is required to fill up the sealed, narrow, and flat space under the action of gravity; on the other hand, very stable SSFSCC is required to avoid the above-mentioned defects.

There are many testing methods developed to evaluate the properties of fresh SCC for the past decades, for example, slump flow, J-ring, V-funnel, L-box, sieve segregation test, settlement column, penetration test for segregation, and U-test, etc. It is generally accepted that SCC needs to have the three key abilities: filling ability, passing ability, and antisegregation ability, which can be evaluated by different methods. Some single test can evaluate two or more abilities. Take the J-ring test, for instance, filling ability by measuring the spread diameter as in the slump-flow test, passing ability by measuring the blocking step BJ, flow-rate by measuring the time to reach 500 mm, and segregation ability can be visually judged. The selection of testing methods for evaluating properties of fresh SCC needs to take the specific conditions and circumstances into account, including reinforcement ratio, formwork, and other special requirements.

3.9.2.2.1 Filling ability

Filling ability is one of the most important properties of fresh SSFSCC. It ensures SSFSSC can fill up the flat, narrow, and sealed space with a large area. Slump-flow value can well describe the flowability of a fresh mixture in unconfined conditions. Visual observations during the test and measurement of the T500 can give additional information on the segregation resistance and uniformity of each delivery. Therefore, the slump-flow test is employed to evaluate the filling ability of SSFSCC. The appropriate slump-flow value and flow rate of fresh SSFSCC are 570−650 mm and 3−7 s based on the results from laboratory experiments and job site applications.

3.9.2.2.2 Passing ability

Passing ability describes the capacity of the fresh mixture to flow through confined spaces and narrow openings without segregation, loss of uniformity, or causing blocking. Passing ability is mainly influenced by the size of confinement gap, maximum aggregate size, and the flowability/filling ability. For the thin platelike filling layer, there is a row steel bar at the center of the sealed-space of the layer, and the confinement gap is only 40 mm. SSFSCC must have a very good passing ability, which makes it smoothly pass the narrow gap and flow on geotextile readily.

The L-box with three bars test is used to assess the passing ability of SSFSCC to flow through tight openings between reinforcing bars without segregation and blocking. Passing ability specifications are set as PA2 according to the European guidelines for self-compacting concrete.

3.9.2.2.3 Stability

Due to density differences among the constituents of SSFSCC, it is easy to suffer from instability. In this paper, the term stability, instead of segregation used in normal SCC, since segregation, bleeding, rising of bubbles, and surface settlement should all be addressed. Segregation and bleeding lead to a very high *w/c* ratio in the top surface of SSFSCC and thus weak interface bonding between the prefabricated slab and SSFSCC layer. Rising of bubbles entrapped in the interface results in a bubble layer; as a consequence, very weak interface bonding in results. Obviously, the surface settlement of SSFSCC layer may result in incomplete contact with the slab, and thus weak bonding. Thus, special attention should be paid to the stability of SSFSCC.

There are several testing methods developed to evaluate the segregation resistance properties of fresh SCC for the past decades, such as sieve segregation test, settlement column, penetration test, etc. These test methods are not enough to secure the good interface bonding between the slab and SSFSCC layer. Three tests, i.e., top surface paste thickness measurement, bleeding test, and expansion rate test, are employed to evaluate the stability of fresh SSFSCC. The bleeding test conducted according to Chinese standard GB/T 50080 is used to evaluate the bleeding resistance and water-holding abilities of fresh SSFSCC. In order to ensure the complete contact between the fresh SSFSCC and slab during hardening, the expansion agent is added to generate a certain expansion rate and to avoid surface settlement or shrinkage of concrete within 24 h. The expansion rate test is conducted according to Chinese standard GB 23439. The top surface paste thickness measurement is a novel and newly developed test method. It is used to assess the segregation degree between paste and aggregate in the fresh mixture. The testing setup of the top surface paste

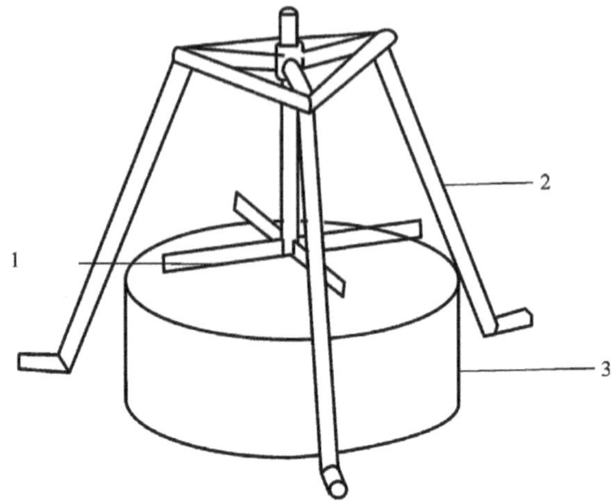

Figure 3.42 Testing setup for top surface paste thickness measurement. Where 1 is steel testing cross; 2 is cylindrical container; 3 is bracket.

thickness is given in Fig. 3.42. It consists of three parts, i.e., steel testing cross, cylindrical container, and bracket. In this test, fresh SSFSCC is placed into a cylindrical container with the height of 110 mm and a diameter of 200 mm, and lets the concrete rest for 15 min. Afterward, the steel testing cross, with the thickness of 1 mm, length of 150 mm, and height of 10 mm, is fixed right on the top of the surface of concrete, followed by releasing the steel testing cross suddenly, then measure the penetration depth of the cross within 30 s. It is believed that the surface paste thickness test is appropriate to evaluate the segregation degree of fresh SSFSCC. The segregation resistance is very good and acceptable when the surface paste thickness is no more than 7 mm, which is about equal to the thickness of the mortar layer enwrapped coarse aggregate. There is no method specially designed for the bubble stability of SSFSCC. However, SSFSCC passed bleeding, and top surface paste thickness tests is quite sticky, and should have good stability of bubble.

The property requirements of fresh SSFSCC for CRTS III slab ballastless track were proposed in Table 3.19 based on comprehensive laboratory tests and full-scale job-site tests.

3.9.3 Mix proportioning of SSFSCC

Raw materials and their fractions in the mix are the key factors influencing the properties of concrete. In past decades, the mix proportioning method has been one of the focuses in SCC research and application fields. More

Table 3.19 Property requirements of fresh SSFSCC for CRTS III slab ballastless track.

Properties	Parameters	Typical value	Testing method
Filling ability	Slump flow/mm	570–650	Slump spread test (workability test GB/T 50080, Chinese standard)
	Flow rate T500/s	3–7	
Passing ability	P A	≥0.8	L-box test
Stability	Top surface paste thickness/mm	≤7	Surface paste thickness test as shown in Fig. 3.42
	Bleeding/%	0	Workability test GB/T 50080
	Expansion rate/%	0–1.0	Expansion test GB 23439(Chinese standard)

than 10 mix design methods for SCC have been proposed by researchers from different countries and regions around the world such as Japan, the Laboratory Central Des Ponts et Chausses (LCPC), the Swedish Cement and Concrete Research Institute (CBI), and research groups in China. Each method has its own characteristics, which basically meets the specific demands in practice. Most of them gave only general guidelines and ranges of quantities of materials to be used in proportioning SCC, in the meantime, the specific conditions and objective properties of SCC in fresh and hardened states had to be taken into account. In addition, these existing methods were normally developed based on experiences and specific experimental results. They may not be applicable to other scenarios with different conditions. Especially for SSFSCC applied in the sealed-space-filling layer of the ballastless slab track system in HSR, some special requirements of SSFSCC in fresh and hardened states have been put forward. In order to realize the optimum properties and minimum cost for SSFSCC, an optimal mix proportioning method has to be established. For a good mix proportion of concrete, some basic requirements have to be taken into consideration collectively:

① Workability in a fresh state.
② Designed mechanical properties.
③ Designed durability when subjecting to a specific environment.
④ Economic and environment-friendly consideration.

Of course, an optimal mix proportioning of concrete should be highly robust, given that the variation in raw materials of SSFSCC along high-speed rail construction sites occurs. The design of SSFSCC should meet the above-mentioned considerations.

3.9.3.1 The key parameters of mix proportion

Generally speaking, concrete can be seen as a binary system consisting of coarse aggregate and mortar, and mortar can also be considered as a binary system including paste and fine aggregate. In the concrete level, concrete can be ideally modeled as a suspension system with two phases of coarse aggregate and mortar as shown in Fig. 3.43A. In this model, the coarse aggregate is idealized as a large spherical particle and uniformly dispersed in the continuous mortar phase. The mortar, as a continuous phase, enwraps the coarse aggregate. tm represents the average space between neighboring coarse aggregates and is equal to the thickness of the mortar layer, enwrapping the surface of the coarse aggregate. In a similar way, mortar can be described by the physical model shown in Fig. 3.43B. The fine aggregate is idealized as a small spherical particle and evenly dispersed in the paste. The paste, as a continuous phase, enwraps fine aggregate. tf represents the average space between neighboring fine aggregates and is equal to the thickness of the paste layer coating the surface of the fine aggregate.

The properties of concrete are determined by the properties and volume fraction of paste and aggregates. The fresh paste contributes to the workability of the mixture. Its consistency and volume fraction are the main factors determining the properties of the fresh mixture. The paste enwraps aggregates, and thus a lubricative paste layer on the surface of each aggregate is formed. The lubrication layer makes aggregates move easily with low friction if the paste has appropriate viscosity. After setting and hardening, the hardened paste binds the dispersed aggregates into a whole

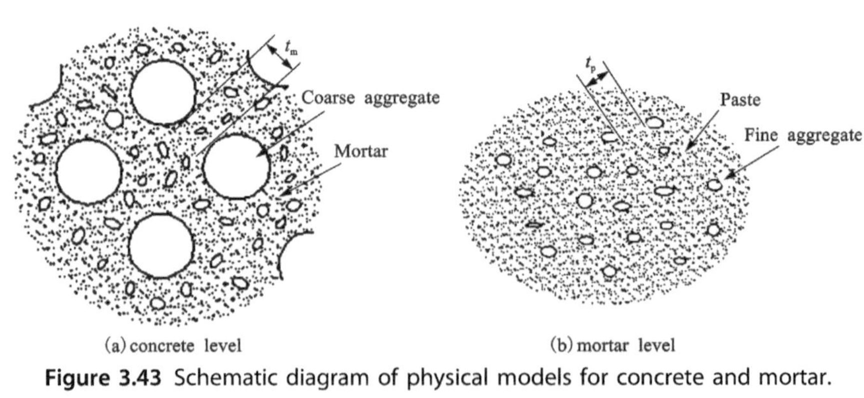

Figure 3.43 Schematic diagram of physical models for concrete and mortar.

body. According to the physical model shown in Fig. 3.43, the average gap between neighboring aggregates in the mixture is one of the key parameters which greatly affect the properties of fresh and hardened concrete. In the case of SSFSCC, a suitable average gap between neighboring aggregates (the thickness of paste layer) is of great importance for ensuring self-compactibility of the fresh mixture and good bonding strength between the SSFSCC filling layer and prefabricated slab. The average gap between aggregates is closely related to the volume fraction of aggregate in unit mixture. Of course, the properties of fresh paste are also the key parameters affecting the workability, mechanical, and durability of SSFSCC. It depends on the water-to-powder ratio, dosage of SP, and compositions of binder powder, etc. Hence, in order to meet the requirements of SSFSSC, the following key mix parameters should be carefully considered:

① The volume fractions of coarse and fine aggregates. They determine the average gap between neighboring aggregates. The volume fraction of aggregates can be easily obtained if the average gaps (t_m, t_p) are determined. Based on the concept of mean free path proposed by Fullman and the aggregate gradation which is assumed to follow Fuller function, the average gaps can be obtained:

$$\text{For concrete system: } t_m = \frac{2(1 - V_{ca})}{3V_{ca}} \times \sqrt{D_{ca,\max} \times D_{ca,\min}} \tag{3.22}$$

$$\text{For mortar system: } t_p = \frac{2(1 - V_{fa})}{3V_{fa}} \times \sqrt{d_{fa,\max} \times d_{fa,\min}} \tag{3.23}$$

where V_{ca} and V_{fa} are the volume fractions of coarse aggregate and fine aggregate in the corresponding system; $D_{ca,\max}$, $D_{ca,\min}$, $d_{fa,\max}$, and $d_{fa,\min}$ are the maximum and minimum diameters of coarse and fine aggregates, respectively.

② Water-to-binder/powder ratio. It is a basic parameter in the mix design of concrete, and closely related to the compressive strength of SSFSCC and the consistency of fresh paste. In order to obtain a high viscosity of paste, a relatively low water-to-binder/powder ratio is adopted.

③ Type and amount of powder. Apart from cement and expansive agent, mineral powders, including cementitious materials and inert powders, are necessary constituents for SSFSCC. They positively affect the fresh and hardened properties, and greenness of concrete. Generally, powder materials applied in SSFSCC include cement, mineral admixture, and expansive agent.

④ Type and dosage of admixtures. SP and viscosity-enhanced agent are both used in SSFSCC. It improves the rheological properties of concrete and also remarkably improves the robustness of fresh SSFSCC.

3.9.3.2 The procedures of mix proportioning of SSFSCC

Based on the above considerations and experiences, the procedures of mix proportioning of SSFSCC are summarized in the following steps.

Step 1: Select raw materials and calculate the volume content of aggregates.

The raw materials used for producing SSFSCC includes crushed limestone with a maximum diameters of 16 mm and continuous gradation, river sand with a fineness modulus ranging from 2.3 to 3.0, Portland cement, fly ash (FA) with a specific area of 450–500 m^2/kg, ground granulated blast furnace slag (GGBS) with a specific area of 450–500 m^2/kg, expansive agent (EA), viscosity-enhanced agent (VEA), and polycarboxylic acid SP. VEA is a compound composing of cellulose polymer and inorganic ultrafine calcareous and siliceous powders.

A reasonable volume of aggregates is crucial to the filling ability, passing ability and segregation resistance of fresh SSFSCC, and the volume stability of hardened SSFSCC. As shown in Fig. 3.38, the volume of aggregate is closely related to the average gap between neighboring aggregates in a mixture. Therefore, it is very important to select the optimal gap between aggregates, according to Eqs. (3.22) and (3.23). The optimal average gap tm between neighboring coarse aggregates in a concrete system is 13–14 mm for coarse aggregate with the diameter ranging from 4.75 to 16 mm, and thus 0.28–0.32 is determined for the volume fraction of coarse aggregate. For the mortar system in concrete, the optimal average gap tp between neighboring fine aggregates is 1.2–1.4 mm for fine aggregate with the diameter ranging from 0.15 to 4.75 mm. 0.29–0.31 is the volume fraction of fine aggregate in unit SSFSCC mixture. The optimal total volume fraction of coarse and fine aggregates ranges from 0.60 to 0.62, which is a reasonable value for good workability in a fresh state and good volume stability in hardened state of SSFSCC. It is worth to mention that the aggregate gradation should meet fuller function. When the fineness modulus of sand is at the low limit, a small volume fraction of sand should be selected.

Step 2: Calculate the volume content of the paste.

Based on the volume fractions of coarse and fine aggregates in the mixture, the volume of paste, V_p, can be easily obtained by Eq. (3.24):

$$V_p = V_c - V_{ca} - V_{fa} - V_{air} \tag{3.24}$$

where V_c is the volume content of the SSFSCC mixture; V_{ca} is the volume content of coarse aggregate; V_{fa} is the volume content of fine aggregate; V_{air} is the volume content of the air bubble in the mixture.

Step 3: Determine the w/b ratio (m_b/m_w).

The relation between w/b ratio and compressive strength of concrete can be described by the classical mathematical equation proposed by Abram or Bolomey. Based on Bolomey equation and experiences, the compressive strength of SSFSCC is correlated with w/b ratio and other factors, as shown in Eqs. (3.25) and (3.26). Eq. 3.26 can be rewritten as Eq. 3.27, and thus the w/b ratio (m_w/m_b) can be determined.

$$f_{cu,0} = f_{cu,k} + 1.645\sigma \tag{3.25}$$

$$f_{cu,0} = 0.42 f_{ce} \frac{m_b(1-\beta) + m_b \cdot \beta \cdot \gamma}{m_w} - 1.2 \tag{3.26}$$

$$m_w/m_b = \frac{0.42 f_{ce}(1 - \beta + \beta \cdot \gamma)}{f_{cu,0} + 1.2} \tag{3.27}$$

where $f_{cu,k}$ is the designed compressive strength of SSFSCC; $f_{cu,0}$ is the required compressive strength of SSFSCC during the production process by considering the fluctuation of strength; σ is the variation value of compressive strength, f_{ce} is the tested compressive strength of cement at the age of 56 d; β is the percentage of mineral admixture (by the mass of total binder); γ is the cementitious coefficient of mineral admixture at the age of 56 d, 0.4 for fly ash and 0.9 for ground granulated blast furnace slag is used.

Step 4: Calculate the content of powder and water.

Given that mineral admixtures can improve performance, reduce cost, and increase greenness of concrete, fly ash and GGBS are both used in the SSFSCC mixture. Meanwhile, an expansive agent (EA) is also used to compensate the shrinkage of concrete. Therefore, the powders used in SSFSCC include cement, fly ash, GGBS, and expansive agent.

According to the volume content of paste in unit mixture, and the density of each powder, the mass content of each powder can be calculated by the following equations:

$$m_b = V_p \Big/ \left(\frac{1 - \beta_{FA} - \beta_{GGBS} - \beta_{EA}}{\rho_c} + \frac{\beta_{FA}}{\rho_{FA}} + \frac{\beta_{GGBS}}{\rho_{GGBS}} + \frac{\beta_{EA}}{\rho_{EA}} + \frac{m_w}{m_b} \right) \tag{3.28}$$

$$m_w = \frac{m_w}{m_b} \times m_b \tag{3.29}$$

$$m_{FA} = \beta_{FA} \times m_b \tag{3.30}$$

$$m_{GGBS} = \beta_{GGBS} \times m_b \tag{3.31}$$

$$m_{EA} = \beta_{EA} \times m_b \tag{3.32}$$

$$m_c = m_b - m_{FA} - m_{GGBS} - m_{EA} \tag{3.33}$$

where ρ_c, ρ_{FA}, ρ_{GGBS}, ρ_{EA}, and ρ_w are the density of cement, fly ash, GGBS, EA, and water, respectively. β_{FA}, β_{GGBS}, and β_{EA} are the amount of fly ash, GGBS, and EA by mass percent of a total binder. m_{FA}, m_{GGBS}, m_{EA}, and m_c are the mass of fly ash, GGBS, expansive agent, and cement in unit mixture.

Step 5: Determine the dosage of SPs and viscosity enhanced agent.

SP is used in combination with viscosity-enhanced agent (VEA). On the one hand, VEA increases the viscosity of the liquid phase and stabilizes the suspension system. On the other hand, SP disperses powder among the liquid phase. Thus, a highly dispersed and stable suspension can be achieved by the combined action of VEA and SP. This is desired for highly flowable cement-based materials. In order to produce SSFSCC with satisfactory workability and robustness, a special VEA, which includes organic and inorganic components, is made. The dosage of VEA is 5%–6% of the total binder by weight. The dosage of SPs is adjusted by the flow spread of SCC.

Step 6: Trial batch.

Based on the above mix proportioning steps, the compositions of a mix can be obtained, and then a trial batch of SSFSCC can be implemented. The workability of fresh mixture, the mechanical properties, durability, and volume stability of hardened concrete are tested according to Tables 3.18 and 3.19. If one of the properties fails to meet the proposed value, adjustments should be made until all properties of SFFSCC satisfy the specified requirements.

3.9.3.2.1 Typical mix for SSFSCC

Based on the proposed mix proportion method, three mixes with different powders were designed, as shown in Table 3.20. The raw materials used include Grade 42.5 Portland cement with a compressive strength of 49.8 MPa at 56 d, GGBS with a specific area of 450 m^2/kg, class F fly ash with a specific area of 460 m^2/kg, sulfoaluminate-based expensive agent (EA), crushed limestone (CL) with continuous gradation and a size ranging

Table 3.20 Mixing proportions of concretes (kg/m³).

Mix	Cement	FA	GGBS	EA	VEA	SP	W	CL	S
1	350	80	100	0	0	5.0	175	810	835
2	308	80	100	42	0	5.0	175	810	835
3	308	50	100	42	30	5.3	175	810	835

from 4.75 to 16 mm, river sand (S) with a size ranging from 0.15 to 4.75 mm and a fineness modulus of 2.6, polycarboxylic acid SP, and compound VEA. The properties of fresh and hardened concrete are given in Table 3.21. It can be found that the properties of the mixes with different powders differ from each other. The compositions of powders in a mixture, especially VEA, have a significant effect on the properties of fresh concrete. The sample without VEA has higher flowability (slump flow), less flow rate, larger surface paste thickness in the fresh state compared to sample with VEA. Mix three is a typical mix of SSFSCC, which satisfies the corresponding properties specified in Tables 3.18 and 3.19. The typical SSFSCC mix has been successfully applied in actual practice. This indicates that the proposed mix proportioning method for SSFSCC is reasonable and feasible.

3.9.4 Construction technology of SSFSCC

With proper design and careful production of SCC, bleeding, segregation, surface settlement, and instability of the air bubble can be reduced to an acceptable level for normal SCC. For a formwork with the upper side open to the atmosphere, slight bleeding, surface settlement, and air bubble floating of normal SCC is acceptable. Furthermore, it can be smoothed by the process of finishing. However, in the case of filling a flat, sealed, and narrow space, no process of finishing is allowed for the top surface of SFFSCC; what's more, the top surface of the SSFSCC filling layer is required to have a strong bonding with the above prefabricated slab. Therefore, the defects on the top surface of the filling layer shown in Fig. 3.44 should be minimized. Obviously, bleeding, segregation, surface settlement, and air bubble floating should be strictly controlled for SSFSCC of CRTS III ballastless slab track. In order to make the SSFSCC filling layer meet the designed requirement, it is also of great significance to develop proper construction technology besides developing an optimal mix proportioning of SSFSCC.

The construction of the CRTS III slab track structure includes a set of steps, as shown in Fig. 3.41. Apart from that, there are several special and

Table 3.21 Properties of fresh and hardened concretes.

Mix	Properties in a fresh state						
	Air content/%	Slump flow/mm	T500/s	PA	Bleeding rate/%	Top surface paste thickness/mm	Expansion rate within 24 h/%
1	3.6	670	3.5	0.86	0	12.0	0
2	4.2	655	3.0	0.89	0	9.5	0.40
3	4.8	610	5.0	0.90	0	7.0	0.50
	Properties in a hardened state						
	Compressive strength/MPa	Flexural strength/MPa	Elastic modulus/GPa	6-h charge passed/c	Freezing–thawing resistance/g·cm-2	Shrinkage/10-6	
1	53.0	6.0	38.0	720	820	395	
2	52.3	5.8	37.2	750	840	335	
3	50.5	6.5	36.0	460	610	310	

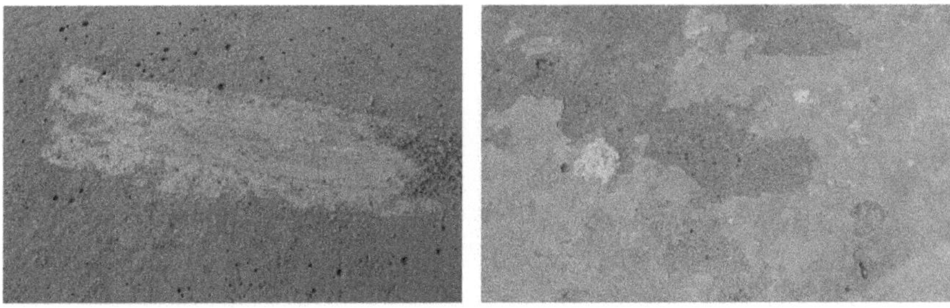

(a) (b)

Figure 3.44 The top surface of the filling layer constructed by SSFSCC with improper workability after removing the prefabricated slab seated above. (A) paste with a very porous structure and low strength on the top surface of the filling layer, (B) a very thin paste layer with a bubble layer beneath on the top surface of the filling layer.

important requirements for the construction of the SSFSSC filling layer of the CRTS III slab track. They are described as follows.

The casting speed of SSFSCC is the first point. SSFSCC flows in the sealed space may be in two different modes, as shown in Fig. 3.45 Mode 1, the SSFSCC is very flowable, or the casting speed of SSFSCC is slow. SCC first spread out on the base plate, that is to say, the profile of SSFSCC has no contact with the top surface at the beginning. The surface of SSFSCC elevates steadily until fresh SSFSCC has full contact with the slab. In this scenario, the air in the sealed space cannot be totally squeezed out, and some big air bubbles are readily entrapped in the interface between the slab and base plate. Mode 2, the SSFSCC is not highly flowable, or the casting speed of SSFSCC is fast; the profile of SSFSCC has contact with upper and bottom faces during the flowing of SSFSCC. The air in the sealed space can be squeezed out to its largest extent, and no air bubble could be entrapped in the interface. Therefore, SSFSCC of the CRTS III slab track is cast in mode 2. To reach the balance of flowability and casting speed of SSFSCC is

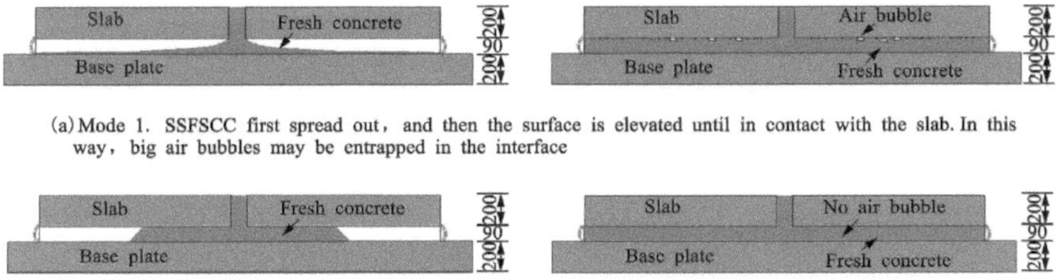

(a) Mode 1. SSFSCC first spread out, and then the surface is elevated until in contact with the slab. In this way, big air bubbles may be entrapped in the interface

(b) Mode 2. SSFSCC has full contact with upper ad bottom surfaces to squeeze out air to its largest extent. In this way, almost no air bubbles may be entrapped in the interface

Figure 3.45 Two different filling modes of SSFSCC.

Figure 3.46 Full-scale field experiment of SSFSCC construction.

crucial for casting SSFSCC. Full-scale tests have been carried out to find the appropriate casting speed of SSFSCC with specified rheological behavior.

Secondly, the slab has been finely adjusted to its design position, and the change of position cannot go beyond 0.5 mm during casting SSFSCC. However, in order to cast SSFSCC in mode 2, the height of SSFSCC in the pouring hole has to be high enough to maintain hydraulic pressure on SSFSCC, and thus the slab can be lifted up easily. Therefore, some clamps are designed to hold the slab still, as shown in Fig. 3.46. However, if the height of SSFSCC in the casting hole is too high, the hydraulic pressure exerted on the slab is high; as a consequence, relatively large deformation of the clamp is inevitable, and the position of the slab may be changed beyond the limit of 0.5 mm. The rheological behavior of SSFSCC, SSFSCC height in the pouring hole, and the clamps jointly affect the position controlling of the slab.

Thirdly, real-time field inspection must be carried out from porthole and exhaust holes during casting. The casting speed and segregation of SSFSCC can be visually judged from the porthole during casting. When the homogenous SSFSCC begins to flow out of the exhaust holes in the corners of the sealed place, as shown in Fig. 3.47, it is believed that the

(a) Exhaust holes (b) Porthole

Figure 3.47 Observation of SSFSCC from exhaust hole and porthole.

sealed space is full of SSFSCC, and the casting of SSFSCC should be stopped to avoid lifting the above prefabricated slab by excess fresh SSFSCC. Besides, the quality of SSFSCC can also be visually judged by the observations from the porthole and exhaust holes.

In addition, it is worth to mention several other important steps during the construction of the SSFSCC filling layer. For example, after the completion of grouting SSFSCC into the sealed space, some fresh concrete in the cylindrical pouring hole should be reserved until the initial setting of concrete to avoid surface settlement. In this way, a little bit hydraulic pressure is maintained to keep SSFSCC in full contact with the above prefabricated slab, and the surface settlement of concrete can be effectively avoided. Prior to casting SSFSCC, the sealed space should be prewetted to avoid water loss of SSFSCC absorbed by geotextile and prefabricated slab.

It is important to note that, although some successful applications of SSFSCC in high-speed rail have been made, there are still some challenges faced. The first challenge is that the air bubble layer on the top surface of the SSFSCC filling layer (see Fig. 3.44) occurs sometimes. Although bleeding, segregation, surface settlement of SCC can be effectively avoided, as long as SSFSCC meets all the requirements listed in Table 3.19. However, the thin paste layer with a bubble layer underneath is hard to be completely avoided. This type of interface has low interface bonding between the slab and SSFSCC layer. The fresh SSFSCC flows in the sealed space in mode 2, both upper and bottom surfaces, provide friction during flowing. It is similar to the flow of pumping SCC. The velocity profile of SCC flows through the pipe is assumed to consist of a plug in the center, a lubrication layer at the wall, and a shear zone. It is believed that the plug is small, and the bulk concrete is sheared. However, conventional vibrated concrete has no shear zone (see Fig. 3.48). When SSFSCC flows in the sealed space, a shear zone may be present. If the SSFSCC is shear thinning under the shear rate, depending on the casting rate, the viscosity of SSFSCC may be reduced, and bubbles become unstable. Floating bubbles gather under the lubrication layer. This is exactly what Fig. 3.44 presents. Due to the presence of the bubble layer, the bonding between the slab and the SSFSCC filling layer is very weak. In practice, the size of the bubble in concrete is decreased by the action of chemical admixture, and the stability of bubbles is verified by small-scale casting test for every new batch of materials prior to application. This helps to prevent the bubbles from rising to the interface in real engineering.

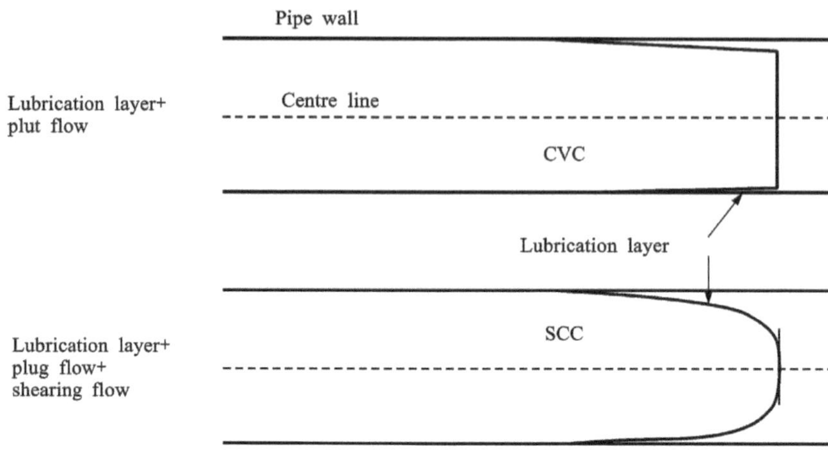

Figure 3.48 Estimated velocity profile during pumping of conventional vibrated concrete (CVC, top) and self-compacting concrete (SCC, bottom), showing that the bulk SCC is sheared while CVC is not.

The second challenge is the quality control of SSFSCC for the CRTS III slab track. The amplitude of quality fluctuation of raw materials is relatively large, because of the long time and large space spans during the construction of an HSR line of CRTS III slab track. Thus, to further improve the robustness of SSFSCC is extremely important.

3.10 Steam-cured concrete

3.10.1 Introduction

Steam curing, which accelerates the early strength development of concrete by means of elevating temperature and providing moisture simultaneously, is commonly used in the production of prefabrication concrete products and elements for high productivity. Prefabrication is a manufacturing process, generally conducted at a specialized facility, in which various elements are joined to form a component part of the final installation. The manufacturing process may be factory prefabrication or off-site prefabrication. Off-site fabrication is often used when both prefabrication and preassembly are integrated. Prefabrication has been identified as the first degree of industrialization, followed by mechanization, automation, robotics, and reproduction.

Off-site fabrication has become popular after the World War II. At that time, various concrete structural elements were developed with the support of the administration to deal with the increasing demand for housing. The demand was at its peak in the 1950s—1970s in Europe for the construction of new towns, and large-scale public housing developments. In the early 1970s, the US government started to use prefabricated building systems. According to the data in 1996, Denmark ranked highest precast levels at 43%, followed by the Netherlands (40%), Sweden and Germany also reached up to 31%. In Asia, the precast levels in Japan and Singapore were about 15% and 8%, respectively. In 2002, prefabricated elements accounted for about 17% of the concrete volume used in Hongkong, China, which are used for façades, staircases, parapets, partition walls (drywall), and semiprecast slabs.

Chinese administrations are even more ambitious in encouraging the prefabrication technology to be used in the construction of the housing sector. For instance, the precast level in Shanghai reached up to 40% in 2018, and the level is still increasing. Other cities are also encouraging to increase the precast level.

During the past decades, high-speed railway infrastructure in China has a rapid development, and the total length of high-speed rail is over 30,000 km in 2019. Precast concrete elements have been widely used in the construction, such as sleepers, track slabs, and prestressed concrete box girders. It is roughly estimated that more than 30,000 box beams of 900 tons and 2,500,000 concrete slabs of 7 tons have been used in the construction of high-speed rail.

To sum up, due to the high productivity, greenness, labor saves, and high quality, governments around the world have been and are encouraging the application of prefabrication in the construction sectors. Since steam-cured concrete is the main materials used for prefabrication, the steam-cured concrete is of paramount significance to the prefabrication technology.

Steam-cured concrete can rapidly improve the strength of concrete at early ages and efficiency of template turnover. However, in comparison with concrete cured at room temperature, some defects or damages can be brought by steam curing, such as concrete brittleness developing, concrete surface layer microcracks increasing, and porosity enlarging. High temperature and the moist steam-cured process can bring about significant differences between surface and inner concrete due to stress gradient, heat-mass transfer, and nonuniform of hydration of cementitious. This has been defined as heat damage. The detailed mechanisms of heat damage have been proposed. Firstly, the formation of a dense shell around cement grains in early age hinders the later hydration, and the rapid early hydration leads to the heterogeneous distribution of hydration products, and hence coarsing the pore structure. Secondly, temperature gradient and the corresponding tensile stress develops during the steam curing process, which may lead to the formation of fine cracks at the surface layer. Thirdly, the concrete surface layer exposed to steam exhibits a more open pore structure than that of the interior parts, reflected by a higher porosity and water absorptivity, which deteriorates the permeability of steam-cured concrete.

3.10.2 Raw materials

Basically, the raw materials of steam-cured concrete have no difference with that of ordinary concrete. In the Chinese code, a special blending material is proposed to be used in the steam-cured concrete for the slab, as shown in Table 3.22.

It is also found that partially replacing cement with supplementary cementitious materials (SCMs) in steam-cured concrete is helpful to mitigate the heat damage as well as reduce the CO_2 footprint. Nevertheless, the use of SCMs can only mitigate the adverse effects to some extent; steam-cured concrete incorporated with SCMs still suffers from the heat-induced problems. In order to reduce the heat damage caused by steam curing, lowering the maximum curing temperature seems to be an efficient way. However, lowering the maximum curing temperature would hinder early strength development, which is of great importance in the precast industry. The use of SCMs with high reactivity in steam-cured concrete may be a promising way to reduce the maximum curing temperature without compromising early strength gain.

Table 3.22 Blending materials for steam-cured concrete.

Items		Requirements
Chloride ion content/%		Should not be greater than 0.02
Loss on ignition/%		≤4.0
SO_3 content/%		≤3.0
Moisture content/%		≤1.0
Water demand ratio/%		≤105
Activity index/%	1 d	≥125
	28 d	≥110

Metakaolin has been studied as an alternative binder for steam-cured concrete due to its excellent pozzolanic properties. Investigations showed that the substitution of 10%–25% cement (by mass) by metakaolin results in increased compressive strength in contrast to concrete or mortar incorporating cement only. The corresponding microstructure tests demonstrated that the incorporation of metakaolin in concrete results in consumption of $Ca(OH)_2$ at early stages, and the formation of secondary C-A-S-H refines the pore structure, leading to an improved early mechanical property. Besides, the addition of metakaolin contributes to enhanced volume stability for normal-cured concrete, and further decreases the autogenous and dry shrinkage to some extent as well.

Limestone powder is often used as an inert filler. Nevertheless, evidence has shown that, within limits, the calcite in limestone can interact with aluminum-bearing phases in cement paste or metakaolin to form mono/hemi-carboaluminate phases which exhibit better thermodynamic stability than ettringite or mono-sulfoaluminate.

A strong synergistic effect between limestone powder and metakaolin has been reported. Steam-cured concrete incorporated with the combination of metakaolin and limestone exhibits improved mechanical properties and sorptivity, refined microstructure in comparison with steam-cured concrete incorporated with ground blast furnace slag and fly ash, which is typically used to produce prefabricated concrete slab, a sleeper for high-speed railway structure in China.

Lightweight aggregates possess unique properties due to their porous characteristics. The incorporation of lightweight aggregate can reduce the bulk density of concrete and affect the mechanical properties of concrete. The water-releasing effect provides an internal curing effect for concrete, which can benefit the improvement of the microstructure of concrete. It is found that the addition of expanded clay sand has a great effect on the compressive strength and elastic modulus of steam-cured concrete. Especially, the compressive strength of steam-cured concrete incorporated with ceramsite sand is even higher than that of reference without ceramsite sand. It is also found that 30% ceramsite sand replacing normal aggregate results in a delay of temperature rise, which is very important for alleviating the damage on microstructure caused by the elevated temperature at an early age and thus results in improving compressive strength of steam-cured concrete.

3.10.3 Curing regime

A typical steam curing of concrete often includes four stages, i.e., delay stage, controlled heating stage, heat treatment stage, and controlled cooling stage, as shown in Fig. 3.49.

Maximum heat treatment temperatures may be in the range of 40–100°C, and the optimum temperature has been found in the range of 60–85°C for ordinary steam-cured concrete. The curing temperature is a compromise between the rate of strength gain and ultimate strength, because the higher the curing temperature, the higher productivity, and the lower the ultimate strength. The main factors determining the behavior of binders subjected to heat treatment are fineness and composition of types of cement, the type and amount of additives used in binders, and curing cycle parameters. For compressive strength development of concrete, the duration of steam curing is also an important parameter as well as temperature. It is obvious that heat treatment application at a lower temperature is more economical and energy-saving. The length of the total curing period must allow for the controlled heating application and cooling of the concrete. Practical curing cycles are chosen as a compromise between the early and late strength requirements. According to ACI code, an appropriate heat cycle would consist of the following: a delay period of 2–5 h, heating at the rate of 22–44°C/h up to a maximum temperature of 60–85°C, then storage at maximum temperature, and finally a cooling period, the total cycle (exclusive of the delay period) should be completed preferably not more than 18 h.

As mentioned above, a large amount of steam-cured concrete slabs and box girders have been successfully used in the construction of high-speed

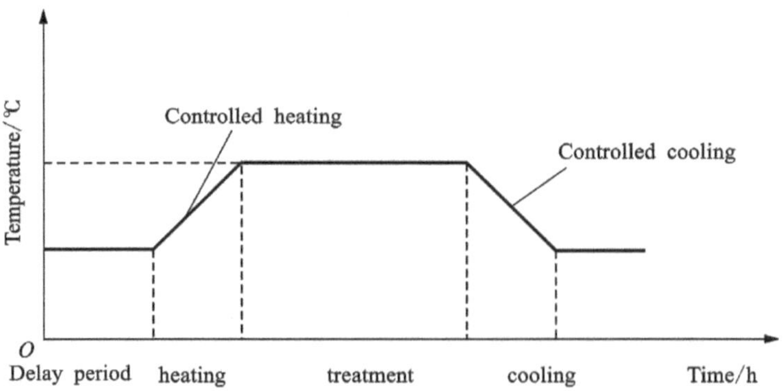

Figure 3.49 Typical steam curing process.

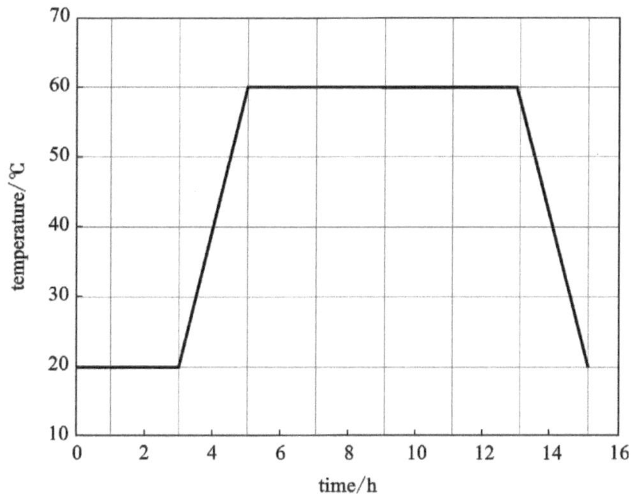

Figure 3.50 Curing regime used for prefabricated elements for the construction of high-speed rail.

rail in China. The curing regimes used are somewhat different from that proposed by ACI, as follows:

① Delay period: at least 3 h after casting.

② Controlled heating stage: heating rate is lower than 15°C/h

③ Heat treatment stage: temperature of the steam is lower than 45°C, the internal temperature of concrete is lower than 55—60°C, heat treatment stage is no longer than 6 h

④ Controlled cooling stage: heating rate is lower than 15°C/h; when demolding, the temperature difference between concrete's surface and ambient is lower than 15°C.

A total heat-curing duration is about 15 h, as shown in Fig. 3.50.

In addition, it was found that the surface permeability and strength development of steam-cured concrete is remarkably influenced by the subsequent curing condition. The subsequent curing, such as long-term water-soaking curing at 20°C or oven-dry curing at 60°C, deteriorates the performances of steam-cured concrete. Their surface permeability indexes are 75.1% and 88.0% higher than that of standard-cured concrete at 90 d, respectively. The continuous dissolution of calcium hydroxide is considered the main reason for the deterioration of the impermeability and strength of steam-cured concrete.

After demolding, the steam-cured concrete should be well cured, such as immersing in water or covered with a plastic sheet. In the case of the

Figure 3.51 Steam-cured concrete slab.

steam-cured concrete slab, the stacking of slabs should follow the specifications to avoid irregular deformation during the period of storage, as shown in Fig. 3.51.

3.10.4 Mechanical properties

3.10.4.1 Compressive strength

Compressive strength is the key parameter for concrete materials. The relationship between compressive strength, curing temperature, and binder composition of steam-cured concrete has been investigated, as shown in Fig. 3.52.

Denote: 100%C is concrete made with pure cement. 100%C + 20%FA + 10%GGBS is concrete made with 70% cement, 20% fly ash, and 10% slag.

In comparison with the sample cured at 20°C, the steam temperature ranging from 45 to 80°C promotes the early strength development of concrete; however, hinders the late strength development after 7 d. In

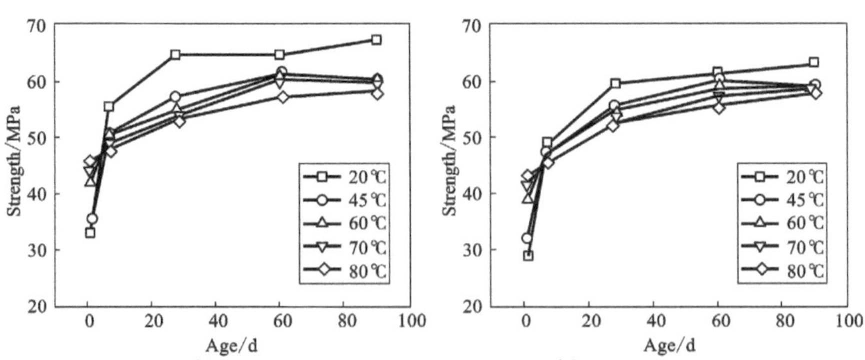

Figure 3.52 Development of compressive strength of concrete with various binders at different cured temperature.

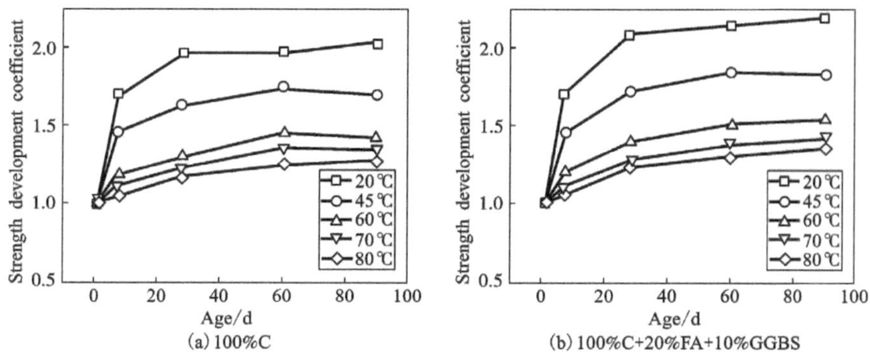

Figure 3.53 Strength development coefficient of steam-cured concrete (in comparison with 1 d strength).

addition, the compressive strength of concrete with supplementary cementing materials is lower than that of concrete with pure cement. This means high curing temperature can cause heat damage on concrete, and thus the compressive can also be reduced.

In order to take a closer look at the effect of temperatures and binder on the strength development, the 1 d compressive strength was taken as reference, and the strength development of specimens can be seen in Fig. 3.53. Although the strength with supplementary cementing materials is lower than concrete with pure cement, the strength development coefficient of the former is higher than that of the later. This means the strength of concrete with supplementary cementing materials can develop better at later age.

From the viewpoint of microstructure, the equilibrium between C—S—H and ettringite, which are the most important hydrates of cement, is affected by the temperature. At high curing temperature, the ettringite becomes more soluble, the sulfate concentration in the solution increases, and the quantity of sulfate bound to the C—S—H also increases. The denser structure of C—S—H and the decrease of the ettringite content at elevated temperature leads to higher capillary porosity and lower compressive strength of concrete.

3.10.4.2 Dynamic mechanical properties

In practice, infrastructure such as railway, highway, and bridge is often subjected to static load as well as dynamic load, which is resulted from mechanical impact and vibration from running vehicle during its service. Therefore, it is of great significance to understand the dynamic mechanical characteristics of concrete material and structure under dynamic load.

Figure 3.54 Pictures of failure modes of specimens under different strain rate conditions.

The split Hopkinson pressure bar (SHPB) test method is one of the most common methods to study the resistance of concrete to impact. SHPB is often used to study the dynamic mechanical properties of concrete. Concrete is a multiphase porous composite material, and contains some initial defects, such as microcracks and pores. The failure of concrete is a process of crack propagation in which the initially isolated defects gradually extend and become a connected macrocrack network. Fig. 3.54 shows the photos of compressive failure modes of concrete under quasi-static (the strain rate is 10^{-5}/s), low strain rate (the strain rate is 10^{-3}/s), and high strain rate (SHPB test) conditions. It can be seen from Fig. 3.54 that the failure modes of a specimen are different from each other at different strain rates condition. The compressive failure mode of the specimen under the quasi-static load presents shear failure, while the one at low strain rate (10^{-3}/s) manifests both shear failure and splitting failure characteristics. The number of cracks on the compression surface obviously increases, and the splitting failure phenomenon of aggregate increases. This indicates that in the range of low strain rate, the crack slightly increases with the increasing loading rate and then spreads freely in hardened paste or aggregate and further connects into several main cracks throughout the whole specimen, which leads to splitting failure of concrete in the end. It is worth noting that under the high strain rate of the SHPB impact load condition, the failure behaviors of concrete are quite different from those under quasi-static and low strain rate loading conditions. Under this impact-load condition, the concrete specimen is immediately broken into small pieces, and the degree of crushing of the concrete specimen becomes more serious with the further increase of strain rate, as shown in Fig. 3.54 C and D. This is mainly because when the strain rate is high, the original defects in the concrete have no time to propagate. At this time, the huge energy from the impact-load with high velocity on concrete is mainly consumed in the form of finer crack generation, resulting in a greater degree of damage or fracture to the concrete.

3.10.5 Durability

According to field engineering experiences, premature deterioration of steam-cured concrete elements takes place quite often. In China, a full investigation of steam-cured concrete sleepers showed that the average life of sleepers was only 14 years. Some sleepers took on seriously deterioration after service of 5–8 years. Cracking, water penetrating, bear capacity decreasing were the main deterioration characters of steam-cured concrete bridges elements in China. In the United States, some steam-cured precast concrete girders used in San Mateo Bridge in San Francisco were found serious deterioration and must be repaired after service of 17 years. In Texas, many drilled-shaft concrete foundations of high-mast illumination poles constructed in the late 1980s were found to have premature concrete deterioration due to alkali-silica reaction (ASR) and delayed ettringite formation (DEF). From 1998 to 2003, a comprehensive investigation was conducted in Texas on the structural performance of in-service bridges with premature concrete deterioration. The premature concrete deterioration was attributed to two expansive distress mechanisms: ASR and DEF. It is also found that the deicer-scaling and corrosion resistances of steam-cured concrete are significantly lower than those of concrete at common curing. The resistances become worse with the increase in the steam temperature and with the reduction of the precuring time.

By measuring the chloride profile after chloride immersion, it was found that chloride penetration resistance of steam-cured concrete is worse than that of standard-cured concrete; this effect only happens at the first 10 mm of the cover layer, as shown in Fig. 3.55. This is induced by the heat

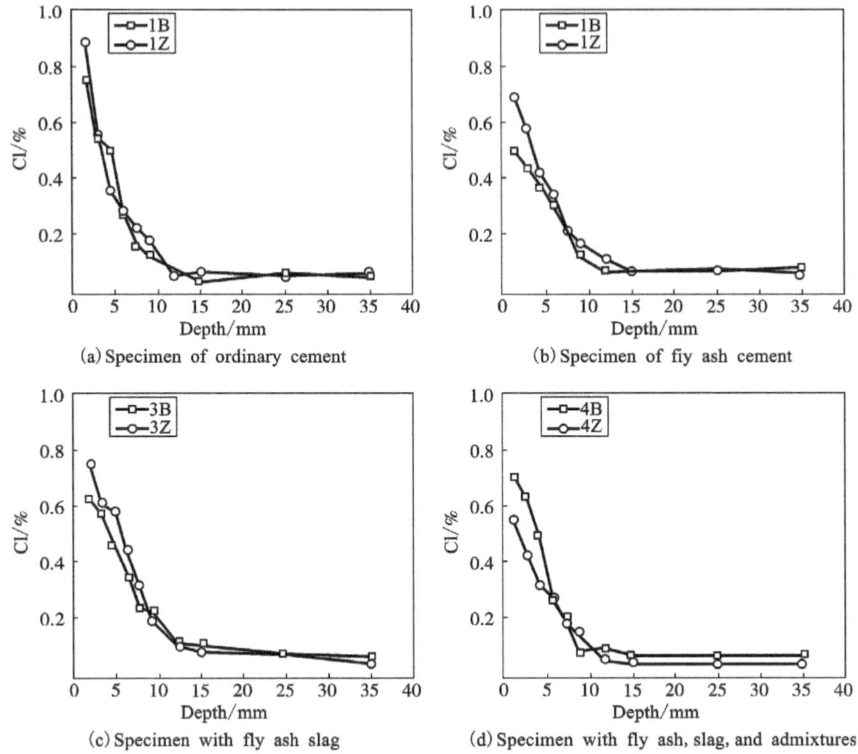

Figure 3.55 Chloride profile of specimens subject to standard curing and steam curing (B is for standard curing, Z is for steam curing).

damage on the surface of steam-cured concrete. A similar phenomenon can be found on the carbonation test. That is to say, the carbonation rate of the first 10 mm of the cover layer of steam-cured concrete is faster than that of standard-cured concrete.

It is found that the anti-FT cycles of steam-cured concrete are worse than that of standard-cured concrete. Appropriate air entrainment can improve the antifrost of steam-cured concrete. Under the coupled effect of static or dynamic loads and freezing, the deterioration rate can be significantly increased. It is reported that the coupled effect of cycled load and FT cycles can cause 10 times deterioration on concrete than the single effect does.

It is worth to mention that delayed ettringite formation (DEF) and alkali–silica reaction (ASR) also were found in the steam-cured concrete slab in the high speed railway, as shown in Fig. 3.56. Microstructural analysis showed that a large amount of ettringite was formed in the steam-cured concrete slab. Also, white substances were found on the surface of the aggregate. SEM and EDS analysis confirmed that it is an ASR reaction, as shown in Fig. 3.57.

Figure 3.56 Map cracking of steam-cured slab.

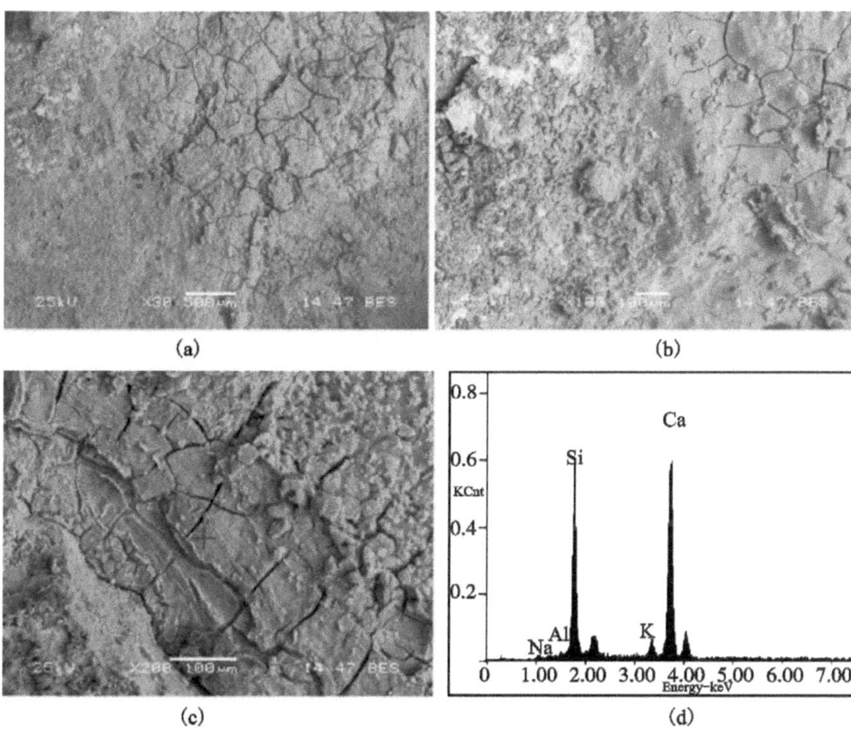

Figure 3.57 SEM pictures of substance around aggregate.

It is generally believed that damage due to the DEF happens at temperatures higher than 70°C. However, it is found that DEF occurs in field where temperatures are far lower than 70°C. This is because DEF was influenced by the curing temperature, as well as various other factors, such as cement composition (alkalis, C_3S, C_3A, SO_3, and MgO), fineness, etc. If an unfavorable combination of these parameters exists, DEF may occur at temperatures less than 70°C. However, the simultaneous occurrence mechanism of the delayed ettringite formation and the alkali–silica reaction in concrete structures are still not well understood.

In summary, heat damage causes initial defects, including the porous structure and microcracks in steam-cured concrete. These initial defects make the antipermeability of steam-cured concrete worse than that of standard-cured concrete. Therefore, the penetration-related durability of steam-cured concrete is bad. Under the action of dynamic fatigue load, the situation may be worse. Thus, the frost-resistance, chloride penetration resistance, carbonation resistance, ASR, and DEF of steam-cured concrete are worse than that of standard-cured concrete (Figs. 3.57 and 3.58).

Exercises

1. Why is concrete the most widely used engineering material? Why are concrete and steel the best combination?
2. Define the terms grading and maximum aggregate size, as used in concrete technology. What considerations control the choice of the maximum aggregate size of aggregate in a concrete? Discuss the reason why grading limits are specified.
3. What is SP of concrete? Explain the action mechanisms of SP, and list its advantages.
4. Explain the following terms and discuss their significance: absorption capacity, saturated-surface-dry condition, damp condition.
5. What are supplementary cementing materials? Name three typical supplementary cementing materials, and their uses in concrete.
6. Explain how retarding and accelerating agents change the hydration of cement.
7. Define the following phenomena, and give their significance and the factors affecting them: slump loss, segregation, and bleeding.

8. Define the initial and the final setting times, and their main affecting factors.
9. Discuss the physical–chemical characteristics of the C–S–H, calcium hydroxide, and calcium sulfoaluminates present in a well-hydrated Portland cement paste.
10. How many types of voids are present in a hydrated cement paste? What are their typical dimensions and their effects on the properties of concrete?
11. How many types of water are associated with a saturated cement paste? Discuss the significance of each and their difference.
12. What is ITZ, and explain the reasons how it forms. Describe the characteristics of ITZ, and its effect on the mechanical properties of concrete.
13. Explain the main factors affecting the compressive strength of concrete.
14. Describe the various stages of microcracking when a concrete specimen is loaded to failure. Explain why the strain–stress curve of concrete is nonlinear.
15. What are drying shrinkage strain and creep strain in concrete? What is their significance? How are the two phenomena similar to each other?
16. From the standpoint of material, describe the strategy to prevent the crack of concrete.
17. What is the AAR? Describe the strategy to prevent the AAR of concrete.
18. What chemical reactions are generally involved in sulfate attack on concrete? What are the physical manifestations of these reactions? How to prevent sulfate attack?
19. Why steel bar does not corrode in concrete? In which situation corrosion of steel can be initiated?
20. Briefly describe the strategy for the control of corrosion of embedded steel in concrete.
21. Discuss the hypothesis of expansion on freezing of a saturated cement paste containing no air. Why is entrainment of air effective in reducing the expansion due to freezing?
22. State the differences between self-compacting concrete and normal concrete.
23. How does the heat-regime affect the microstructure and properties of heat-curing concrete?

24. Use the following parameters to make a preliminary concrete mix design

Desired strength:	**C30**		
Desired slump:	55–60 mm		
Raw materials	Cement	River sand	Crushed aggregate
Strength	36 MPa		
Density	3.1 g/cm^3	2.65 g/cm^3	2.72 g/cm^3
Maximum size			20 mm
Water content		1.4%	0.5%

CHAPTER 4
Metal

4.1 Introduction

Metals are described as either ferrous, containing a substantial proportion of iron (Fe), or as nonferrous.

Ferrous metals include mild steel, carbon steel, stainless steel, cast iron, and wrought iron. Structural steel is the most useful of the ferrous metals. The steels are used to produce the various sections (shapes, plates, and bars) from which structural members (beams, girders, columns, struts, ties, and hangers) are fabricated. The primary purpose of such members is to support loads and/or resist forces acting on structures such as bridges; industrial, institutional, commercial, and residential buildings; rails for railroad and transportation, and other construction equipment. On the other hand, the steel is also used as a reinforcement material (bars, wires, and strands) in Portland cement concrete structural, to resist tensile stresses which could lead to the failure of the brittle concrete. The mechanical properties of construction steel, the cold working and strengthening of steel, and the standards and selection of steel will be focused on in this chapter.

Nonferrous metals include aluminum, brass, copper, nickel, tin, lead, and zinc, as well as precious metals like gold and silver. Nonferrous metals are much more malleable than ferrous metals. Nonferrous metals are also much lighter, making them well-suited for use where strength is needed, but weight is a factor, such as in the aircraft or canning industries. Nonferrous metals are widely used for gutters, water pipes, roofing, and road signs due to a higher resistance to rust and corrosion. They are also nonmagnetic, which makes them perfect for use in small electronics and as electrical wiring.

Civil Engineering Materials
ISBN 978-0-12-822865-4
https://doi.org/10.1016/B978-0-12-822865-4.00004-0

4.2 Structural steel

4.2.1 Chemical composition

The chemical composition of steel is very important since it has a significant effect on the microstructure of the material and hence on its mechanical behavior and properties.

Steels are alloys that have iron content ranging from more than 99% to less than 60%, with carbon and other metallic elements to harden the alloys or give them other properties. Iron by itself is relatively soft and prone to rusting. Steel is hard and corrosion resistant.

Each element added to the basic constituent of iron has some effects on the properties of the material. Following is a list of the elements commonly added to iron:

4.2.1.1 Carbon

Carbon is the most important element in steel and can be present up to 2% (although most welded steels have less than 0.5%). The carbon can exist either dissolved in the iron or a combined form, such as iron carbide (Fe_3C). The influence of carbon on the mechanical properties of carbon is shown in Fig. 4.1. With the increase of carbon content, the rigidity and the strength of steel will increase, and its plasticity and toughness will decrease. If the carbon content is more than 1%, the ultimate strength of the steel begins to fall. Also, if the carbon content is too high, the brittleness and aging sensitivity of the steel will rise, which reduces its ability to resist the corrosion of the atmosphere and weld ability.

4.2.1.2 Manganese

Steels usually contain at least 0.3% manganese, which acts in a threefold manner: it assists in DE oxidation of the steel, prevents the formation of iron sulfide inclusions, and promotes greater strength by increasing the hardenability of the steel. Manganese appears in structural steel grades in amounts ranging from about 0.50 to 1.70%. Manganese is a necessity for the process of hot rolling of steel by its combination with oxygen and sulfur.

4.2.1.3 Aluminum

Aluminum is one of the most important deoxidizers in the material, and also helps form a more fine-grained crystalline microstructure. It is usually used in combination with silicon to obtain a semi- or fully killed steel.

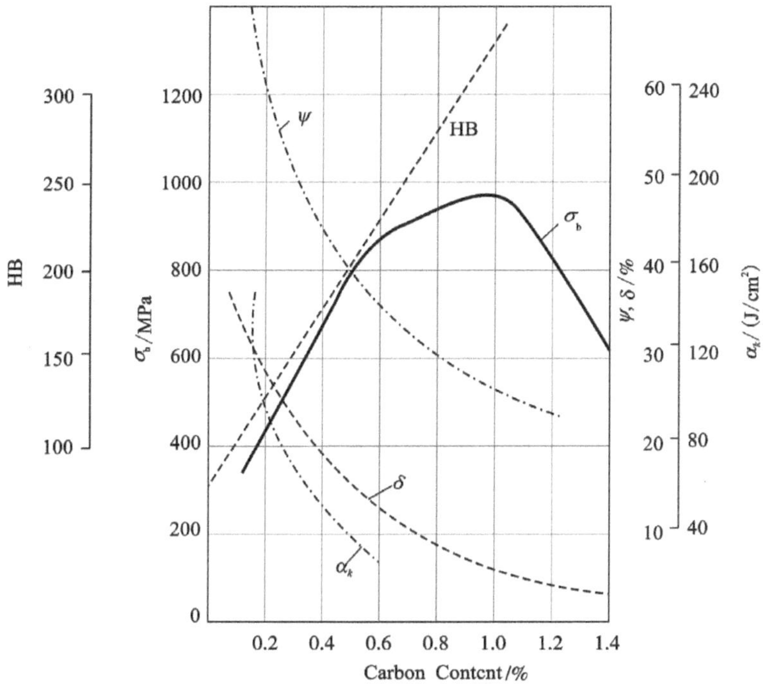

Figure 4.1 Influences of carbon content on properties of hot-rolled carbon steel (*d*, Elonyationl; *HB*, Hardness; *j*, Shrinkage of Cross section; *sb*, Tensile Strength; *SK*, Impact Toughness).

4.2.1.4 Silicon

Usually only small amounts (less than 0.40%) are present in rolled steel when silicon is used as one of the principal deoxidizers along with aluminum. Silicon is the element that is most commonly used to produce semi- and fully killed steels, dissolves in iron, and tends to strengthen it. The resulting decrease in ductility could present cracking problems in some situations.

4.2.1.5 Phosphorus and sulfur

Phosphorus and sulfur are considered to be undesirable impurities in the steel. Both act to reduce the ductility of the material. All steel grade specifications, therefore, place severe restrictions on the amount of P and S that are allowed, basically holding them to less than about 0.04–0.05%.

Phosphorus tends to cause embrittlement. In low-alloy, high-strength steels, phosphorus can be added in amounts up to 0.10% to improve both strength and corrosion resistance.

In amounts exceeding 0.05%, sulfur tends to cause brittleness and reduce weld ability. Additions of sulfur in amounts from 0.1% to 0.3% will improve the machinability of steel but impair weld ability.

4.2.1.6 Chromium, molybdenum, and nickel

The addition of these elements is generally considered to increase the hardenability of steel. Chromium can markedly improve the corrosion resistance of iron and steel in oxidizing types of media. Molybdenum can elevate temperature strength. Nickel is often used to improve the toughness and ductility of the steel at low temperatures.

4.2.2 Strengthening mechanisms

The ability of a metal to deform plastically depends on the ability of dislocations to move. Hardness and strength are related to how easily a metal plastically deforms, impeding the movement of dislocations will therefore result in the strengthening of the material. There are several ways to impede dislocation movement, which include:
① Controlling the grain size (reducing continuity of atomic planes).
② Strain hardening (creating and tangling dislocations).
③ Alloying (introducing point defects and more grains to pin dislocation).

4.2.2.1 Controlling the grain size

The size of the grains has a significant effect on the strength of the material. The boundary between grains acts as a barrier to the dislocation movement and the resulting slip because adjacent grains have different orientations with the different atom alignments and discontinuous slip planes. The smaller the grains, the shorter the distance atoms can move along a particular slip plane (shown in Fig. 4.2). Therefore, smaller grains improve the strength of a material. The size and number of grains are controlled by the rate of solidification from the liquid phase.

4.2.2.2 Strain hardening (cold working)

Strain hardening is a process to promote the metal harder and stronger due to plastic deformation. The dislocations are generated when plastic deformation occurs in the metal. The dislocations will interact and become pinned or tangled (shown in Fig. 4.3). This can block the further movement of the dislocations and promote the strength of the material. This type of strengthening is commonly called work-hardening or cold working

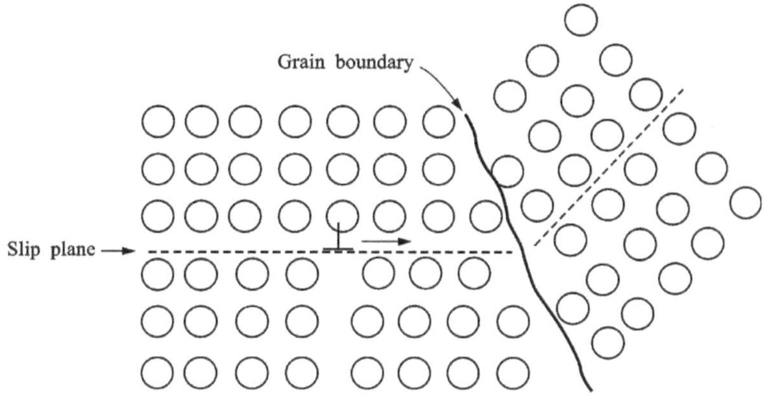

Figure 4.2 A slip plane.

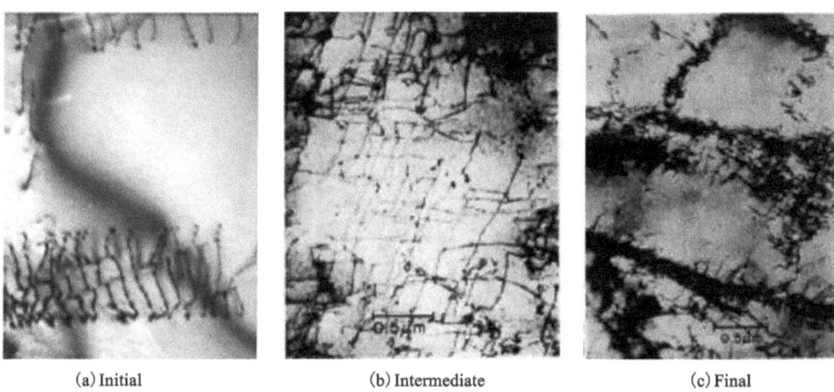

Figure 4.3 Dislocation tangles produced by plastic deformation in iron.

because the plastic deformation occurs at a temperature low enough that atoms cannot rearrange themselves.

The common forming operations of cold working include rolling, forging, drawing, and extrusion (shown in Fig. 4.4).

With the extension of time, if the strength and the rigidity of steel increase and the plasticity and the rigidity of steel decrease, it is called aging. The aging process of steel under the natural state is very slow. If the steel often suffers vibrating and impact loads in cold working or use, the aging will develop fast. After cold working, the yield strength, tensile strength, and rigidity of steel will increase but the plasticity and the toughness keep decreasing if the steel stays at room temperature for 15—20 days or is heated

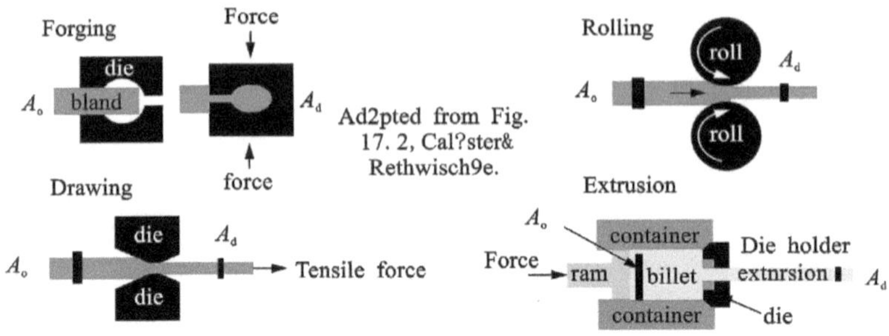

Figure 4.4 Common forming operations.

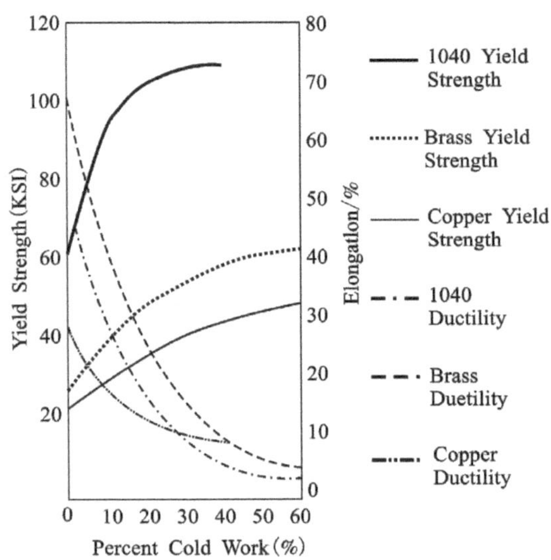

Figure 4.5 Effects of cold work on the different strengths of steel.

to 100–200°C for 2 h. The former is called natural aging, and the latter is called artificial aging. It should be understood, however, that increasing the strength by cold working will also result in a reduction in ductility. Fig. 4.5 gives the effect of cold working on different strengths. It should be noted that a small amount of cold working results in a significant reduction in ductility.

However, strengthening due to strain hardening is not always desirable, especially since the ductility will be lowered. Heat treatment can be used to remove the effects of strain hardening. Recovery, recrystallization, and grain growth things can occur during heat treatment.

4.2.2.3 Heat treatment

The heat treatment includes heating, heat preservation, and cooling.

The steel material could be heated below the critical point or above a critical point. The former heating way can stabilize the structure and eliminate the residual stress. The latter way can make material austenitizing. Austenitizing is to heat steel metal over its critical temperature long time enough, so it could be transformed. If a quenching followed after austenitizing, then the material will be hardened. Quenching will take fast enough to transform austenite into marten site. Once reached austenitizing temperature, suitable microstructure, and full hardness, the steel pipe material will be attained in further heat treatment processes.

The purpose of heat preservation is to uniform the heating temperature of steel material, then it will get a reasonable heating organization.

The cooling process is the key process in heat treatment; it determines the mechanical properties of steel after the cooling process.

The heat treatment processes for the steel include normalizing, annealing, tempering, quenching, and other processes (shown in Fig. 4.6).

4.2.2.3.1 Normalizing

Heating the steel above the critical temperature, and cooled in the air. Through normalizing, the steel material stress could be relieved, and thus improves ductility and toughness for the cold-working process. Normalizing is usually applied for carbon and low alloy steel material. It will produce different metal structures, pearlite, binate, some marten site. This brings harder and stronger steel material, and less ductility than full annealing material.

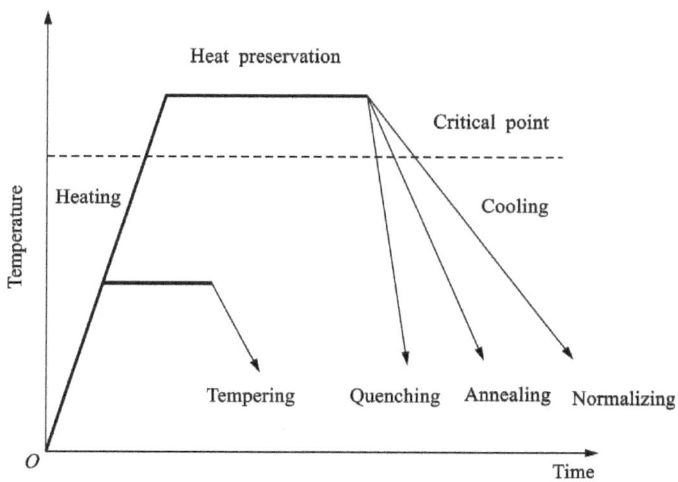

Figure 4.6 The heat treatment processes.

4.2.2.3.2 Annealing

Heating the material to above its critical temperature long enough until microstructure transforms to austenite. Then slow cooled in the furnace, get the maximum transformation of ferrite and pearlite. Annealing will eliminate defects, uniform the chemical composition, and fine grains. This process is usually applied for the high carbon, low alloy, and alloy steel need to reduce their hardness and strength, refine the crystal structure, and improve the plasticity, ductility, toughness, and machinability.

4.2.2.3.3 Quenching

Heating the steel material to critical temperature until microstructure transformation is done, cooling it at a rapid rate. Quenching's purpose is to produce thermal stress and tissue stress. It can eliminate and improve through tempering. The combination of quenching and tempering can make comprehensive performance improved.

4.2.2.3.4 Tempering

Heating the steel material to a precise temperature below the critical point, and often done in the air, vacuum, or the inert atmospheres. There are low temperature (150°C~250°C), medium-temperature (350°C~500°C), and high-temperature (500°C~650°C).

The purpose of tempering is to increase the toughness of steel and alloy steel pipes. Before tempering, this steel is very hard but too brittle for most applications. After process can improve the plasticity and toughness of steel pipe, reduce or eliminate the residual stress, and stabilize the steel pipe's size. Brings good comprehensive mechanical properties, so that it does not change in service.

4.2.2.4 Alloying

Only a few elements are widely used commercially in their pure form. Generally, other elements are present to produce greater strength, to improve corrosion resistance, or simply as impurities leftover from the refining process. The addition of other elements into metal is called alloying and the resulting metal is called an alloy. When a second element is added, two different structural changes are possible:

① Solid solution strengthening occurs when the atoms of the new element form a solid solution with the original element, but there is still only one phase. Recall that the term "phase" refers to that region of space occupied by a physically homogeneous material.

② The atoms of the new elements form a new second phase. The entire microstructure may change to this new phase or two phases may be present.

Solid solution strengthening involves the addition of other metallic elements that will dissolve in the parent lattice and cause distortions because of the difference in atom size between the parent metal and the solute metal. Recall from the section on crystal point defects that it is possible to have substitution AL impurity atoms and interstitial impurity atoms. A substitution AL impurity atom is an atom of a different type than the bulk atoms, which has replaced one of the bulk atoms in the lattice. Substitution AL impurity atoms are usually close in size (within approximately 15%) to the bulk atom. Interstitial impurity atoms are much smaller than the atoms in the bulk matrix. Interstitial impurity atoms fit into the open space between the bulk atoms of the lattice structure. Since the impurity atoms are smaller or larger than the surrounding atoms, they introduce tensile or compressive lattice strains. They disrupt the regular arrangement of ions and make it more difficult for the layers to slide over each other. This makes the alloy stronger and less ductile than the pure metal.

Still another method of strengthening the metal is adding elements that have no or partial solubility in the parent metal. This will result in the appearance of a second phase distributed throughout the crystal or between crystals. These secondary phases can raise or reduce the strength of an alloy. The properties of a polyphase (two or more phase) material depend on the nature, amount, size, shape, distribution, and orientation of the phases. Greek letters are commonly used to distinguish the different solid phases in a given alloy.

4.2.3 Mechanical properties

4.2.3.1 Stress—strain behavior: tensile test

A typical stress—strain curve of the tensile test coupon is shown in Fig. 4.7 in which a sharp change in yield point followed by plastic strain is observed (the elastic stage). After a certain amount of plastic deformation of the material (the yield stage), due to reorientation of the crystal structure, an increase in load is observed with an increase in strain (the reinforcement stage). After a little increase in load, the specimen eventually fractures (the necking stage).

At the yield stage, the corresponding stress of the highest point on the hackle is called the upper yield point; the corresponding stress of the lowest point is called the lower yield point. Because the yield points are unstable,

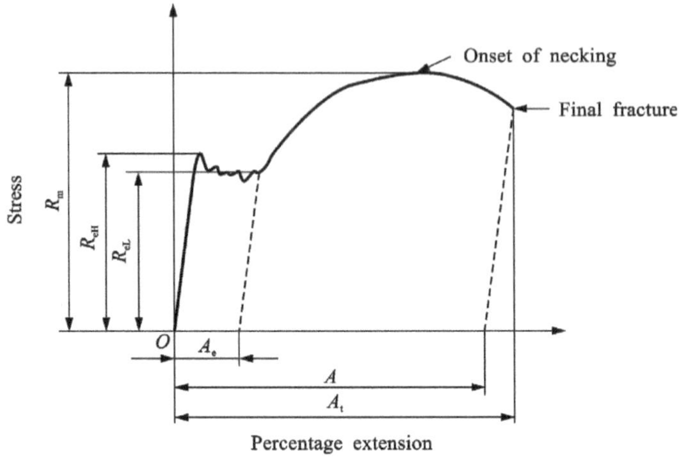

Figure 4.7 Stretching of low carbon steel Q–E (R_m—tensile strength; R_{eH}—upper yield strength; R_{eL}—lower yield strength; A—Percentage yield point extension; A_t—percentage total elongation after fracture).

the Chinese Standard regulates that the stress of the lower yield point is the yield strength of the steel, expressed by σ_s. Medium carbon steel and high carbon steel have no obvious yield points, so 0.2% of the stress of the residual deformation is the yield strength. Yield strength is very important to the use of steel. When the actual stress of a structure reaches the yield point, there will be irretrievable deformation which is not allowed in constructions. Thus, yield strength is the main base to determine the allowable stress of the steel.

Ultimate tensile strength (or tensile strength) is the ultimate tensile stress that the steel can bear under the role of tension.

Under the role of alternating loads, steel will be damaged suddenly when the stress is far below the yield strength, and this damage is called fatigue failure. The value of stress at which failure occurs is called fatigue strength, or fatigue limit. The fatigue strength is the highest value of the stress at which the failure never occurs. Generally, the biggest stress that the steel bears alternating loads for 106–107 times and no failure occurs is called the fatigue strength.

4.2.3.2 Elasticity

Fig. 4.7 show that the steel deformation property in the elastic stage is called elasticity. At this stage, the ratio of the stress to the strain is the modulus of elasticity, that is, $E = \sigma/\varepsilon$ MPa. The bigger E is, the higher the stress that causes its deformation is; and under the certain stress, the smaller the elastic deformation will be. In projects, the modulus of elasticity reflects the rigidity of the steel which is an important value to calculate the deformation of a structure under stress. The elastic modulus of Q235, the carbon structural steel commonly used in constructions, is calculated as follows: E = (2.0 − 2.1) × 105 MPa.

4.2.3.3 Plasticity

The construction steel should have good plasticity. In projects, the plasticity of the steel is usually expressed by the elongation (or the reduction of cross-section area) and cold bending.

Elongation refers to the ratio of the increment of the gauge length to the original gauge length when the specimen is stretched off, expressed by δ (%), shown in Fig. 4.8. Elongation is an important index to measure the plasticity of steel. The bigger the elongation is, the better the plasticity of steel is. The elongation is related to the gauge length, and usually δ5, and δ10, are used to express the elongation when $l_0 = 5d$ and $l_0 = 10d$, respectively. Reduction of cross-section area is the percentage of the cross-section shrinkage quantity of the neck-shrinking part to the original cross-section area when the specimen is stretched off.

Cold bending is the property that the steel bears the bending deformation under normal conditions. The cold bending is tested by checking whether there are cracks, layers, squamous drops, and ruptures on the bending part after the specimen goes through the regulated bending. Generally, it is expressed by the ratio of the bending angle σ and the diameter of the bending heart d to the thickness of the steel or the diameter of the steel a. Fig. 4.9 shows that the bigger the bending angle is, the smaller the ratio of d to a is, and the better the cold bending property is. Cold

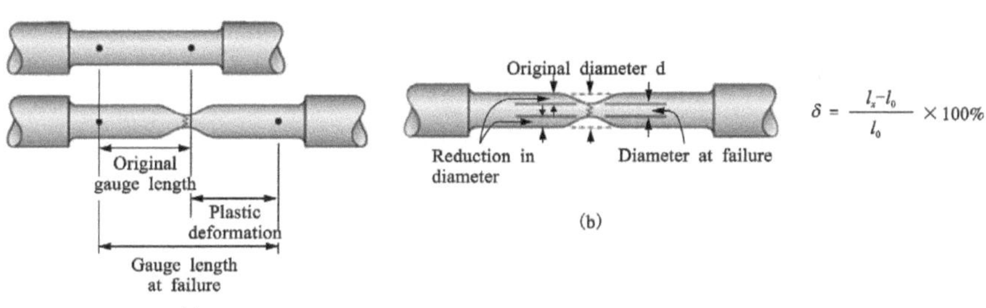

Figure 4.8 Elongation of steel (l_0-Original gauge length; l_x-Gauge length at failure; d-original diameter).

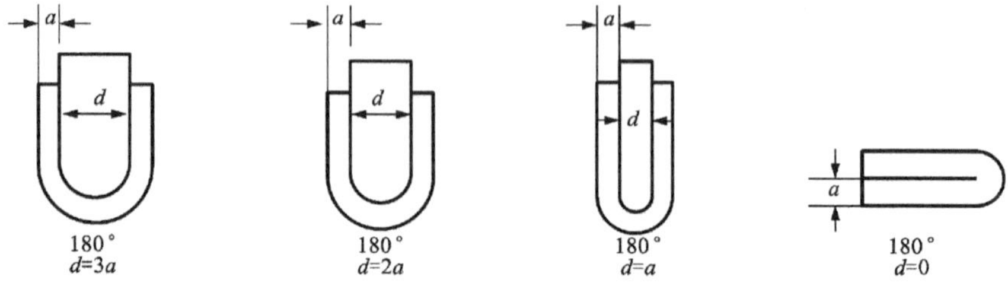

Figure 4.9 Cold bending test of steel (d—diameter of the bending heart; a—the thickness or the diameter of the specimen).

bending is a method to check the plasticity of steel and is related to elongation. The steel with bigger elongation has better cold bending property. But the cold bending test for the steel is more sensitive and strict than the tension test. The cold bending test is helpful to expose some defects of steel, such as pores, impurities, and cracks.

4.2.3.4 Impact toughness

The Charpy test method (shown in Fig. 4.10) determines the toughness or impact strength of the material in the presence of a flaw or notch and fast loading conditions. This destructive test involves fracturing notched impact test specimens at a series of temperatures with a swinging pendulum. The amount of energy absorbed by the material during fracture is measured. Charpy V notch or U notch test specimens (10 mm × 10 mm × 55 mm) are used. The impact toughness of the steel is related to its chemical

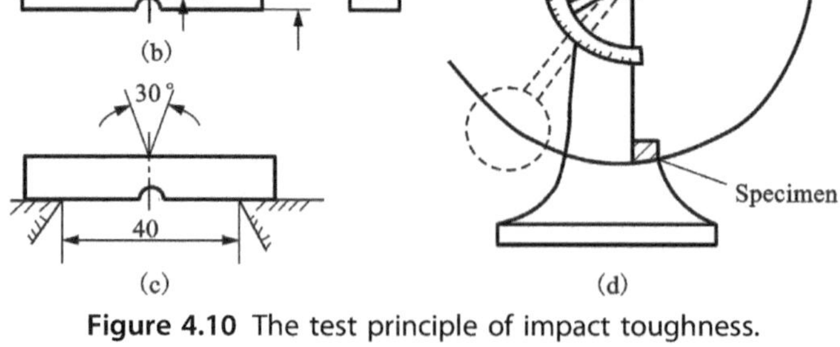

Figure 4.10 The test principle of impact toughness.

elements, smelting, and processing. Generally, P and S contents in steel are high, and impurities and the tiny cracks forming in smelting will lower the impact toughness.

Besides, the impact toughness of the steel can be influenced by temperature and time. At room temperature, the impact toughness will decline little with the temperature falling, and the damaged steel structure reveals the ductile fracture (shown in Fig. 4.11); if the temperature falls into a range, the steel reveals the brittle fracture, and the temperature is very low when a cold brittle fracture occurs. The critical temperature of its brittle fracture should be lower than the lowest temperature of the place.

4.2.3.5 Rigidity

Rigidity is the property to resist the plastic deformation when there is a hard object press into the steel within the partial volume of the surface, often related to the tensile strength. Recently, there are various methods to measure the rigidity of the steel, and the most common one is Brinell hardness, expressed by HB.

The yield strength, tensile strength, elongation, cold bending, and impact toughness of the steel are usually used as the base for the evaluation mark.

4.2.4 Classifications of steel

4.2.4.1 According to composition

Most commercial steel can be classified into two groups: carbon steel and alloy steel.

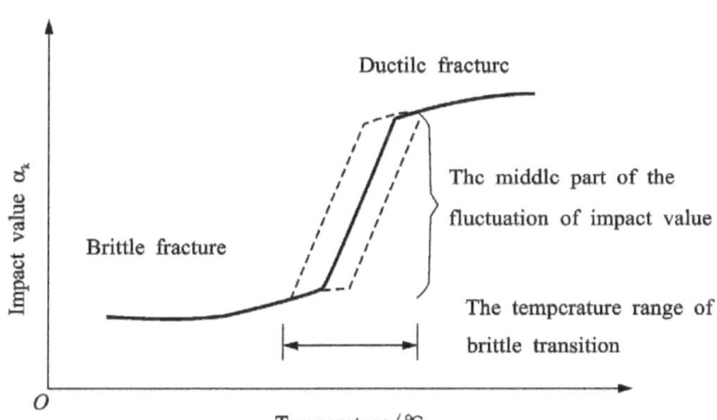

Figure 4.11 The effect of temperature on impact toughness.

The carbon steel has less than 1% of carbon and traces of manganese, sulfur, silicon, and phosphorus. The properties and characteristics of this type of steel are according to the carbon content present in it and there is a minor influence on this type of carbon due to the alloying and residual materials. Plain carbon steel is subdivided into four groups:

① Low carbon steel: the carbon content is less than 0.30% and is the most commonly used grade. They can be machined and welded nicely and their ductility is greater than high carbon steel.

② Medium carbon steel: the carbon content is from 0.30% to 0.45% carbon. Due to increased carbon content, there is an increase in hardness and tensile strength and a decrease in ductility. And its machining and welding are difficult than low carbon steel due to the increased content of carbon.

③ High carbon steel: the carbon content is between 0.45% and 0.75%. And it is a challenge for welding and machining this type of steel heating is necessary to produce acceptable welds and is also used to control the mechanical properties of steel after welding.

④ Very high carbon steel: the carbon content is up to 1.50%. This type of steel requires heat before, during, and after welding to control its mechanical properties. This type of steel is used for hard steel products such as metal cutting tools and truck springs.

⑤ Alloy steel is a type of steel in which one or more elements other than carbon have been intentionally added, to produce a desired physical property or characteristic. Common elements that are added to make alloy steel are molybdenum, manganese, nickel, silicon, boron, chromium, boron, and vanadium. There are two types of alloy steel:

⑥ Low alloy steel: the carbon content is below 0.25% and often 0.15% for especially welding applications, and the common alloying materials are manganese, nickel, chromium, molybdenum, silicon, vanadium, and boron and the less common alloying elements are aluminum, cobalt, copper, titanium, tungsten, tin, and zirconium. Mostly low alloy steel is used to achieve better hardenability and is increased corrosion resistance in a certain environment. Low alloy steel is difficult to weld. By lowering the carbon content to 0.10% along with other alloying materials increase the strength of the material.

⑦ High alloy steel: there are alloying elements of more than 8% by weight of total other than carbon and iron is classified as high alloy steel. High alloy steel consists of at least two chemical elements and the properties of this type of steel depend on the percentage of the chemical elements present in it. If the percentage is high, then its properties depend on that chemical element with a high percentage. High alloy steel is highly corrosion resistant with high reliability.

Low carbon steel and low alloy steel are widely used in civil engineering.

4.2.4.2 According to the application

According to application steel can be classified into two types:

Stainless steel: this is a steel alloy with a minimum of 10% chromium content. Stainless steel is more resistant to stains, corrosion, and rust than ordinary steel.

Tool and die steel: tool and die steels are high carbon steels with alloying elements of chromium, tungsten, vanadium, manganese, and molybdenum.

4.2.4.3 According to deoxidation practice

Deoxidation of steel is a steel-making technological operation, in which the concentration of oxygen dissolved in molten steel is reduced to a required level. Base steel in descending order of oxygen removal can be divided into killed, semikilled, and rimmed steels.

① Killed steel: killed steels are completely deoxidized steel and the commonly used deoxidizing elements are silicon and aluminum. Their solidification does not cause the formation of carbon monoxide (CO). Ingots and castings of killed steel have a homogeneous structure and no gas porosity (blowholes). Killed steel is used for forging, carburizing, heat treatment, and other applications.

② Semikilled steel: semikilled steels are incompletely deoxidized steels and have characteristics between rimmed and killed steel. Structural steels contain 0.15%–0.25% carbon and some amount of excess oxygen, which forms carbon monoxide during the last stages of solidification. This type of steel is suitable for drawing operations (except severe drawing).

③ Rimmed steel: rimmed steels are also known as drawing quality steel. Rimmed steels are low carbon steel that is partially deoxidized or non-oxidized. Carbon content is less than 0.25% and manganese content is less than 0.6% in rimmed steel. Rimmed steels evolve a sufficient amount of carbon monoxide during solidification. These steel are ideal for rolling, a large number of applications, and is adapted to cold-bending, cold-forming, and cold header applications.

4.2.4.4 According to shape

Steel can be formed into many products with different shapes (shown in Fig. 4.12).

4.2.4.5 According to press-working modes

The press-working modes include hot working and cold working.

4.3 Standards and selection of building steel

4.3.1 The steel used for steel structures

4.3.1.1 Carbon structural steel

4.3.1.1.1 Designation system

The Chinese National Standard of Carbon Structural Steel (GB700-2006) regulates the different grades consisting of the letter of yield point, the value of yield point, the quality level, and the DE oxidation method.

① "Q" represents the yield point; the value of the yield point includes 195, 215, 235, and 275 MPa.

② The quality level is expressed by the content of sulfur and phosphor: A, B, C, and D. Steel grade D and grade C with lower sulfur and phosphor contents are better than steel grade B and grade A:

③ F represents rimmed steel, B represents semikilled steel, Z and TZ represent fully killed steel and special fully killed steel, and Z and TZ

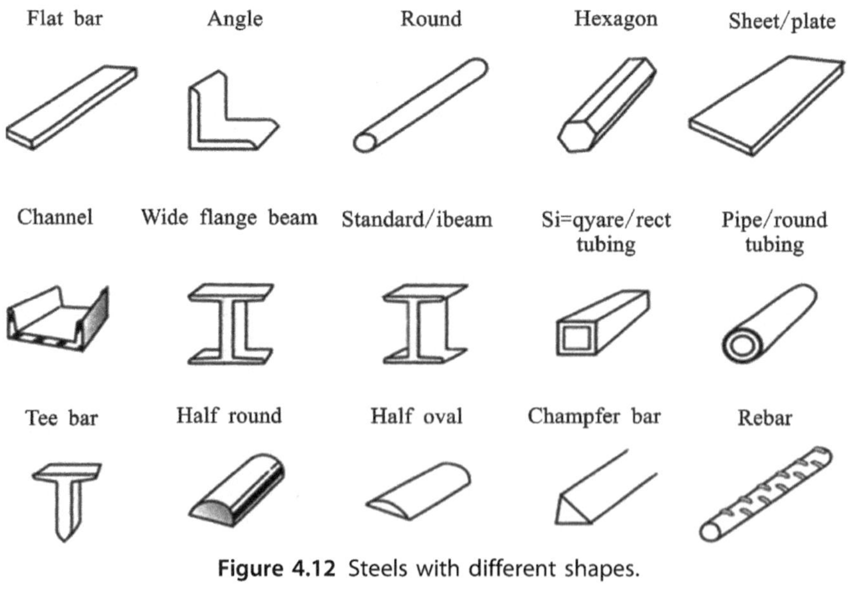

Figure 4.12 Steels with different shapes.

can be omitted in the grades of steel. Special fully killed steel and fully killed steel are better than rimmed steel.

For example, Q235-A.F represents A-grade rimmed steel with a yield point of 235 MPa.

4.3.1.1.2 Technical requirements

The mechanical properties and technological characteristics should be in line with Tables 4.1 and 4.2.

4.3.1.1.3 Selection of carbon structural steel

The selection of steel depends on the quality, properties, and the corresponding standards of steel on one side; on the other side, it depends on the requirements of the projects for the properties of steel.

The load types, welding, and temperatures of the projects have different requirements for the properties of steel. And the demands should be met in the selection of steel. Usually, the rimmed steel is restricted under the following conditions: it is a welding structure directly bearing the dynamic loads; it is a nonwelding structure and the calculating temperature is equal to or lower than −20°C: it is a welding structure bearing static loads and indirect dynamic loads and the calculating temperature is equal to or lower than −30°C.

In the construction of steel structures, carbon steel Q235 is mainly used, namely various profiles, steel boards, and coffins made of Q235. Steel Q235 has good strength, toughness, plasticity, and process ability, and is easy to be smelted, and has a lower cost. Because Q235-D has enough elements to form fine-particle structures and controls the contents of sulfur and phosphor strictly, it has better impact toughness to resist vibrating and impact loads than other grades, especially at negative temperature. Steel grade A is often used for structures bearing static loads.

Steel Q215 has low strength and high plasticity and deforms a lot under stress. It can replace Q235 after cold working. Steel Q275 has high strength but low plasticity, and sometimes is rolled to ribbed bars used in concrete.

4.3.1.2 High strength low alloy structural steels

According to the national standard of high strength low alloy structural steel (GB 1591—2018), there are 8 grades of Q345, Q390, Q420, Q460, Q500, Q550, Q620, and Q690. The alloy elements include manganese, silicon, barium, titanium, niobium, chromium, nickel, and lanthanum. The representation of grades consists of the letter of the yield point, the value of

Table 4.1 Mechanical properties of carbon structural steel (GB700).

Grade	Level	Tensile test												Impact test	
		Yield point R_{eH}(N/mm^2)						Tensile strength R_m(N/mm^2)	Elongation A/%						
		Thickness of steel (Diameter)/mm							Thickness of steel (Diameter)/mm						
		≤ 16	> 16 – 40	> 40 – 60	> 60 – 100	> 100 – 150	> 150 – 200		≤ 40	> 40 – 60	> 60 – 100	> 100 – 150	> 150 – 200	T(°C)	V imapct work (vertical) (J)
		≥							≥						≥
Q195	—	195	185	—	—	—	—	315–430	33	—	—	—	—	—	—
Q215	A	215	205	195	185	175	165	335–450	31	30	29	27	26	—	—
	B													+20	27
Q235	A	235	225	215	215	195	185	370–500	26	25	24	22	21	—	—
	B													+20	27
	C													0	
	D													−20	
Q275	A	275	265	255	245	225	215	410–540	22	21	20	18	17	—	—
	B													+20	27
	C													0	
	D													−20	

Table 4.2 Technological characteristics of carbon structural steel (GB700).

Grade	Direction of samples	Cold bending test 180 degrees B = 2a	
		Thickness of Steel(Diameter)/mm	
		≤60	>60–100
		Diameter of bending heart d	
Q195	Vertical	0	—
	Horizontal	0.5a	
Q215	Vertical	0.5a	1.5a
	Horizontal	a	2a
Q235	Vertical	a	2a
	Horizontal	1.5a	2.5a
Q275	Vertical	1.5a	2.5a
	Horizontal	2a	3a

Note: *a*, The thickness (diameter) of samples; *B*, The width of samples.

the yield point, and the quality level (including B, C, D, E, F, the five levels).

The addition of alloy elements into the steel can modify the organization and properties of steel. If 18Nb or 16Mn (the yield point is 345 MPa) with a similar carbon content (0.14%–0.22%) is compared with Q235 (the yield point is 235 MPa), the yield point is improved by 32%, and it has good plasticity, impact toughness, and weld ability and can resist low temperature and corrosion; and under the same conditions, it can make the carbon structural steel save steel consumption by 20%–30%. The ore or the original alloy elements in steel waste, such as niobium and chromium, are often used for the alloying of steel; or some cheap alloy elements, such as silicon and manganese, are added; if there is a special requirement, a little number of alloy elements, such as titanium and vanadium, can be used. The smelting equipment is the same as the equipment to produce carbon steel, so the cost increases a little.

The adoption of low-alloy structural steel will reduce the weight of structures and extend the useful time, and the high-strength low-alloy structural steel is especially used in the large-span or large column-grid structures for better technical and economic effects.

4.3.2 Steel for the reinforcement of concrete

4.3.2.1 Hot-rolled reinforced bars

The hot-rolled reinforced bars used for concrete structures should have high strength, certain plasticity, toughness, cold bending, and weld ability. The hot-rolled reinforced bars include the hot-rolled plain round steel bar (HPB) and the hot-rolled rib round bar (HRB), H represents "hot-rolled" P represents plain, R presents "ribbed" and B represents "bar."

There is a grade of HPBQ300, the chemical compositions are shown in Table 4.3, the technical requirements are shown in Table 4.4.

There are 9 grades of hot-rolled ribbed bar: HRB400, HRB500, HRB600, HRB400E, HRB500E, HRBF400, HRBF500, HRBF600, HRBF400E, and HRBF500E, where E means earthquake, F means fine crystals. The chemical compositions are shown in Table 4.5, the technical requirements are shown in Table 4.6.

4.3.2.2 Cold-rolled ribbed reinforced bars

The cold-rolled ribbed bar is the bar made by cold drawing or cold rolling the ordinary low-carbon steel, the quality carbon steel, or the low-alloy hot-rolled coiled bar to reduce the diameter and form crescent cross ribs on three faces or two faces of the bar. The base metal of the cold-rolled ribbed bar should be in line with the existing national standard Cold-rolled Ribbed Bur (GB 13788—2017). The grades of CRB550, CRB600H, and CRB680Hb are used for the reinforced concrete structure. The grades of CRB650, CRB800H, and CRB800Hb are special for the reinforced prestressed concrete structure.

The cold-rolled ribbed steel bars have high strength, good plasticity, and high cohesion force with concrete and stable quality. Grade 550 steel bars are mainly used for reinforced concrete structures, especially the main load-bearing bars of slab members and the non-prestressed steel bars in pre-stressed concrete structures.

Table 4.3 Chemical compositions of hot-rolled plain round bars (GB1499.1).

	Chemical compositions ($\leq$wt%)				
Grade	**C**	**Si**	**Mn**	**P**	**S**
HPB300	0.25	0.55	1.50	0.045	0.050

Table 4.4 Technical requirements for hot-rolled plain round bars (GB1499.1—2017).

Grade	Yield strength/ MPa	Tensile strength/ Mpa	Elongation/%	Total elongation under maximum force/%	Cold bending 180° Diameter of bending heart, a-nominal diameter of bar
			≤		
HPB300	300	420	25.0	10.0	d = a

Table 4.5 Chemical compositions of hot-rolled bibbed bars (GB1499.2).

Grade	Chemical compositions (≤wt%)				
	C	Si	Mn	P	S
HRB400, HRBF400, HRB400E, HRBF400E HRBF400, HRBF500, HRBF400E, HRBF500E	0.25	0.80	1.60	0.045	0.045
HRB600	0.28				

4.3.3 Prestressed steel wire for concrete or steel strain

They are the special products made by cold working, rebackfiring, cold rolling, or crossing the high-quality carbon structural steel, also called high-quality carbon steel wire or steel strain. The prestressed steel wire for concrete can be divided by processing way: cold-drawn steel wire and stress-relieved wire, the two types. The stress-relieved wire can be divided into low loose plain round wire, spiral rib steel wire, and deformed steel wire, three types.

For the prestressed steel wires for concrete, the mark of the products should contain the following content: prestressed steel wire, nominal diameter, tensile strength grade; code of processing state, code of appearance, and standard code. Example 1: The mark of the cold-drawn plain and round wire with a diameter of 4.00 mm and tensile strength of 1670 MPa should be prestressed steel wire Example 2: The mark of the low loose spiral rib steel wire with a diameter of 7.00 mm and tensile strength of 1570 MPa should be prestressed steel wire Steel strand is made by seven steel wires undertaking crossing hot treatment.

Table 4.6 Technical requirements for hot-rolled ribbed bars (GB1499.2).

Grade	R_{eL} lower yield point strength/MPa	R_m tensile strength/Mpa	Elongation/%	Total elongation under maximum force/%	$R^o{}_m/R^o{}_{eL}$	$R^o{}_m/R_{eL}$
			≥			≤
HRB400 HPBF400	400	540	16	7.5	—	—
HRB400E HPBF400E			—	9.0	1.25	1.30
HRB500 HPBF500	500	630	15	7.5	—	—
HRB500E HPBF500E			—	9.0	1.25	1.30
HRB600	600	730	14	7.5	—	—

Note: $R^o{}_{eL}$, measured lower yield strength; $R^o{}_m$, Measured tensile strength.

4.3.4 Steel for bridge

Railway and highway bridges bear not only dead load but also dynamic load directly, and some parts of which also bear alternating stress. Exposed to the air, some bridges are in rainy and humid areas, some are in ice and snow areas. They have to endure the baptism of climate change and corrosive medium under long-term stress. Therefore, compared with general structural steel, bridge structural steel must possess good plasticity, toughness, weld ability, and higher fatigue strength. Considering the effect of low temperature and long-term reliability, it is also required to have lower cold brittleness and aging sensitivity, in the case of brittle fracture accidents.

4.3.4.1 Codes for representing steel types

According to the Chinese National Standard of Carbon Structural Steel (GB/T 714), the code consists of four parts, namely, letter Q that representing yield strength, specified minimum yield strength value, Chinese Pinyin letter Q which means bridge, and quality level symbol. For example, Q420qC refers to the bridge structural steel with the specified minimum yield strength value of 420 MPa and quality level of C. Besides, when the steel plate, delivered in a thermo mechanical controlled process, possesses characteristics of weather resistance and through-thickness, the symbol of weather resistance (NH) and through-thickness (Z) shall be added after above-specified code, like Q420qDNHZ15.

4.3.4.2 Technical requirements

According to the specified minimum yield strength value, bridge structural steel mainly includes eight types, which are Q345q, Q370q, Q420q, Q460q, Q500q, Q550q, Q620q, and Q690q. There are four quality levels which are C, D, E, and F, divided by the content of P and S, among which content P is lower than 0.030%, 0.025%, 0.020%, and 0.015%, respectively, and content S is lower than 0.025%, 0.020%, 0.010%, and 0.006%, respectively.

There shall be no defects on the surface of the structural steel for the bridge, such as fractures, bubbles, scabs, inclusions, folds, mill scales, and visible laminations. In terms of steel plate with a thickness of more than 20 mm, it should be inspected by ultrasonic flaw detection. Impact toughness is another important performance of structural steel for the bridge. For level C, D, and E, the shock test temperatures are 0°C, −20°C, and −40°C respectively. At that temperature, the energy from Charpy (V) Impact Test should be no less than 120 J; and at −60° C, the energy from the Charpy (V) Impact Test of level F should be a less than 47 J.

In the 180°bending test, to the specimen whose thickness is less than or equal to 16 mm, its diameter of bending pressure head should be twice of the thickness; to the specimen whose thickness is greater than 16 mm, the diameter should be three times of the thickness. Based on chemical composition, the atmospheric corrosion resistance index is used to manifest its corrosion resistance.

4.3.4.3 Characteristics and applications

The mechanical properties of each structural steel type, such as yield strength and tensile strength, are shown in Table 4.7. Through completely deoxidizing, Q345q, Q370q, Q420q, and Q460q have had good comprehensive properties with fewer impurities. Not only do they have high strength but they can also perform excellently in plasticity, toughness, and weld ability. Codes like Q370q and Q420q are the basic steel types for building the main structure of the steel plate girder bridge. And Q500q, Q550q, Q620, and Q690q also have high strength, as well as good plasticity and toughness. They are mainly used for bars that putting stress on the bridge steel structure. High-strength structural steels are generally not used for main structures. Even if their strength is greatly improved, the elastic modulus does not change much. Therefore, although they can bear higher stress which ensures the bearing capacity of bridge structures, they cannot meet the design requirements of structural stiffness with low elastic modulus.

4.3.5 Rail steel

4.3.5.1 Properties

Under the effect of wheel pressure, impact, and wear, the rail steel should not only possess a higher strength to bear higher pressure and peel-off strength but also have higher hardness, abrasive resistance, impact toughness, and fatigue strength. And with the development of continuously welded rail track, it should have good weld ability also. Simultaneously, rail steel, operated in rainy and humid areas, saline-alkali areas, and tunnels, is often subjected to the corrosion of various environmental media, so good atmospheric corrosion resistance should also be taken into account.

Table 4.7 Technical requirements for steel for bridge (GB/T 714).

Grade	Level	R_{eL} lower yield point strength/MPa, Thickness/mm ≥, ≤50	>50−100	R_m tensile strength/Mpa ≥	Elongation/% ≥
Q345q	C, D, E	345	335	490	20
Q370q	C, D, E	370	360	510	20
Q420q	C, D, E	420	410	540	19
Q460q	C, D, E	460	450	570	18
Q500q	C, D, E	500	480	630	18
Q550q	C, D, E	550	530	660	16
Q620q	C, D, E	620	580	720	15
Q690q	C, D, E	690	650	770	14

To meet the above needs, the rails should be smelted by basic oxygen furnace or electric furnace process, and continuously casting rolled by the killed steel with higher carbon contents (high-carbon steel) which has been disposed of via secondary refining and vacuum DE oxidation. There are five types, which are U71Mn, U75V, U77MnCr, U78CrV, and U76CrRE with content C of 0.65%−0.81% which also contains alloy elements such as Mn, Cr, V, and Si. Besides, their content P and S are no more than 0.030%, content O is no more than 0.0030%, content N is no more than 0.0090%. It can be seen that harmful elements are low in rail steel. The microstructures of the full cross-section rail should be pearlite and a little iron element instead of marten site, bainite, and grain boundary cementite. And the uniform bending should not exceed 0.5% of the total length of the rail. Besides, defects like fractures, lines, folds, transverse scratches, and laminations are not supposed to exist on the surface.

The wheel rail in the rail joint makes a big impact force. Thus, the quenching treatment on the rail top should be carried out within the range of 30−70 mm at both ends of the rail, to enhance the wear resistance.

Table 4.8 Mechanical properties of steel rail (TB/T2344).

Grade	Tensile strength/ Mpa	Elongation/%	Hardness in the centerline of top surface of rail(JBW 10/3000)
U71Mu	≥880	≥10	260−600
U75V	≥980	≥10	280−320
U70MnCr	≥980	≥9	290−330
U78CrV	≥1080	≥9	310−360
U76CrRE	≥1080	≥9	310−360

The quenching depth is between 8 and 12 mm. The mechanical properties of hot-rolled steel are shown in Table 4.8.

The impact toughness of the rail should be assessed by drop test, which requires no fractures between two supports of the specimen after one hit. At −20°C, the minimum and average values of fracture toughness KIC are 26 MPa·m1/2 and 29 MPa·m1/2, respectively. And the fatigue life should be greater than 5 × 106 times.

Rail types can be demonstrated by the average quality per meter. Rail types in China mainly include 75, 50, and 43 kg/m, with standard lengths of 12.5 m, 25 m, 75 mm, and 100 mm. With the rapid development of heavy-haul and high-speed tracks, rail needs to be heavy. At present, the heaviest rail in the world has reached 77.5 kg/m, and the performance and quality requirements of rail are getting higher and higher.

High-speed railway and railway passenger dedicated line, whose speed is 200−350 km/h, mainly consist of ballast less track structure and seamless rail. So they have higher requirements for rail technical conditions. Their additional requirements include high cleanliness of the steel inside (strictly control the harmful elements of P and S in rails), no original defects on the rail surface, the depth of decarbonized zone, residual tensile stress index, fracture toughness, the technical conditions of fatigue crack growth rate, the use of long specified-length rail, and the reduction of welded joints.

4.3.5.2 Rail grinding

Rail grinding is the most effective way to maintain steel rail. Known from the railway track working principle, the railway provides a continuous and elastic foundation for train operation. As the main part of the railway track system, steel rail directly attaches wheel bear a lot load from the train. A part of the load is absorbed by the steel rail itself, and the rest of the load produces the rail wears like surface fatigue, crack, scratch, deformation, etc.

Figure 4.13 Rail wears.

(shown in Fig. 4.13). Rail wears can result in reducing steel rail service life, more maintenance work, more maintenance cost, and damaging railway running. So that rail grinding becomes vital processing to the railway track.

Steel rail grinding is applied widely because of its high efficiency in the world. Generally, rail grinding is realized by rail grinding wagon. There are four strategies for different rail wears, such as corrective grinding, transitional grinding, preventive grinding, and particular grinding.

① Corrective grinding.
② Transitional grinding.
③ Preventive grinding.
④ Particular grinding.

The advantages of rail grinding are shown as follows:

① Increase 50%−100% useful life of steel rail (5−8 years).
② Decrease the risk of rail failure.
③ Make steel rail and wheel have correct contact, reduce the snake movement of the train.
④ Decrease the deterioration rate of the wheel, railway fasteners, and track geometry.
⑤ Give the possibility of higher train operation speed.
⑥ Reduce rolling noise and vibration and control rail corrugation.

⑦ Save the maintenance cost of railway fasteners, railway track system.
⑧ Reduce the locomotive fuel consumption and cost of wheel repair.

4.4 Corrosion and prevention of steel

When the surface of steel contacts the surrounding environment under a certain condition, it will be corroded. The corrosion will reduce the load-bearing cross-section of steel, the uneven surface will lead to the convergence of stress, which will lower the load-bearing ability of steel; also, the corrosion will lower the fatigue strength greatly, especially the impact toughness of steel, which will result in the brittle fracture of steel. If the steel bars in concrete are corroded, there will be an expansion of volume, which makes the concrete crack along bars. Thus, the measures to resist corrosion should be adopted to prevent the corrosion of steel in working.

4.4.1 Reasons for corrosion of steel

4.4.1.1 Chemical corrosion

It is pure chemical corrosion caused by the electrolyte solution or various dry gases (such as O_2, CO_2, and SO_2. etc.), without any electric current. Usually this kind of corrosion will generate loose oxide on the surface of steel by oxidation, and it is very slow under dry conditions, but it will be very fast under high temperature and humidity.

4.4.1.2 Electrochemical corrosion

When steel contacts with electrolyte solution and generates an electric current, there will be electrochemical corrosion caused by the role of the primary battery. The steel contains ferrite, cementite, and nonmetal impurities, and all of these components have different electrodes and potentials, which means their activity is diversified; if there is an electrolyte, it will be easy to form two poles of the primary battery. When the steel contacts with humid media, like air, water, and earth, a layer of water film will cover its surface and various ions coming from the air dissolves in water, which forms electrolyte. At first, the ferrite in steel loses its electron, that is, $Fe \rightarrow Fe^{2+} + 2e$, to become anode and cementite becomes the cathode. In acidic electrolyte, H^+ obtains electron to become H_2 and runs away; in neutral media, water gets OH^- due to the DE oxidation of oxygen and generates insoluble $Fe(OH)_2$ (Figure); it can be oxidized into $Fe(OH)_3$ and its dehydration product $Fe_2(OH)_3$ which is the major component for bronze rust (Fig. 4.14).

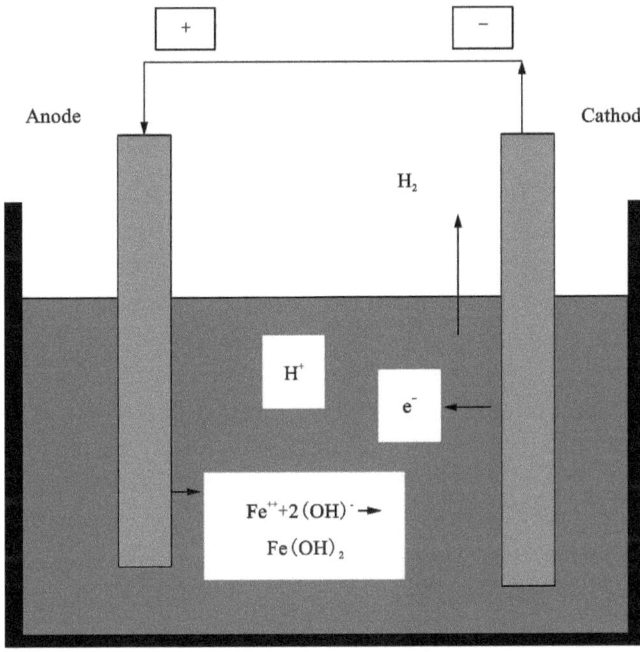

Figure 4.14 Schematic of electrochemical corrosion.

4.4.2 Corrosion prevention of steel

4.4.2.1 Protective film

This method is to isolate the steel from the surrounding media with the protective film to prevent or delay the damage caused by the corrosion of external corrosive media. For example, paint coatings, enamel, or plastic on the surface of steel; or use the metal coating as the protective film, such as zinc, tin, and chrome.

4.4.2.2 Electrochemical protection

Current-free protection is to connect a piece of metal, such as zinc and magnesium, more active than steel to the steel structure, and because zinc and magnesium have lower potentials than steel, the anodes of the corrosion cells coming from zinc and magnesium have been destroyed, but the steel structure will be protected. This method can be used for the places which are difficult or impossible to be covered with a protective layer, such as a steam boiler, the shell of the steamboat, underground pipe, maritime structure, and bridge. Impressed current protection is to emplace some waste steel or other refractory metals around the steel structure, such as high silicon iron and silver-lead alloy, and to connect the cathode of the impressed direct current to the protected steel structure and the anode to the refractory metals and the refractory metals become the anode to be covered and the structure becomes the cathode to be protected.

4.4.2.3 Alloying

The addition of alloy elements into carbon steel to produce various alloy steel will improve its anticorrosion, such as nickel, chrome, titanium, and copper. The above method can be adopted to prevent the corrosion of the steel bars in concrete, but the most economic and effective way is to improve the density and the alkalinity of concrete and make sure that the steel bars are thick enough. In the hydration products of the cement, there is about 1/5 $Ca(OH)_2$, and when the pH value of the media reaches about 13, there is a passive film on the surface of steel bars, so the bars in concrete are difficult to generate rust, But when CO_2 in the air diffuses into the concrete and reacts with $Ca(OH)_2$ to neutralize the concrete. When pH value falls to 11.5 or below, the passive film will be destroyed and the steel surface reveals an active state; and if it is humid and oxygen condition, the electrochemical corrosion will start on the surface of steel bars; because the volume of rust is 2–4 times than steel, it will lead to the cracking of concrete along bars. CO_2 diffuses into the concrete and carries the carbonization, so the improvement of the density of concrete will effectively delay the carbonization process. Because CL- will destroy the passive film, the consumption of chloride should be controlled in the preparation of reinforced concrete.

4.5 Nonferrous metals

The first cost of nonferrous metals is usually much greater than that of ordinary ferrous metals, but the difference is often offset by their superior working properties and resistance to corrosion.

4.5.1 Copper

The Copper Development Association, Mutton Lane, Potters Bar, Hertfordshire, provides technical advice and many excellent publications dealing with all aspects of the properties and uses of copper and its alloys.

Usually the three grades of copper used in the building are

① Deoxidized copper: this is used for domestic plumbing tubes where welding is to be carried out and for general engineering purposes.

② Fire refined tough pitch copper: this contains oxygen and is stronger than deoxidized copper. It also has higher thermal and electrical conductivity and higher resistance to atmospheric corrosion than deoxidized copper. It is used as a sheet for fully supported roof coverings.

③ Electrolytic tough pitch high conductivity copper: this metal is similar to fire refined tough pitch copper but contains fewer impurities. It is largely used for electrical conductors.

The salmon-red color of clean copper is well known. Its alloys vary from red, gold, and pale yellow to soft silver. In ordinary atmospheres and waters, copper develops protective skin. In certain environments, a green patina slowly develops; an effect which can be obtained more rapidly by chemical methods, although in the case of copper roofing success depends very much on the climatic conditions prevailing at the time of treatment. Washings from copper may stain adjacent materials and inhibit the growth of lichen and they may give rise to the corrosion of other metals inhibit the growth of lichen and they may give rise to the corrosion of other metals.

Copper is, in general, very resistant to corrosive agents, particularly to seawater, but it is attacked by strong mineral acids and ammonia. Water containing a high proportion of free carbon dioxide is cuprosolvent. Up to 1.5 mg/l copper can be tolerated in drinking water, but low concentrations of copper greatly accelerate the pitting corrosion of galvanized steel. Pitting corrosion of copper tubes has been caused by the cathodic scale deposited by soft moorland waters containing manganese salts, occasionally by certain hard well waters, and by the scale of carbon produced in the process of extrusion. However, the formation of a sound protective film in copper cylinders is assured by the installation of aluminum protector rods, which control the electrochemical potential of the copper. Copper is supplied in the fully annealed dead/soft, half-hard, and full hard conditions.

Copper in the annealed or hot worked condition is relatively strong, and it is extremely ductile. Its strength characteristics and hardness are increased by cold working.

Rod is rolled from continuously cast sections. Wire of 0.025—5 mm diameter is drawn from the rod. Tube from about 1.5 to 610 mm diameter drawn from cylindrical billets or cold drawn from hollow extrusions. Plate, i.e., flat material larger than 305 × 10 mm up to about 3.7 m long. Sheet and strip from 0.5 to 10 mm thick sheet wider than 450 mm. Foil with 0.15 mm maximum thickness, any width. Plate, sheet, strip, and foil are either hot rolled from slabs or cakes or formed by electrodeposition.

Unalloyed copper can be hot rolled, forged, and extruded. In the annealed condition it is eminently suitable for site working although work hardening may necessitate further annealing by heating to a dull red heat. Copper and also alloys can be joined by welding, brazing, and soldering.

4.5.2 Aluminum

Aluminum is the third most common element, so its price tends to be stable. Economical use is aided by the ease of forming a high strength/weight ratio and good resistance to corrosion and possible avoidance of the costs of painting.

About four tones of bauxite clay and 17,000-kilowatt hours of electricity are needed to produce one tons of aluminum. The metal can be classified as being either pure, containing 99% or more aluminum, or as alloys.

The density of aluminum is low, 2700 kg/m^3, about the same as granite and about one-third that of steel, 7850 kg/m^3. Aluminum is second only to copper in thermal conductivity. Electrical conductivity is about 60% that of copper. Aluminum has long been used for electrical conductors on the grid system and its use for building installations is being developed.

In ordinary atmospheres, a thin, but dense, whitish film of oxide forms almost instantaneously, and under damp conditions of external exposure roughening of the surface may follow if it is not kept clean. It is not recommended for exposure to marine atmospheres. The products of corrosion do not stain adjacent materials and are not toxic to animals or plants. Aluminum may suffer electrolytic corrosion in damp or wet conditions and contact must be avoided with copper, copper alloys such as brass, and to a less extent with bare mild steel. Contact with zinc, stainless steel, and lead is normally safe, but the glazing bars should not have lead wings where there may be salt spray. Metallic salts such as those originating from water that has passed through copper pipes or over copper roof coverings must not be allowed to come into contact with aluminum.

Acids, such as those which may be used to clean building materials or which arise from decaying vegetable growths on tiled roofs, attack aluminum.

Free alkalis such as those in wet Portland cement also attack aluminum. The metal should be protected from fresh concrete or mortar by bitumen of the solution type.

Aluminum, either with original mill or cast, or later anodized finish, is very liable to disfigurement by scratches, knocks, or stains. Cement and lime adhere strongly and leave marks where they are removed. Components should be carefully handled and protected during construction by removal wrappings. Preferably they are fixed after all wet construction and finishes are completed.

Aluminum is being used for rainwater gutters and pipes. They are light, strong, and do not need to be painted. They are made in sheets or are extruded in long lengths. Aluminum is not recommended for pipes conveying drinking water or for waste pipes and traps which would be attacked by bleach, strong detergents, or soda.

For building purposes, the alloys most commonly used may contain magnesium, manganese, and silicon, together with several minor additions that increase the strength of pure aluminum. Strength is further improved by cold working or by heat treatments which are described later.

Alloying reduces thermal and electrical conductivities, but they remain high. Thus, aluminum alloys are used for corrugated and troughed roof sheeting. A roofing system of interlocking roll-formed sections can be used for roof pitches down to 1°C. "Secret" sliding fixings accommodate thermal movement so that transverse laps are needed only at 30 m centers. Window, door, and shop-fitting sections, Venetian blind slats, and for structural members. Alloys, particularly those containing copper, are less resistant to corrosion than pure aluminum and may become unsightly and cause sliding windows to "stick." However, in "breathable" air, corrosion is unlikely to affect strength. Sheet and strip are sometimes clad with pure aluminum to improve their corrosion resistance.

4.5.3 Magnesium

Magnesium is mainly produced by electrolysis from brine. This is a very energy-intensive process. Among all lightweight metals, magnesium is attractive due to its low density, good mechanical properties and damping capacity, high thermal conductivity, good cast ability, and electromagnetic shield property. Magnesium alloys are comparable to engineering metals, such as iron-based and aluminum-based alloys, with an excellent high strength-to-weight ratio. It is less dense, that is, one-third that of aluminum, and has huge potential in the transportation industry for saving fuel. However, the widespread use of magnesium is hindered due to its poor corrosion resistance. Impurities such as iron, nickel, and copper form secondary-phase particles due to their low solubility limits. Since these secondary-phase particles are generally electrochemically cathodic to the matrix, they can induce magnesium corrosion. Alloying elements such as aluminum and rare-earth metals have been shown to improve the corrosion resistance of magnesium. Magnesium alloys are finding applications in the automobile and aircraft industry, as well as in electronic products and sports equipment. Magnesium is also used as an alloying element for making other metal alloys, such as aluminum alloys.

Exercises

1. What is steel? What is construction steel? What are the properties of steel?
2. From what aspects is steel divided? How many subdivisions of each aspect? How the construction in steel divided?
3. What are the technical properties of construction steel? How to express each property? What is the actual significance? How to determine?
4. In the figure of the stress—strain curve of low-carbon steel, how many stages are there? What are the characteristics and indexes of each stage?
5. What is the yield ratio? What is the actual significance of projects?
6. What is the basic organization of steel? What are their characteristics? What kinds of impact do the chemical components of steel have on the properties?
7. What is cold working and aging? How does the property of steel change after cold working and aging?
8. How to express the grade of carbon structural steel and low alloy?
9. What kinds of corrosions does construction steel have? How to resist?

CHAPTER 5

Wood

5.1 Introduction

Wood is one of the oldest building materials used by humans. Due to its wide availability and strength, it has been an important building material since humankind began building houses and bridges. The term wood is generally used to refer to a hard and fibrous material that forms the trunk, branches, and roots of a tree. In Commonwealth countries, wood that has been harvested and sawn to suitable dimensions for use in carpentry and building construction is referred to as "timber." Usually, lumber is used extensively as the structural members for building residential houses, commercial buildings, bridges, and other structures. Due to their strength, versatility, and construction efficiency, lightweight lumber frames and lumber roof trusses are widely used for domestic houses in many parts of the world. Laminated veneer lumber, glued laminated wood (glulam), and other engineered lumber products are used in large-scale structures, such as industrial and commercial buildings, bridges, and so on. Since long before the advent of steel and concrete, lumber has been used as a chief building material around the world due to its wide availability and esthetics. Lumber has also been extensively used in building construction around the world. Lumber is also extensively used as decorative nonstructural members, such as flooring, windows and doors, staircases, railings, and partition walls.

This chapter mainly focuses on the structure and the physics characteristics of the wood.

It's well known, wood has several advantages as follows:

① Specific intense strength;
② Lightweight and high strength;
③ Great elasticity and tenacity;
④ Low thermal conductivity and good thermal isolation;
⑤ Be easy to process;
⑥ Beautiful-grained, mild-toned, elegant-styled, and well-affected in decoration.

Civil Engineering Materials
ISBN 978-0-12-822865-4
https://doi.org/10.1016/B978-0-12-822865-4.00005-2

And for sure, the wood also has the following disadvantages:
① Anisotropy;
② Expanding with wetness and shrinking with dryness;
③ Be liable to crack or warp;
④ Be corrupted or mildew and rot or even eaten by worms;
⑤ Poor fireproof.

The wood is made of trees, and trees are mainly classified into two species (as is shown in Table 5.1).

5.2 Structure and composition

Wood is a highly orthotropic material because of how trees grow in nature. The macrostructure of wood is shown in Fig. 5.1A. The trunks are cut into three different sections:
① Transverse section: the section that is vertical against the trunk axis (longitudinal direction);
② Radial section: the section that passes the trunk axis;
③ Tangential section: the section that parallels with the trunk axis and tangent with the annual ring.

Table 5.1 Classification and characteristics of trees.

Classification	Characteristics	Usage	Examples
Conifer	The leaves are long and needle-like, the trunks are straight and tall, and ligneous tissue is soft, liable to process. Of superior strength, apparent density is low, and shrinkage deformation is low.	Used for load-carrying members, doors or windows, etc.	Pine, juniper, cypress, etc.
Broadleaf	Leaves are broad and shape in sheets, most of which are hardwood. The straight parts of the trunks are short, and ligneous tissue is hard, not easy to process. The apparent density is high, and the shrinkage deformation is high, easy to crack or warp.	Used for the minor load-carrying member in interior decoration or veneer, etc.	Elm, birch, manchurian ash, etc.

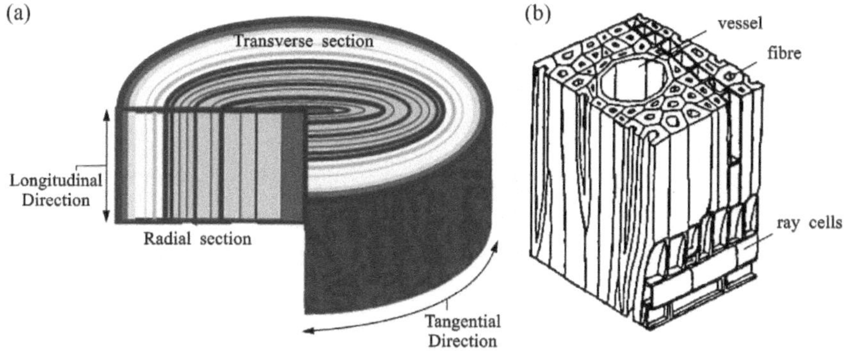

Figure 5.1 Structure of wood: (A) macrostructure, (B) microstructure.

The microstructure of wood is shown in Fig. 5.1B. Every cell can be classified into two parts: the cell wall and the vessel. The cell wall is composed of fibrils. The longitudinal combination is firmer than the transverse combination.

5.3 Engineering properties

5.3.1 Relative density

As shown in the microstructure, the void space (porosity) between cells and the vessel leads to large differences in relative densities and large changes in mechanical property. For all the species of wood, the relative density (specific density) of the cell wall material itself is about 1.5. However, wood's density or specific gravity depends on wood species, cell size, the thickness of the cell wall, cell density, the log cut, and moisture content. Freshly cut green wood has a higher density than dry wood. The volume and hence the density of wood varies with its moisture content. The dry density of wood ranges from 300 to 700 kg/m^3. The strength and stiffness of wood increase with the increase in specific gravity.

5.3.2 Moisture in wood

The water in the wood can be classified into two states:

① As free water (or water vapor) lies in intercellular space (cell cavities), the free water is relatively accessible.

② As physically absorbed water, bound water that lies inside the cell walls (fibers).

Moisture content (MC) is defined as the weight of water in wood given as a percentage of oven-dry weight:

$$MC = \frac{m_s - m_0}{m_0} \tag{5.1}$$

where m_s is the mass of wet wood; m_0 is the mass of wood after oven-drying at 100–105°C.

Depending on the species and type of wood, the moisture content of living wood ranges from approximately 30% to more than 250% (two-and-a-half times the weight of the solid wood material). In most species, the moisture content of the sapwood is higher than that of the heartwood.

Fiber saturation is an important benchmark for both shrinkages and decay. When fibers absorb water, it first is held in the cell walls themselves. When the cell walls are full, any additional water absorbed by the wood will now go to fill up the cavities of these tubular cells. Fiber saturation is the level of moisture content where the cell walls are holding as much water as they can.

Two important MC numbers to remember are 19% and 28%. We tend to call a piece of wood dry if it is at 19% or less moisture content. Fiber saturation averages around 28%.

Wood placed in an environment with stable temperature and relative humidity will eventually reach a moisture content that yields no vapor pressure difference between the wood and the surrounding air. In other words, its moisture content will stabilize at a point called the equilibrium moisture content (EMC). The wood used indoors will eventually stabilize at 8%–14% moisture content; outdoors at 12%–18%. The different moisture contents in the wood are schematized in Fig. 5.2.

5.3.3 Dimensional stability

Wood is dimensionally stable when the moisture content is above the fiber saturation point. Below the fiber saturation point, wood shrinks when moisture is lost and swells when moisture is gained. This susceptibility to dimensional change is one of the few wood properties that exhibits significant differences for the three orthotropic axes. In the longitudinal direction, average shrinkage values from green to oven-dry conditions are between 0.1% and 0.2%, which is generally of no practical concern. In the tangential and radial directions, however, shrinkage is much more pronounced (Fig. 5.3).

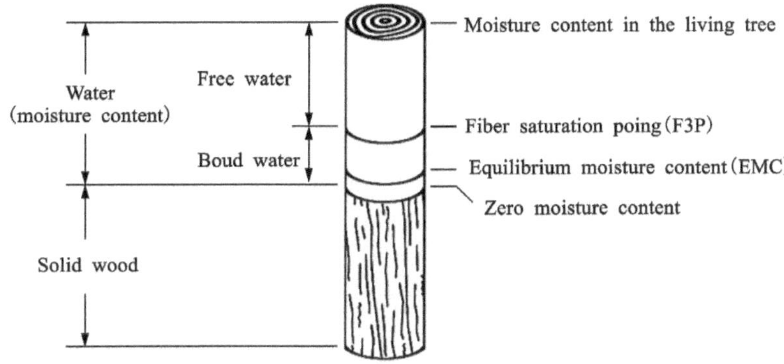

Figure 5.2 Schematic illustration of wood moisture content.

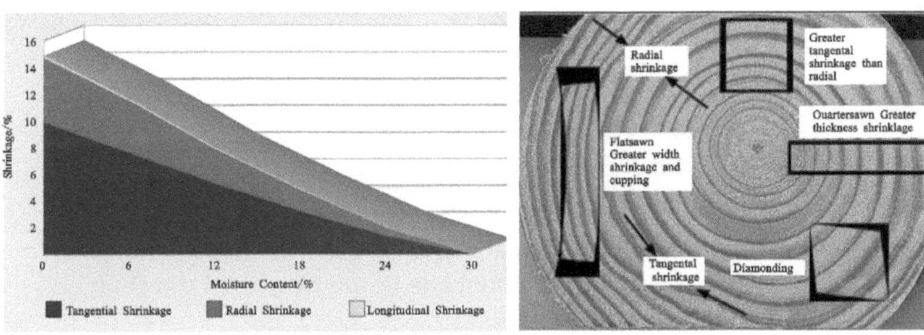

Figure 5.3 Dimension distortion.

As the piece dries, it develops a moisture gradient across its section (dry on the outside, wet on the inside). The dry outer shell wants to shrink as it dries below the fiber saturation point, however, the wetter core constrains the shell. This can cause checks to form on the surface (Fig. 5.4). The shell is now set in its dimension, although the core is still drying and will in turn want to shrink. But the fixed shell constrains the core and checks can thus form in the core. Most checks are not of structural significance; however, when checking extends from one surface to the opposite or adjoining surface (through-checks), the strength and other properties of the piece may be affected.

5.3.4 Mechanical properties

Wood is an anisotropic material. Anisotropy in wood is due to the tubular nature of its cells. The wood cell structure is analogous to a bundle of parallel straws bound together by weak glue (shown in Fig. 5.5). The straw

Figure 5.4 Checks are lengthwise separations of the wood, perpendicular to the growth rings, caused by uncontrolled shrinkage in the tangential direction.

Figure 5.5 Straw is bundled parallel to and perpendicular to its longitudinal axes.

bundle is the strongest parallel to its longitudinal axes and the straw bundle is the weakest perpendicular to its longitudinal axes (shown in Fig. 5.5). Straws here represent the fibers or grain of the wood.

Compared to other materials discussed in this book, the properties of wood display a very high degree of variability. It is not uncommon to find coefficients of variation above 30% for many properties.

5.3.4.1 Elastic properties

Elastic properties relate a material's resistance to deformation under applied stress to the ability of the material to regain its original dimensions when the stress is removed. For an ideally elastic material loaded below the proportional (elastic) limit, all deformation is recoverable, and the body returns to its original shape when the stress is removed. Wood is not ideally elastic, in that some deformation from loading is not immediately recovered when the load is removed; however, residual deformations are generally recoverable over a while. Although wood is technically considered a viscoelastic material, it is usually assumed to behave as an elastic material for most engineering applications.

For an isotropic material with equal properties in all directions, elastic properties are described by three elastic constants: modulus of elasticity (E), shear modulus (G), and Poisson's ratio (μ). Because wood is orthotropic, 12 constants are required to describe elastic behavior: 3 moduli of elasticity, 3 moduli of rigidity, and 6 Poisson's ratios. These elastic constants vary among species and with moisture content and specific gravity. The only constant that has been extensively derived from test data, or is required in most bridge applications, is the modulus of elasticity in the longitudinal direction. Other constants may be available from limited test data but are most frequently developed from material relationships or by regression equations that predict behavior as a function of density.

5.3.4.2 Compression strength

Wood can be subjected to compression parallel to the grain, perpendicular to the grain, or at an angle to the grain When compression is applied parallel to the grain, it produces stress that deforms (shortens) the wood cells along their longitudinal axis (Fig. 5.6). When compression is applied perpendicular to the grain, it produces stress that deforms the wood cells

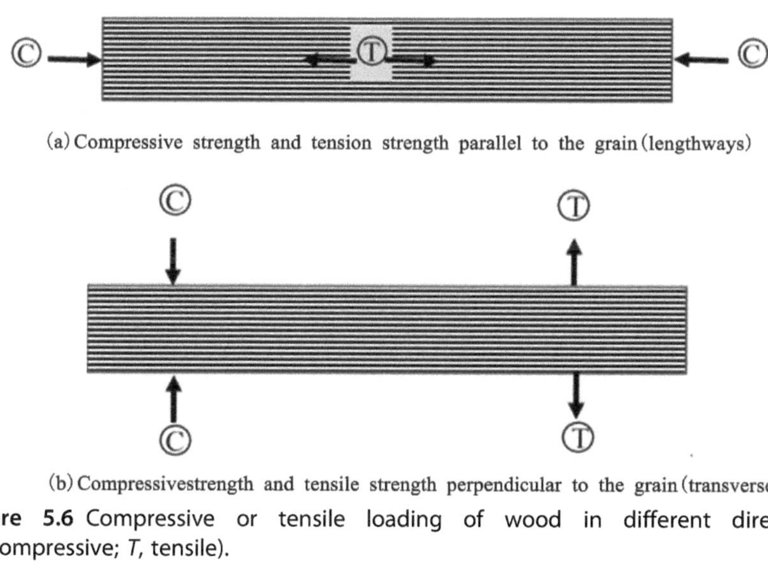

(a) Compressive strength and tension strength parallel to the grain (lengthways)

(b) Compressivestrength and tensile strength perpendicular to the grain (transverse)

Figure 5.6 Compressive or tensile loading of wood in different directions (*C*, compressive; *T*, tensile).

perpendicular to their length (Fig. 5.6). Wood cells collapse at relatively low-stress levels when loads are applied in this direction. However, once the hollow cell cavities are collapsed, wood is quite strong in this mode because no void space exists. Compression applied at an angle to grain produces stress acting both parallel and perpendicular to the grain.

5.3.4.3 Tension strength

Tension strengths parallel to grain and perpendicular to grain differ substantially (Fig. 5.6). Parallel to its grain, wood is relatively strong in tension. In contrast to tension parallel to the grain, wood is very weak in tension perpendicular to the grain.

5.3.4.4 Shear strength

There are three types of shear that act on wood: vertical, horizontal, and rolling (Fig. 5.7). Vertical shear is normally not considered because other failures, such as compression perpendicular to the grain, almost always occur before cell walls break in vertical shear. In most cases, the most important shear in wood is horizontal shear, acting parallel to the grain. It produces a tendency for the upper portion of the specimen to slide concerning the lower portion by breaking intercellular bonds and deforming the wood cell structure.

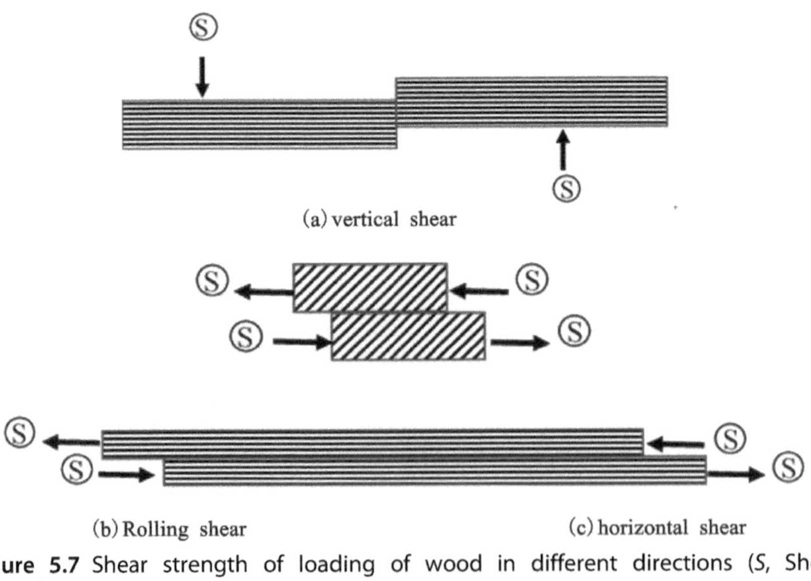

Figure 5.7 Shear strength of loading of wood in different directions (*S*, Shear strength).

Rolling shear is caused by loads acting perpendicular to the cell length in a plane parallel with the grain. The stress produces a tendency for the wood cells to roll over one another. Wood has low resistance to rolling shear, and failure is usually preceded by large deformations in the cell cross sections. Wood has low resistance to rolling shear.

5.3.4.5 Bending strength

When wood specimens are loaded in bending, the portion of the wood on one side of the neutral axis is stressed in tension parallel to the grain, while the other side is stressed in compression parallel to the grain. Bending also produces horizontal shear parallel to grain and compression perpendicular to grain at the supports. A common failure sequence in simple bending is the formation of minute compression failures followed by the development of macroscopic compression wrinkles. This effectively results in a sectional increase in the compression zone and a section decrease in the tension zone, which is eventually followed by tensile failure.

In summary, the relationship between the strengths of wood is shown in Table 5.2.

5.3.5 Factors affecting the wood strength

5.3.5.1 Moisture content

An empirical relationship between moisture content and mechanical properties, written in a logarithmic form, is as follows:

$$\log P = \log P_{12} + \left[(M - 12) / \left(M_p - 12\right)\log(P_R / P_{12})\right] \qquad (5.2)$$

where P is the property of interest; P_{12} is the value of the property at a moisture content of 12%; M is the moisture content; M_P is the moisture content below which property changes due to drying are first observed (this may be slightly less than the fiber saturation point and is often taken to be 25%); P_g. is the value of the property for moisture contents above M_P.

Below the saturation point, the absorbed water increases, and the cell walls expand, losing the structure, and lower the strength. When the moisture exceeds the fiber saturation point, the increase of water has little effect on the strength (shown in Fig. 5.8). However, care should be exercised when adjusting properties below 12% moisture. Although most properties will continue to increase when wood is dried to very low moisture content levels, for most species some properties may reach a maximum value and then decrease with further drying.

Table 5.2 The relationships between strengths of wood.

Compression strength		Tensile strength		Bending strength	Shear strength	
Parallel grain	Perpendicular grain	Parallel grain	Perpendicular grain		Parallel grain	Perpendicular grain
1	1/10–1/3	2–3	1/20–1/3	1.5–2	1/7–1/3	1/2–1

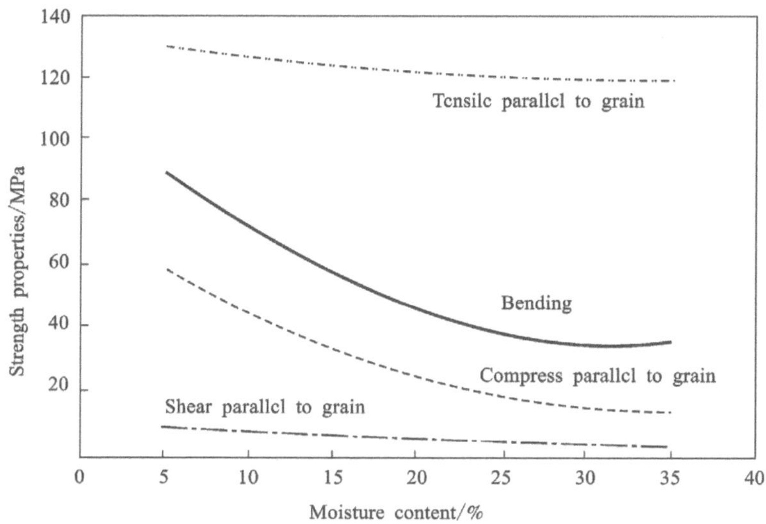

Figure 5.8 Effect of moisture content on the strengths of wood.

5.3.5.2 Environment temperature

In general, the mechanical properties of wood decrease when heated and increase when cooled. At a constant moisture content and below approximately 150°C, mechanical properties are approximately linearly related to temperature. The change in properties occurs when the wood is quickly heated or cooled.

The temperature has a direct effect on the wood strength. when the temperature rises from 25 to 50°C, the wood compression strength will be reduced by 20%–40% and the wood sharing strength will be reduced by 12%–20% because the collide among wood fibers is softened. Also, if the wood is in hot and dry surroundings, it may become fragile. The boiling method is often employed to reduce its strength contemporarily to meet the needs of processing (such as the production of plywood).

5.3.5.3 Time under load

Mechanical properties values are usually referred to as static strength values. Static strength tests are typically conducted at a rate of loading or rate of deformation to attain maximum load in about 5 min. Higher values of strength are obtained for wood loaded at a more rapid rate and lower values are obtained at a slower rate. For example, the load required to produce failure in a wood member in 1 s is approximately 10% higher than that obtained in a standard static strength test.

When initially loaded, a wood member deforms elastically. If the load is maintained, additional time-dependent deformation occurs. This is called creep. Creep occurs at even low stresses, and it will continue over years. For sufficiently high stresses, failure eventually occurs. This failure phenomenon is called the duration of the load (or rupture).

The strength of wood will be reduced with the lasting loading time since the plastic-flow deformation will occur to wood, and the rupture strength of wood may be only 50%—60% of the limited strength of wood (as shown in Fig. 5.9).

5.3.5.4 Defects

During the growing, cutting, and processing process of wood, there may be such defects as knots, splits, and worm rot. The influence of a defect on the mechanical properties of a wood member is due to the interruption of continuity and change in the direction of wood fibers associated with the defect. The influence of knots depends on their size, location, shape, and soundness; the attendant local slope of grain, and the type of stress to which the wood member is subjected.

5.4 Wood-based composites

Structural timber does have several natural limitations: ① The size of sawn timbers, in terms of both length and cross-sectional dimensions, is limited by the size of a tree; and ② the presence of knots and other imperfections,

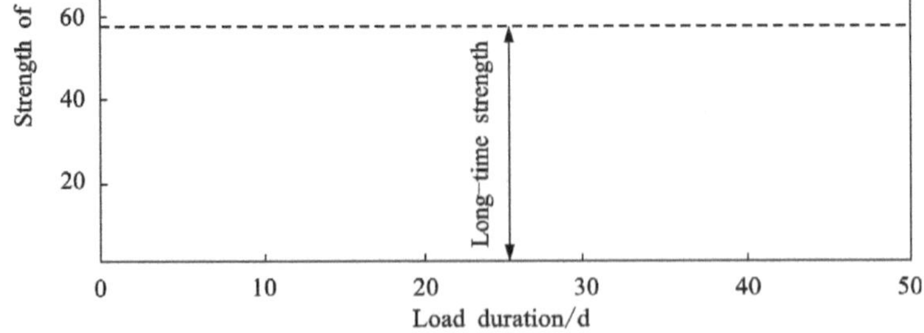

Figure 5.9 The creep rupture strength of wood.

particularly in larger timber, puts a severe limit on the allowable stress level. To improve the utilization of wood in construction, a wide variety of wood-based composites have been produced with the particular aims of

(a) Producing different sizes and shapes, ranging from panel products to very large beams, or curved beams;
(b) Producing materials with better mechanical properties than those of structural timber;
(c) Producing materials without shrinkage and warp;
(d) For economic considerations, finding a use for more of the volume of the tree and wood wastes and scrapped wood.

Therefore, wood-based composites encompass a range of products, from fiberboard to laminated beams, and can be used for nonstructural and structural applications in product lines ranging from panels for interior covering purposes to panels for exterior uses and in furniture and support structures in buildings. Fig. 5.9 shows the different composite products (Fig. 5.10).

5.4.1 Composition and manufacture

The term composite used here means that any wood material is adhesively bonded together. The main compositions of wood-based composites include the following as given below(Fig. 5.10).

Figure 5.10 The different composite products: (A) Laminated Veneer Lumber(LVL), (B) Glued Laminated Lumber (Glulam), (C) Cross-Laminated Timber (CLT), (D) Oriented strand board (OSB), and (E) Plywood.

5.4.1.1 Elements

The primary component of wood-based composites is the wood element, often 94% or more by mass. Common elements for conventional wood-based composites include veneers, strands, particles, and fibers. Properties of composite materials can be changed by changing the size and geometry of the elements and by combining, reorganizing, or stratifying elements.

5.4.1.2 Adhesives

Bonding is the most conventional wood-based composites provided by thermosetting (heat-curing) adhesive resins. Commonly used resin binder systems include phenol-formaldehyde (PF), urea-formaldehyde (UF), melamine-formaldehyde (MF), polymeric methylene di-isocyanate (pMDI), and bio-based adhesives.

5.4.1.3 Additives

Several additives are used in the production of conventional composite products. The most common additive is wax, which is used to provide products with some resistance to liquid water absorption. In flake-, particle-, and fiberboard products, wax emulsions provide limited-term water resistance and dimensional stability when the board is wetted. Even small amounts (0.5%−1%) act to retard the rate of liquid water pick up for limited periods. Other additives used for specialty products include preservatives, mildewcides, and fire retardants.

5.4.1.4 Manufacturer

The manufacturing process steps generally include element preparation, classification and drying, adhesive application, mat formation, pressing, and finishing. Fig. 5.11 shows a typical manufacturing process of an oriented strand board.

5.4.2 Plywood

Plywood is a panel product built up wholly or primarily of sheets of veneer called plies. It is constructed with an odd number of layers with the grain direction of adjacent layers oriented perpendicular to one another. A layer can consist of a single-ply or two or more plies laminated with their grain direction parallel. A panel can contain an odd or even number of plies but always an odd number of layers.

Plywood panels have significant bending strength both along with the panel and across the panel, and the differences in strength and stiffness along

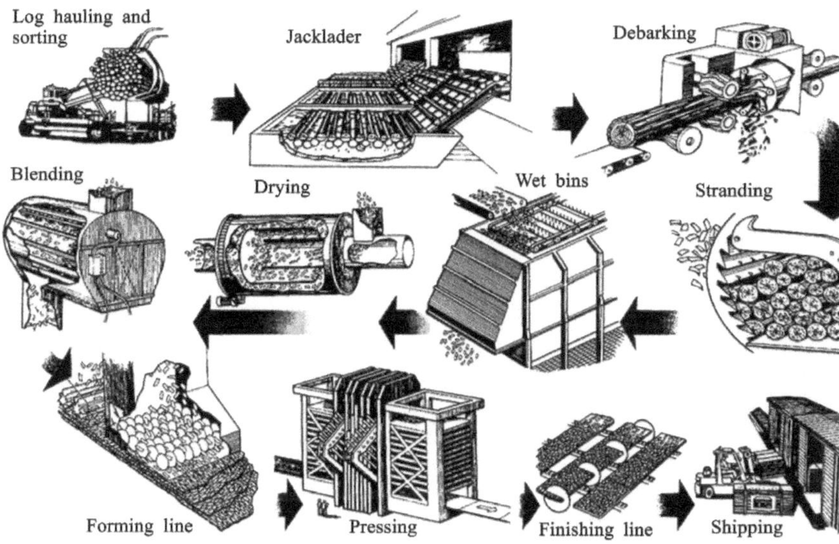

Figure 5.11 The manufacturing process of oriented strand board.

the panel length versus across the panel are much smaller than those differences in solid wood. Plywood also has excellent dimensional stability along its length and across its width. Minimal edge-swelling makes plywood a good choice for adhesive-bonded tongue-and-groove joints, even where some wetting is expected. Therefore, plywood panels can be used in various applications, including construction sheathing, furniture, and cabinet panels. Plywood is also used as a component in other engineered wood products and systems in applications such as prefabricated I-joists, box beams, stressed-skin panels, and panelized roofs. Two classes of plywood are commonly available: ① construction and industrial plywood and ② hardwood and decorative plywood.

The bulk of construction and industrial plywood is used where performance is more important than appearance. However, some grades of construction and industrial plywood are made with faces selected primarily for appearance and are used either with clear natural finishes or lightly pigmented finishes.

Hardwood plywood is normally used in applications including decorative wall panels and furniture and cabinet panels where appearance is more important than strength. Most of the production is intended for interior or protected uses, although a very small proportion is made with adhesives suitable for exterior services, such as in marine applications.

Hardwood and decorative plywood are categorized by species and characteristics of face veneer, bond durability, and composition of center layers.

The glue species, characteristics, and using the range of normal plywood are shown in Table 5.3.

5.4.3 Oriented strand board

Oriented strand board (OSB) is an engineered structural-use panel manufactured from thin wood strands bonded together with water-resistant resin, typically PF or pMDI. It is used extensively for the roof, wall, and floor sheathing in residential and commercial construction. The wood strands typically have an aspect ratio (strand length divided by width) of at least 3. OSB panels are usually made up of three layers of strands, the outer faces having longer strands aligned in the long direction of the panel and a core layer that is counter-aligned or laid randomly using the smaller strands or fines. The orientation of different layers of aligned strands gives OSB its unique characteristics, including greater bending strength and stiffness in the oriented or aligned direction. Control of strand size, orientation, and layered construction allows OSB to be engineered to suit different uses.

To manufacture OSB, debarked logs are sliced into long, thin wood elements called strands. The strands are dried, blended with resin and wax, and formed into thick, loosely consolidated mats that are pressed under heat and pressure into large panels.

5.4.4 Particleboard

Particleboard is produced by mechanically reducing the wood raw material into small particles, applying adhesive to the particles, and consolidating a loose mat of the particles with heat and pressure into a panel product. The particleboard industry initially used cut flakes as a raw material. However, economic concerns prompted the development of the ability to use sawdust, planer shavings, and to a lesser extent, mill residues and other waste materials. To manufacture particleboard with good strength, smooth surfaces, and equal swelling, manufacturers ideally use a homogeneous raw material.

Particleboard is typically made in layers. But unlike OSB, the faces of particleboard usually consist of fine wood particles, and the core is made of coarser material. The result is a smoother surface for laminating, overlaying, painting, or veneering. Particleboard is readily made from virtually any wood material and a variety of agricultural residues. Low-density insulating

Table 5.3 Species, characteristics, and applications of plywood.

Kind	Species	Name	Glue species	Characteristics	Applications
Normal broadleaf plywood	Species I	NQF (climate tolerant plywood)	Phenolic resin	Tolerant of age, boiling and steam, dry and heat, fungi tolerant	Outdoor engineering
	Species II	NS (water tolerant plywood)	Urea-formaldehyde resin	Tolerant of cool water and hot water immersion, but not tolerant of boiling	Outdoor engineering
	Species III	NC (moisture tolerant plywood)	Blood glue, urea-formaldehyde resin with many other ingredients, and other glues of a similar capacity	Tolerant of cool water immersion in a short period	Indoor engineering in normal circumstances
	Species IV	BNS (moisture intolerant plywood)	Soy-adhesive or other glues of a similar capacity	Certain agglutinative strength, but intolerant of water	Indoor engineering in normal circumstances
Normal pine plywood	Species I	Species I plywood	Phenolic resin or other synthesis resin of a similar capacity	Tolerant of age, heat, and fungi	Enduring outdoor engineering
	Species II	Species II plywood	Under hydrated or other synthesis resin of a similar capacity	Tolerant of water and fungi	Engineering in damp circumstances
	Species III	Species III plywood	Blood glue or urea-formaldehyde resin with a litter ingredient	Tolerant of damp	Indoor engineering
	Species IV	Species IV plywood	Soy-adhesive and urea-formaldehyde resin with plenty of ingredients	Intolerant of water and damp	Indoor engineering in dry circumstances

or sound-absorbing particleboard can be made from kenaf core or jute stick. Low-, medium-, and high-density panels can be produced with cereal straw. Rice husks are commercially manufactured into medium- and high-density products.

5.4.5 Fiberboard

The term fiberboard includes hardboard, medium-density fiberboard (MDF), and cellulosic fiberboard. Several things differentiate fiberboard from particleboard, most notably the physical configuration of the wood element. Because wood is fibrous by nature, fiberboard exploits the inherent strength of wood to a greater extent than does particleboard.

To make fibers for composites, bonds between the wood fibers must be broken. Attrition milling, or refining, is the easiest way to accomplish this. During refining, the material is fed between two disks with radial grooves. As the material is forced through the preset gap between the disks, it is sheared, cut, and abraded into fibers and fiber bundles. Refiners are available with single- or double-rotating disks, as well as steam-pressurized and unpressurized configurations.

5.4.6 Specialty composite materials

Special-purpose composite materials are produced to obtain enhanced performance properties such as water resistance, mechanical strength, acidity control, and fire, decay, and insect resistance.

5.4.6.1 Water-repellant composites

Sizing agents are used to increase the water repellency of wood-based composites. Sizing agents cover the surface of fibers, reduce surface energy, and increase fiber hydrophobicity. Sizing agents can be applied in two ways. In the first method, water is used as a medium to ensure thorough mixing of sizing and fiber. The sizing is precipitated from the water and is fixed to the fiber surface. In the second method, the sizing is applied directly to the fibers. Common sizing agents include rosin, wax, and asphalt.

5.4.6.2 Flame-retardant composites

Two general application methods are available for improving the fire performance of composites with fire-retardant chemicals: ① pressure impregnating the wood with water-borne or organic solvent-borne fire-retardant chemicals, and ② applying fire-retardant chemical coatings to the wood surface. The pressure impregnation method is usually more effective and longer-lasting; however, this technique is standardized only for plywood. It is not generally used with structural flake, particle, or fiber composites, because it can cause swelling that permanently damages the wood adhesive bonds in the composite and results in the degradation of some physical and mechanical properties of the composite. For wood in existing constructions, surface application of fire-retardant paints or other finishes offers a possible method to reduce flame spread.

5.4.6.3 Preservative-treated composites

Composites can be protected from attack by decay fungi and harmful insects by applying selected chemicals as wood preservatives. The degree of protection obtained depends on the kind of preservative used and the ability to achieve proper penetration and retention of the chemicals. Wood preservatives can be applied using pressurized or nonpressurized processes. As in the application of fire-retardant chemicals, the pressurized application of wood preservatives is generally performed after manufacture and is standardized for plywood. Postmanufacture pressurized treatments are not standardized for all types of flake, particle, or fiber composite due to the potential for swelling. Preservatives can be added during the composite manufacturing process, but the preservative must be resistant to vaporization during hot pressing. Proprietary flake board and fiberboard products with incorporated nonvolatile preservatives have been commercialized. Common preservative treatments include ammonia Cal copper quat (ACQ), copper azole (CA), and boron compounds.

5.5 Durability

5.5.1 Moisture

The durability of wood is often a function of water, but that does not mean wood can never get wet. Wood can safely absorb large quantities of water before reaching moisture content levels that will be inviting for decay fungi. Normally, decay can generally only get started if the moisture content of the wood is above fiber saturation.

5.5.2 Decay

A variety of fungi, insects, and marine borers can break down the complex polymers which make up the wood structure. The wood-inhabiting fungi can be separated into molds, strainers, soft-rot fungi, and wood-rotting basidiomycetes. The molds and strainers can discolor the wood; however, they do not significantly damage the wood structurally. Soft-rot fungi and wood-rotting basidiomycetes can cause strength loss in the wood, with the basidiomycetes the ones responsible for decay problems in buildings.

Wood-rotting fungi require wood as their food source, an equable temperature, oxygen, and water. Water is normally the only one of these factors that we can easily manage. This may be made more difficult by some fungi, which can transport water to otherwise dry wood. It can also be difficult to control moisture once decay has started since the fungi produce water as a result of the decay process.

5.5.3 Termites

Termites, sometimes called "white ants" are social insects, related to cockroaches rather than ants. They can be distinguished from ants by the absence of a narrow waist on the body and their typically white color. Under a hand lens, termite antennae are straight, whereas those of ants have an elbow. Flying reproductive termites can be distinguished from flying ants by the equal size of all four termite wings.

5.5.4 Preservative treatments

Protective coatings such as paint, stains, and water repellents are commonly applied to wood members to protect them from weathering. These protective coatings help inhibit water penetration into the wood and also reduce the weathering due to ultraviolet (UV) radiation from the sun. Protective coatings also help to stabilize wood members against shrinkage and expansion of wood members caused by fluctuation in wood moisture content.

A termite attack on wood can be prevented by proper consideration during design and construction of the structures, application of physical barriers such as stainless steel mesh beneath the concrete foundations supporting wood structures, minimizing direct contact of wood with soil, and chemical barriers such as application of chlorpyrifos and bifenthrin.

At the industrial scale, chemical treatments are used to prevent and inhibit the degradation of wood and hence enhance the durability and service life of wood products. Preservative treatment of wood consists of impregnating wood cells with chemicals that protect the wood from organisms like fungi and insects, fire damage, and weathering. Pressure and vacuum treatments are often used in preservative treatment to inject chemicals into wood cells. The most commonly used chemicals for preservative treatments are soil-borne preservatives (such as creosote), water-borne preservatives (such as chromate copper arsenate (CCA), ammonia Cal copper quaternary, copper azole, and boron-based compounds), and pentachlorophenol (Penta).

Chemicals, such as monoammonium phosphate, diammonium phosphate, borax, and sodium fluoride, are often used to improve the fire resistance of wood.

Exercises

1. What are the main components of wood according to its general structure?
2. What are the fiber saturation point, equilibrium water content, and standard water content of the wood? What practical meanings do they have?
3. How can the change of wood moisture affect the capacities of wood?
4. What are the factors affecting the strength of wood? How to influence?
5. Briefly list the reason for wood rot and the preservative measures.

CHAPTER 6
Polymers

Polymeric materials are used mainly for nonstructural applications in the construction industry, such as adhesives, sealants, restoration, repair, interior finishing, and components in the envelope of buildings. Their main advantage is that for various applications, they can be customized for various properties. This topic is further discussed in this chapter, with the aim of providing the general background required from the perspective of the use of these materials in the construction industry.

Wood and thatch may be the first materials containing polymers that are used to build shelters or houses. Natural polymers such as cellulose and lignin, among others, make up wood and thatch. Thatch is still used as a "functional material" such as roofs and coverings, for dividers, and for "structural material," wood is still used in main walls, flooring, beams, pillars, and rafters. For Western Europe, wood represents 7.6% of the total (weight) of materials consumption in building and construction industry now.

Polymers, including lignin and cellulose, are macromolecules composed of many small molecules through chemical bonds. Monomer means "one part," which comes from Greek "μονος + μερῶς." Polymer comes from "πολύς + μερῶς" (polys + meros), meaning "many parts."

Cellulose and lignin are natural or biological polymers that are synthesized by living plants to perform one or more given biological functions. By combining a scarce number of chemical units, living organisms developed the ability to produce an enormous range of different macromolecules. More than 10 million different veneers can be constructed with 20 kinds of amino acids, and hundreds of existing polysaccharides or glycans can be constructed by more than 20 kinds of monosaccharides. One ribose or deoxyribose, one phosphate group, and five different amino groups are enough to make up a nucleic acid.

About two centuries ago, man successfully synthesized polymers in chemical plants. The first few polymers were made from natural molecules. In 1839, the first kind of rubber, vulcanized natural rubber, was successfully synthesized. It was obtained by vulcanizing polyisoprene extracted from the

Civil Engineering Materials
ISBN 978-0-12-822865-4
https://doi.org/10.1016/B978-0-12-822865-4.00006-4

resin of rubber tree (rubber tree) of Brazil. In 1863, celluloid, the first thermoplastic plastic, was invented in the search for alternatives to ivory; the raw materials for celluloid were nitrocellulose and camphor. Elephants gratefully acknowledged the invention of celluloid. Celluloid is also used in photography, film, and table tennis. The search for substitutes for natural silk began circa 1880: the first one was rayon (regenerated cellulose fibers) made from viscose (cellulose xanthate) and cellophane.

Phenolic resin was the first polymer to be fully synthesized. As a thermoset polymer, Bakelite is made by cross-linking a phenol formaldehyde resin. It was developed by Leo Baekeland in 1907 and produced since 1910.

It is interesting that thermoplastic oligomers were synthesized at the research laboratory of Charles Wurtz, in Paris, and this was 50 years before the beginning of Bakelite production. Between 1859 and 1861, a series of oligomers were synthesized by Agostinho V. Lourenco (1822–93), as so-called ethers compose du glycol ("composed ethers of glycol") and alcohols polyethyleniques ("polyethylenic alcohols"). Today, they are named as polyethylene glycol or poly(oxyethylene) and the high-molecular-weight counterparts are known as poly (ethylene oxide).

As substitutes for more expensive materials, polymer materials started to be manufactured. They were established quickly on the market, and became the preferred material for a large number of applications. The main benefits are as follows:

① Lightweighting—They make it possible to design and build-up of lighter structures and objects, with improved performance at lower costs (per unit of weight).

② Processability—In general, plastics can be heated and molded at lower temperatures (50–250°C), relative to metals and ceramics.

③ Durability—Polymer has good resistance to harsh environment and is not subject to electrochemical corrosion, it can keep its performance unchanged for a long time.

④ Versatility—It is easy to change polymer properties according to specific requirements.

⑤ Economy—Generally speaking, the use of polymer materials can reduce the cost of raw materials, manufacturing (processing), maintenance, and operation.

⑥ Competitiveness—Plastic products are usually cheaper for the same application and similar properties.

⑦ Environmental advantages—When evaluating the environmental costs related to the whole life cycle of products, polymer materials often present the best cost—benefit ratio. Polymers are also estimated to help reduce environmental costs of transportation (45%), power consumption (42%), raw material consumption (4%), and others (9%).

⑧ Recyclability—In general, polymers can be recycled. In the past 20—30 years, polymer recycling has been gradually integrated into the current polymer industry activities. For the polymer materials used in specific environment (such as automobile plastic fuel tank, PET bottle, etc.), the closed-loop recycling has been realized and expanded. In addition, some new raw materials and innovative products are currently being developed for recycling from polymers.

6.1 Engineering plastics

6.1.1 Introduction

Plastics are a class of materials made from polymers, which consist of a large number of monomers. Chemical reactions occur between monomers to form extended molecular chains containing hundreds to thousands of monomer units. Polymers are formed from organic compounds as monomers, characterized by carbon chains as main chains (Fig. 6.1). The properties of materials are controlled by the molecular structure of very large molecular units. The main properties of plastics can be changed by

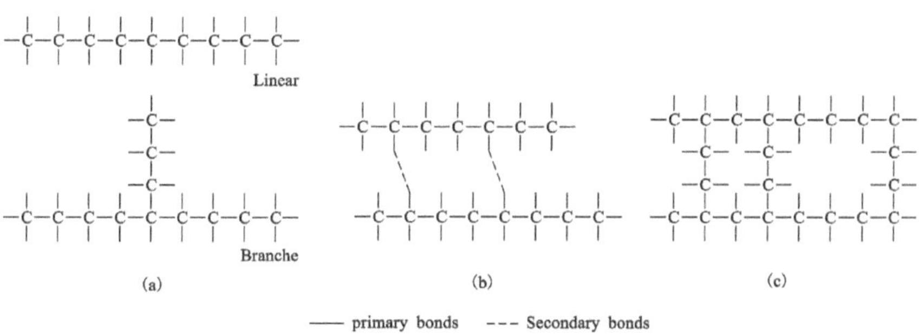

Figure 6.1 Schematic representation of the structure of polymer: (A) individual chain of carbon backbone, linear or branched; (B) thermoplastic polymer with secondary bonds between polymeric molecules, and (C) thermoset polymer with chemical bonds cross-linking between polymeric molecules. *(After Tang Lei, Study on the relationship between molecular structure and solution structure of modified nano-SiO$_2$ hyperbranched polymer, Southwest Petroleum University, Sichuan Province, 2018.)*

changing the stiffness, density, and regularity of internal chains and the interaction between molecular chains. Thermoplastic and thermosets are two major classes of polymers that are used. The main difference between the two is the secondary van der Waals force in thermoplastics and the chemical cross-linking in thermosets (Fig. 6.1). This leads to considerable differences in properties. Despite this, there are sufficient similarities that make it useful scientifically to consider the two classes together as a single class of materials. This section discusses the properties of polymer bonds and the structure of polymers. The production of polymeric materials, including forming plastics and polymerizing monomers, is beyond the scope of this book.

6.1.2 The polymeric molecule

An important feature of the carbon backbone is the tetrahedral bonding structure (Fig. 6.2A), which allows one part of the carbon backbone to rotate freely relative to another (Fig. 6.2B and C). This characteristic is due to the fact that the stress between C—C bonds does not change due to

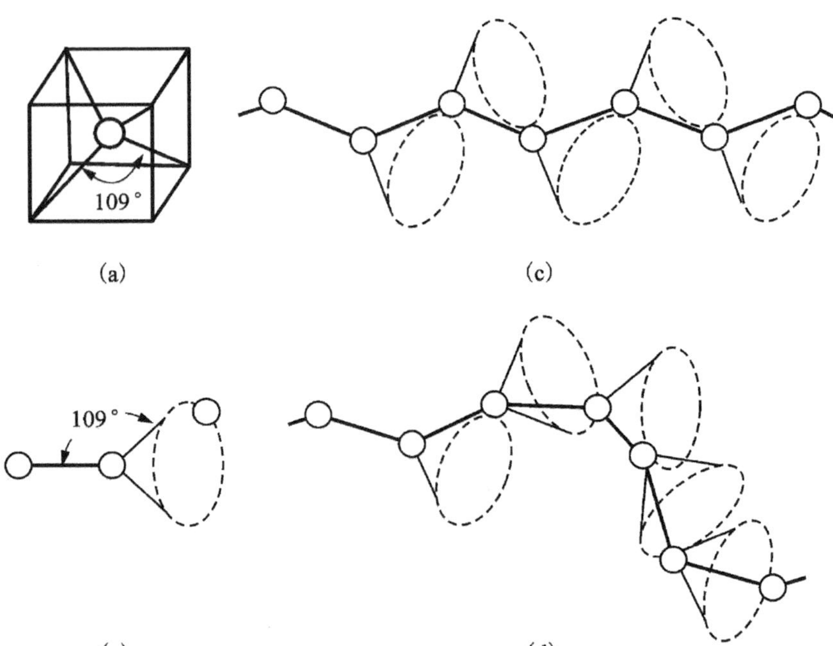

Figure 6.2 The geometrical shape of the polymeric molecule: (A) tetrahedral structure of the bonds; (B) rotation between adjacent segments: (C) straight, extended conformation; and (D) coiled conformation. *(After D. R Askeland. The Science and Engineering of Materials, PWS-Kent Publishing Company, 1985, pp. 336, 340.)*

rotation, because the distance between adjacent atoms and bond angle remains unchanged. As a result, molecules can take on different conformations in which the chains are distorted (Fig. 6.2D). At one end is an extended long chain, and at the other end is a very coiled chain (Fig. 6.3). In fact, the material consists of a series of chains of different winding structures (such as a plate of spaghetti, "snake nest" or tangled string), with some distribution around an average conformation. Molecules are not static; they change conformation constantly.

One of the most important properties of polymer molecules is the coiling nature. This fundamental property of elastomer makes it more ductile than conventional solids with limited deformation. The difference is that in traditional solids, primary chemical bonds must be emphasized, while in elastic polymers, the free rotation allows a large change in molecular size (end-to-end distance). As long as the rotation between segments is basically unimpeded when the load is removed, the extended molecules roll back to the average conformation.

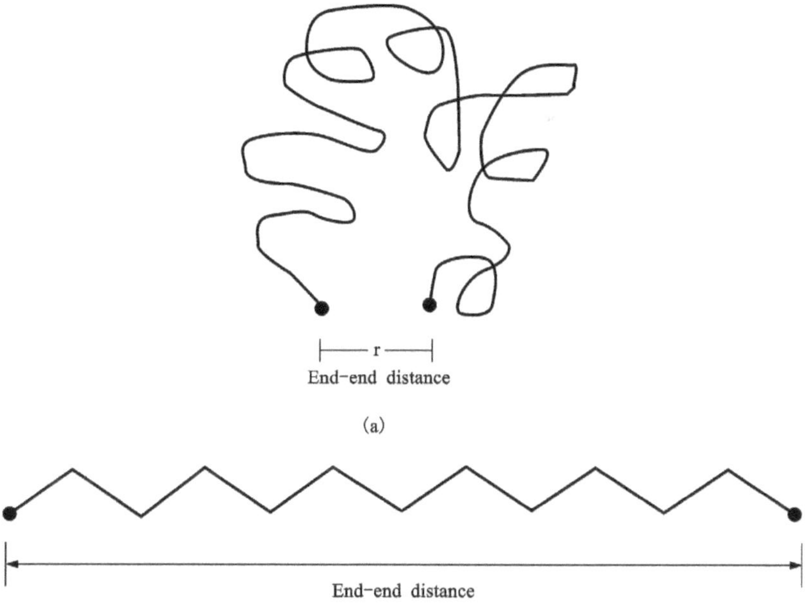

Figure 6.3 Schematic presentation of polymer chain in the extended (A) and coiled conformations (B) with the end-to-end distance, *r*. *(After Hu Cheng, Synthesis and properties of hydrophobically associating polymer P (AM-AMPS-SA), Southwest Petroleum University, Sichuan Province, 2014.)*

When discussing the behavior of a single polymer chain, it must be based on the fact that the segments between carbon atoms can rotate almost without any hindrance. However, there will be differences between the actual situation and the theory. Several of the most important factors affecting the flexibility of the chain are as follows:

① The interaction between the bulk or polar side group in the chain segment rotation will produce steric hindrance
② Chemical bonds between chains (cross-linking)
③ van der Waals bonds between neighboring chains
④ Change in the bonding within the chains.

So in practice, it may be that the polymer material will behave more like a rigid solid, because the flexibility of the individual chain may not be obtained.

6.1.3 Thermoplastic polymers

In thermoplastic materials, flexibility of polymers is hindered due to space constraints and van der Waals bonding. In order to maintain elasticity, the segments of the polymer chain need to obtain enough internal energy to overcome these rotational barriers. This can be achieved by thermal activation, following the concept of thermal activation process. Therefore, the overall behavior of thermoplastic polymers will depend on the balance between these effects. There is enough thermal energy to enable unhindered rotation of the individual chains if the temperature is sufficiently high. Because the bond rotation of the IHC segment can easily break the van der Waals bond, this can have two effects: the flexibility of the chain itself, and the ease with which one chain can move through the other. In this case, the polymer behaves as a viscous liquid.

A single van der Waals bond between adjacent atoms in different chains may be weak. However, because polymer chains are long and contain a large number of atoms, the cumulative effect of a single van der Waals gravity can provide a strong intermolecular bond sufficient to inhibit the rotation and flexibility of the chain at room temperature, thus create a solid material. Some polymer precursors have short chain segments, so the intermolecular bonding is not strong enough to form solid with mechanical properties, such as semisolid such as waxes and greases. The melting point of polymers varies from 70 to 120°C due to insufficient intermolecular bonds to completely block rotation and flexibility (even if the polymer chain is long).

As mentioned above, melting temperature (T_m) and glass transition temperature (T_g) can quantitatively describe the temperature dependence of polymer behavior. When the temperature is above T_m, the chain segment of polymer is flexible, and can coil and slide through each other, so it behaves like a viscous liquid. When the temperature drops below TM, the van der Waals force between molecules is large enough to prevent the large-scale winding and sliding of molecules. To do this, the molecules must be linear and have a regular structure in order to fill in orderly. In addition, when a greater intermolecular attraction is generated than usual, for example in molecules where hydrogen bonds can be developed, the crystallization is enhanced. The ordered structure maximizes the intermolecular bonding, thus eliminating rotation and obtaining rigid solids. If the polymer chains are irregular in shape, they cannot be packed into a regular structure and the cooling below T_m does not lead to any drastic change in the arrangement of the chains (they remain in the coiled and entangled structure without any repeatable order; i.e., they will he amorphous). Although the polymer is greatly affected, it cannot be considered as a rigid solid; the flexibility of the chain is reduced, but local chain rotation is still feasible. The material behaves like a supercooled liquid with a high viscosity. When cooled below T_g, the properties of the material change dramatically because the structure is now "frozen" in the amorphous state (Fig. 6.4A) and the chain segments lose the ability to rotate. It appears as a relatively brittle, rigid solid. These structural and performance changes can be easily quantified by monitoring changes in specific volume (Fig. 6.4B) and modulus of elasticity as a function of temperature.

Because the glass transition only involves a secondary transition, that is, from a highly deformable state to a harder state, rather than from an amorphous state to a crystal structure, it is a change in molecular fluidity. Monitoring T_g is a good method to evaluate molecular mobility and its influencing factors. Although the lack of fluidity of the chain below T_g causes the material to become a brittle solid, it can be regarded as glass because of its amorphous structure. The preservation of the amorphous structure shows that the glass transition only involves the rate of specific volume changing with temperature, while the specific volume decreases significantly during the crystallization process (Fig. 6.4B). The decrease of chain flexibility and the enhancement of chain bonding parameters will lead to the increase of T_g.

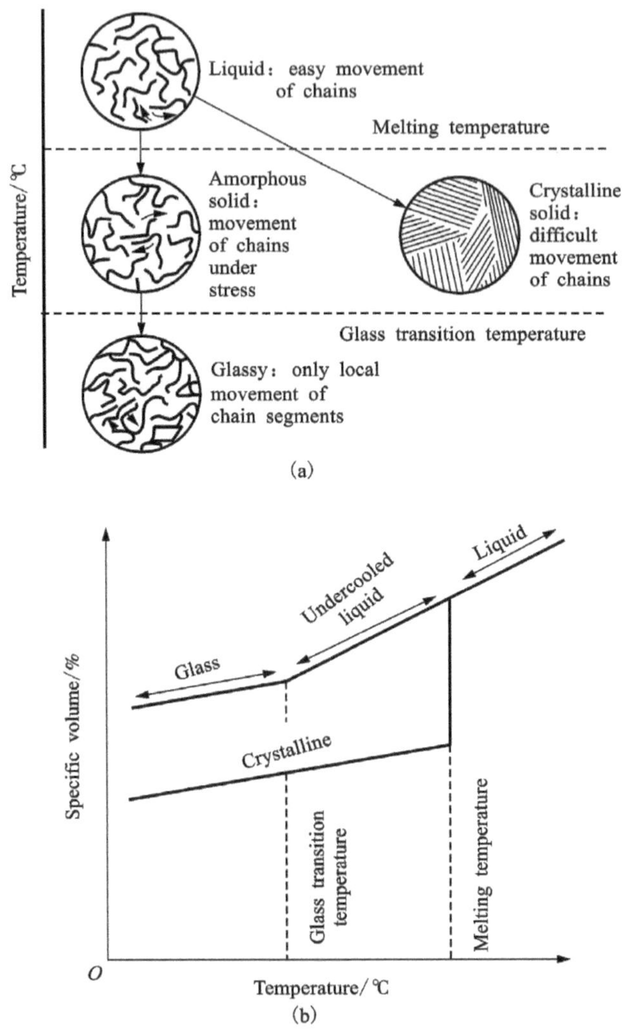

Figure 6.4 The effect of temperature on the structure and behavior of thermoplastic polymer (A) structure and (B) specific volume. *(After D. R. Askeland, The Science and Engineering of Materials, PWS-Kent Publishing Company, 1985, pp. 345, 347.)*

Therefore, T_g values increase with the presence of bulky and charged substituents, as shown in Table 6.1 for polyvinyl chloride and polystyrene (vinyl polymers have the highest glass transition temperature).

Both in the structure of the crystals and in the degree of crystallinity that can be achieved, crystallinity in polymeric materials is different from that in

Table 6.1 Composition and transition temperatures of different thermoplastic polymers.

Class	Composition	T_m(°C)	T_g(°C)	Remarks
Vinyl with one substituent	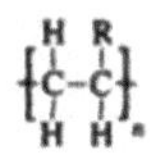			
	R=H Polyethylene	135	-120	-80% Crystalline
		115	-120	-50% Crystalline
	R=CH_3 Polypropylene	176	-27	CH_3 group inhibits crystallization
	R=OH Polyvinylalcohol	—	85	Hydrogen bonding Between chains
	R=Cl Polyvmykhloride	212	87	PVC
	R=C_6H_5 Polystyrene	—	100	Benzene ring prevents crystallization
Vinyl with two substituents	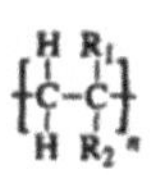			
	R_1=CH_3Polyisoprene	28	-73	cis-Natural rubber trans-Gutta Perch
	R_2=vinyl	28	-73	
	R_1=CH_3 Polymethyl- R_2=CO_2CH_3methacrylate		100	
Fully substituted vinyl	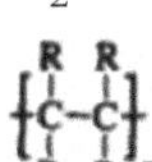 Polytetrafluoroethylene (PTFE)R-F	327		E.g., Teflon, all hydrogen atoms replaced by fluorine
Polyamide	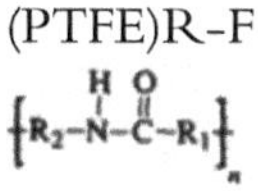	265	50	Nylon 66 R_1=R_2=C_6H_{12}
Polyester	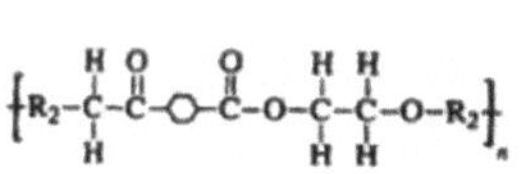	270	80	Dacron
Polycarbonate	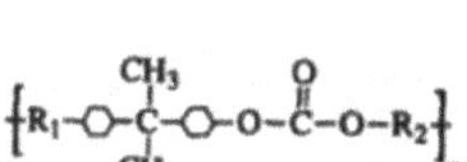	—	150	Polycarbonate

low-molecular-weight compounds. Due to the large size of polymer molecules, when they are cooled through T_m, a considerable amount of movement B is required to arrange into regular crystal structures. As a result, a lamellar crystal structure b is formed by regular folding of the chain (Fig. 6.5); a part of the polymer chain is always in the amorphous state, separating the lamellar microcrystals. The thickness of lamellar crystal corresponds to the fold period, which is about 10 nm. Therefore, crystalline phase and amorphous phase are closely combined to form crystalline polymer, which can present various forms in a wide range of crystallinity (−20%−80%). Materials of this kind can show both crystalline (melting) and glass transitions, as demonstrated in Table 6.1 for several polymers.

In contrast, crystalline polymers usually have better mechanical properties and chemical stability than amorphous polymers. This means that higher intermolecular bonds can be formed in ordered structures, in which crystal segments can be tightly packed together. The regularity of molecular structure and the volume of substituents in the molecular chain determine the difficulty of crystallization. In vinyl-based polymers, where the repeating unit is

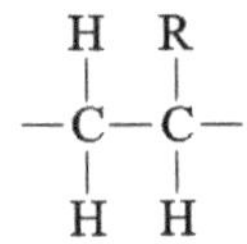

R is a substituting group (substitutes for H), and more than one of the H atoms can be replaced by similar or different groups. The nature of this substitution is important. For example, polyisobutylene {R=$(CH_3)_2$} is more difficult to crystallize than polypropylene (R=CH_3); while polyvinyl

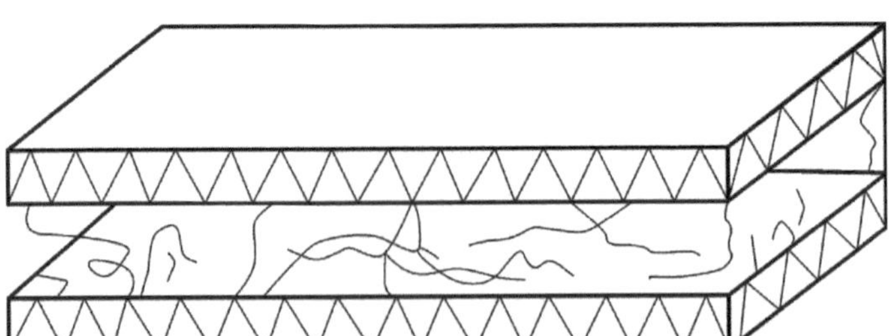

Figure 6.5 Crystalline structure of a polymer, showing the lamellar morphology of the crystalline phase and the amorphous nature of the chain segments between the crystallites. *(After Zhai Ximin, Preparation and interfacial structure evolution of polymer based composite electrolyte for all solid state battery, Harbin Institute of Technology, Heilongjiang Province, 2019.)*

chloride (PVC) (R=Cl) is always amorphous. The size of the latter substituent has a negative effect on the regular packing of the chain. These characteristics are clearly shown in the difference of the transition temperature values in Table 6.1. If the substituents can be arranged according to the main chain (isotactic structure) rather than random (atactic structure), then they will have a positive effect on crystallization.

If the functional groups that can form hydrogen bonds are placed regularly, their existence can also promote crystallization; examples arc esters and amides. Therefore, polyester and centroid used in the production of fibers have high T_m values (Table 6.1), and the stable oriented crystals in these fibers can produce excellent properties. Another requirement for crystallization is the linearity of the polymer chain. Due to the interference of the side branches in the branched chain, the crystal structure cannot be formed. A notable example is polyethylene: crystalline polyethylene is linear, while branched chain polyethylene is still amorphous. Therefore, two forms of polyethylene are widely used: (1) high-density polyethylene, which is crystalline, is a hard solid with good thermal stability; and (2) low-density polyethylene, which is amorphous and therefore soft and elastic.

6.1.4 Thermosetting polymers

The characteristic of thermosetting polymers is the dense cross-linking between polymer molecules. Strictly speaking, the toughest material is made up of a single three-dimensional polymer molecule. This structure results in a hard, rigid, inflexible solid which generally starts to decompose before melting. Therefore, these materials do not exhibit the different response to temperature observed in thermoplastic. However, even if the bonding strength between the chains is very strong, even if there is close cross-linking, there is still a space for small-scale winding. Like steel and ceramics, no chain slippage is allowed, but the modulus of elasticity is smaller than that of conventional materials.

6.2 Sealants

The principle of sealant is to form a barrier between the external liquid or gas and the interior. Unlike adhesives, sealants can only be used for insulation, not for structural purposes. The design of sealants requires strict control over the function of viscoelastic materials, because the final performance evaluation is based on deformability (flexibility), modulus (or mechanical sensitivity), and elastic recovery. For example, a sealant for glass should have good elastic recovery (>40%). The selection of sealants for other unions depends on the type of joint considered and the expected load history: joints with less or higher mobility, joints subject to cyclic loading, joints involving different substrates, etc.

6.3 Adhesive

Adhesive is a kind of adhesive properties, and the surface of two objects can be tightly bonded together. There are many kinds of adhesives with different properties. This section mainly introduces the polymer adhesive used in civil engineering.

6.3.1 Composition and type of adhesive

6.3.1.1 Composition and function of adhesive

Various macromolecular gels are mainly composed of gel or substrate, fillers, and auxiliaries. Auxiliaries include diluents or solvents, curing agents and accelerants, plasticizers or toughening agents, and other antiaging additives.

① Gel: The main components of polymer adhesives are polymer gels with strong adhesive properties, also known as substrates, which give physical and mechanical properties such as adhesive strength and durability. Polymer adhesives with type groups or side groups in macromolecular chains generally have strong adhesion properties, such as bone gel, starch, shellac, and other natural resins: polyurethane, epoxy, phenolic, silicone, polyvinyl alcohol, polyacrylic acid, and other synthetic resins; Nitrile butadiene rubber, neoprene, and other synthetic rubber.

② Diluent or solvent: Diluents are generally gel solvents, divided into inert diluents and reactive diluents. Adjusting the viscosity of the adhesive is the main function of the inert diluent, so as to facilitate the construction, improve the permeability of the adhesive surface, and improve the bonding strength, such as xylene, acetone, alcohol, water, and so on. Active diluents are low viscosity liquid substances, such as low-molecular-weight epoxy compounds and acrylate compounds, which can participate in the curing reaction of thermosetting adhesives.

③ Curing agent: Curing agents such as amines and anhydrides are the curing agents of polymers in adhesives.

④ Plasticizers and toughening agents: Plasticizers or toughening agents such as dioctyl phthalate and low-molecular-weight polyamide resin can improve the toughness of the cured adhesive.

⑤ Fillers: The purpose of adding active or inert powder filler is to reduce the shrinkage of adhesive during curing process and improve the strength and heat resistance of adhesive. The commonly used fillers are silica fume, quartz powder, asbestos powder, and metal powder.

⑥ Other additives: The purpose of adding antiaging agents is to increase the antiaging properties of adhesives. In addition, additives such as bactericide and preservative can make the adhesive have some special functions.

6.3.1.2 Type of adhesive

① Adhesives can be divided into structural adhesive, nonstructural adhesive, and special adhesive.

After curing, it can bear large load and has the ability of heat resistance, cold resistance, and chemical corrosion resistance, which is called structural adhesive. Common structural adhesives such as epoxy resin, phenol dibutyronitrile, and epoxy nylon are mainly used for bonding or strengthening structural parts.

However, nonstructural adhesives such as polyvinyl acetate, polyvinyl alcohol, carboxymethyl cellulose, and rubber cannot bear large load.

The adhesive with high temperature resistance, ultralow temperature resistance, electric conductivity, heat conduction, photosensitivity, and strain resistance can be used as special adhesives. These special adhesives are mainly used for bonding parts with specific functional requirements.

② According to physical classification, adhesives are water emulsion, water soluble, solvent free, solvent based, paste and putty type, solid type, and so on.

6.3.2 Adhesion of adhesive

6.3.2.1 Bond force

Adhesive can produce mechanical bite force, adsorption force, and chemical bonding force between objects, so that objects can be firmly bonded together.

① mechanical bite force

The mechanical biting force of adhesive comes from the tiny anchoring point in the surface space of rough porous object after curing.

② physical adsorption force

The secondary bonding force comes from the mutual adsorption between the adhesive molecules and the surface molecules.

③ chemical bonding force

The adhesive molecules react with the surface molecules to form a strong chemical bond force.

6.3.2.2 The main factors affecting the bonding strength

① Composition and molecular weight of the adhesive

Epoxy resin, chloroprene rubber, and other polymer adhesives, the polymer chain contains more polar groups. It has stronger bonding strength, especially for objects on polar surfaces. The polymer adhesive has the characteristics of small molecular weight, low cohesion, low viscosity, and low adhesion. In the case of high molecular weight, it is difficult to dissolve and disperse in solvent. Only when the molecular weight of the adhesive is appropriate, the adhesive has a higher bonding strength.

② Wettability of adhesives on the surface of object

If the adhesive can produce good wettability on the surface of the object, then its bond strength will increase. Therefore, it is necessary to select the appropriate adhesive according to different objects. Polar adhesives should be used for polar objects and nonpolar adhesives for nonpolar objects. If solvents in adhesives can dissolve the surface of the adhesives, the bonding strength will be higher.

③ Fillers

Appropriate fillers can reduce the cost, save materials, and improve the bond strength. The shear failure of adhesive interface is due to the linear expansion coefficient which is different from the object. The expansion coefficient can be adjusted by adding appropriate fillers to avoid shear failure.

④ Thickness of adhesive film

The bond strength increases with the decrease of the thickness of the adhesive film. If the thickness of the adhesive layer is too thin or the layer is not uniform, it will produce defects at the interface and reduce the bond strength. The thickness of the adhesive film should be controlled between 0.05 and 0.25 mm.

⑤ Bonding technology factors and construction conditions

The following are the main factors affecting the bond strength: surface treatment of adhesive, viscosity and use time of adhesive, standing time after coating, ambient temperature, and hardening time in hardening process.

6.3.2.3 Basic requirements for adhesives

In order to produce enough bonding strength on the surface of the objects, the adhesives should meet the following basic requirements:

① It can produce enough fluidity and wettability on the surface of the object to ensure the full penetration of the surface.

② Choose the binder which is easy to adjust and control the curing speed and viscosity.
③ Choose the adhesive with small expansion deformation and good stability.
④ It has the properties of antiaging, high temperature resistance, low temperature resistance, and chemical resistance.
⑤ The bonding strength is high.

6.3.3 Types and properties of common adhesives

6.3.3.1 Synthetic resin adhesives

6.3.3.1.1 Polyvinyl acetate

The adhesive is made from polyvinyl acetate emulsion. It has the properties of nontoxic, tasteless and high bond strength, but its water resistance and heat resistance are poor. It can be used alone or mixed with cement, gypsum, carboxymethyl cellulose, etc. As a nonstructural adhesive, it is often used in plastic wallpaper, bonding wood, ceramic decorative materials, gypsum board, and other uses.

6.3.3.1.2 Polyvinyl alcohol and polyvinyl acetal adhesives

Polyvinyl alcohol adhesive can be made by dissolving polyvinyl alcohol resin in water. It is a kind of nonstructural adhesives. It is widely used in building wallpaper or ceramic board adhesion, can also be mixed into the cement mortar, to improve the adhesion of mortar.

6.3.3.1.3 Epoxy resin adhesive

The epoxy resin binder supported by epoxy resin is generally two-component. Epoxy resin adhesive and curing agent, toughening agent, and other additives are packed separately. When required, the two components are mixed together to form an epoxy adhesive.

Epoxy resin and other general adhesives have strong adhesion with wood, metal, rubber, plastic, and cement-based materials. It can be cured with different curing systems. The cured film has the properties of low shrinkage (about 2%), high bonding strength, heat and humidity resistance, good volume stability, and corrosion resistance. In the construction industry, it is mainly used for the reinforcement of concrete structure, the bonding of metal and concrete, the repair of cracks, and the adhesion of glass, natural stone, and ceramics.

6.3.3.1.4 Polyurethane adhesive

It can be cured at room temperature. It has good adhesion to plastic, metal, glass, and other materials, and has solvent resistance, acid resistance, oil resistance, impact resistance, and other properties, suitable for waterproof and anticorrosion engineering.

6.3.3.2 Rubber adhesives

This kind of adhesive is made of rubber. The adhesive prepared with natural rubber or synthetic rubber has excellent flexibility and impact resistance, good adhesion but poor heat resistance. It is mainly used for the bonding of plastics, rubber, metal, and nonmetallic materials.

6.4 Fiber reinforced polymer

6.4.1 Introduction

Throughout history, the use of new materials has often led to major changes in civil engineering. In the 18th and 19th centuries, the application of cast iron and wrought iron led to the development of the industrial revolution. In the second-half of the 20th century, the application of concrete also greatly accelerated the reconstruction after the Second World War.

At present, the maintenance and repair costs of structures constructed with traditional materials have become a major challenge for civil engineering, while the use of materials such as steel, reinforced concrete, and wood has increased dramatically in the past few years. As an example, in the United States of America (USA), according to the 2013 American Society of Civil Engineers (ASCE) report for America's infrastructure, almost 25% of the 600,000 bridges are either structurally deficient (in most cases due to corrosion-related anomalies) or functionally obsolete. It is estimated that about 40% of the bridges built after 1945 in the United States need to be replaced before the end of this century.

In the current design, the durability of traditional materials and the existing problems in the current specifications should be considered. The increase in functional requirements and the increase in building speed have encouraged the development of innovative structural solutions, including the development of new materials with high mechanical properties but low specific gravity and corrosion resistance, which require less maintenance during their service life.

Within these new materials, fiber reinforced polymers (FRPs) are presently assuming a particular relevance. Polymer matrix (such as polyester, vinyl ester, or epoxy resin) and fiber reinforced materials (such as glass, carbon, or aramid) embedded in the matrix constitute these composites. FRP materials were first developed in the Navy and aerospace industries in 1940, and then extended to several other industries including petrochemical and automotive industries. In these fields, FRP materials have been proved to have excellent mechanical properties, light weight, and good durability even in relatively harsh environments. Since 1980, FRP materials have been widely used in the construction industry, which makes them more and more interested in FRP materials.

6.4.2 General properties of FRP materials

Composite materials result from the combination of two or more materials which, when used separately, may not present adequate properties to be used as construction materials but, when combined and maintaining an identifiable interface surface, may constitute a new material that symbiotically merges the best properties of the original materials.

FRP material is mainly composed of two parts: ① the fiber which provides most of the strength and stiffness of the material, which is mainly responsible for the mechanical properties of the material; ② the polymer matrix as the adhesive of composite material ensures the load transfer between the fibers and between the applied load and the composite itself. The polymer matrix of FRP materials usually contains fillers and additives in addition to resins. These fillers and additives can improve the manufacturing process, reduce the production cost, and improve the performance of the final product.

6.4.2.1 Constituent materials

6.4.2.1.1 Fibers

The main function of fiber reinforced materials is to provide sufficient strength and stiffness along the development direction of components and support the mechanical traction of structural components. Glass fiber, carbon fiber, and aramid fiber are the most commonly used fiber types in commercial applications. The physical, mechanical, and thermal properties of these fibers are presented in Table 6.2.

Glass fiber is divided into glass fiber reinforced plastic (GFRP) pultruded profiles, reinforcement and sandwich panel skin, which is the most

Table 6.2 Typical characteristics of main fiber reinforcements.

Property	Test standard	E-Glass	Carbon	Aramid
Tensile strength/MPa	ISO 5079, ISO 11566,	2350–4600	2600–3600	2800–4100
Elasticity modulus in tension/GPa	ASTM C 1557, ASTM D 2343,	73–88	200–400	70–190
Maximum strain in tension/%	ASTM D 3379	2.5–4.5	0.6–1.5	2.0–4.0
Density/g/cm^3	ISO 1889, ISO 10119, ASTM D 1577	2.6	1.7–1.9	1.4
Thermal expansion coefficient/ ($\times 10^{-6}$/K)	ISO 7991	5.0–6.0	Axial: −1.3 to −0.1 Radial: 18.0	−3.5
Fibers diameter/μm	ISO 1888, ISO 11567	3–13	6–7	12
Fibers structure	–	Isotropic	Anisotropic	Anisotropic

commonly used fiber reinforced material in the construction industry. Their main advantages are relatively low cost and high strength. However, the main disadvantages of glass fiber are the decrease of long-term strength (stress fracture), low elastic modulus, moisture resistance, and alkali resistance. In different types of glass fibers (code e, s, AR, c), they all have the same elastic modulus, but have different mechanical strength and durability. The traditional E-glass fiber has good electrical insulation performance, which is the most used up to now, equivalent to 80%–90% of commercial products. Type S fiber is basically used in the aerospace industry because it has better mechanical resistance, but is much more expensive (3–4 times) than Type E glass fiber. Type AR fiber can work better in alkaline environment after increasing zirconia content, so it can be used in cement-based composite materials, such as GRC (glass fiber reinforced concrete); Type C fiber has better corrosion resistance, but it is rarely used in civil engineering.

There are several forms of carbon fiber used in reinforcement applications: carbon fiber cloth, sheet, reinforcement, and prestressing tendons. They have high tensile strength and elastic modulus, low self-weight, high fatigue, and creep resistance and excellent chemical resistance. They have low self-weight, high tensile strength and elastic modulus, fatigue and creep resistance, and excellent chemical resistance. Their main disadvantages are high production cost and high energy consumption. There are several grades of carbon fiber, including standard modulus (conventional grades listed in Table 11.1), and several improved grades, including medium modulus, high modulus, and ultrahigh modulus.

Aramid fibers are stronger than glass fibers and present an elasticity modulus that is 50% higher. In addition, the excellent toughness of these fibers makes them have great potential in industrial applications requiring high energy absorption, such as bulletproof vests and helmets, automobile crash attenuators. The potential benefits of civil engineering applications are weakened by the relatively low compressive strength (500−1000 MPa), the susceptibility to stress fracture (long-term strength reduction), and high sensitivity to ultraviolet radiation.

These different types of fibers can be used to reinforce different forms of FRP materials, i.e., short fibers (chopped), usually 3−50 mm in length, and nearly parallel continuous long tow, untwisted (roving) or twisted (yarn). These fiber reinforced forms can be further processed into textiles with different reinforcement directions. Several products can be developed, either random oriented fibers (chopped strand mats) or continuous fibers (continuous fiber mats), or directional reinforcing bars (such as woven and nonwoven fabrics, stitched fabrics, grids and grids), which can be biaxial (0°/90 degrees or + 45 degrees/−45degrees) or triaxial (0°/+45 degrees/ −45 degrees). Finally, all these forms can be further combined into textile products with continuous oriented fibers, continuous oriented or random oriented short fibers. Figs. 6.6 and 6.7 show examples of available forms of fiber reinforcement.

(a) (b) (c)

Figure 6.6 Examples of available forms of fiber reinforcements: (a) rovings, (b) chopped strand mat, (c) biaxial woven fabric.

Figure 6.7 Different types of mats: (a) rovings, (b) yarns, (c) continuous strand mat (random fiber orientation), (d) biaxial woven fabric (0/90), (e) complex mat (0/90 weave + random fiber orientation), (f) bidirectional complex mat (0/45/90 weave + random fiber orientation).

Surface treatment of steel bars can prevent damage during processing and help fibers stick together. This surface treated component is called sizing (usually made of starch, oil, or wax). Because of the coupling agent in the sizings, it can provide lubrication and antistatic function, and promote the fiber matrix peeling. When sizing is applied to certain types of fibers, such as glass, durability during service life can be improved by providing a means to prevent moisture degradation.

6.4.2.1.2 Polymeric matrices

The mechanical performance of FRP composites relies mostly on the fiber reinforcement. However, polymer matrix is also responsible for carrying some types of loads, especially those related to transverse stress and interlaminar shear stress. In addition, the polymer matrix shall meet the following requirements: ① holding the fibers in their intended positions; ② ensuring load transfer/distribution between fibers; ③ preventing fibers from buckling when subjected to compressive stresses; ④ protecting fibers from environmental degradation agents such as moisture (similar to the steel protection provided by the concrete cover in reinforced concrete structures).

The basic resins that make up the polymer matrix of FRP materials usually contain auxiliary components, which are used to improve material processing, induce polymerization reaction, reduce material cost, and adjust the properties of final FRP products. These additional constituents can be divided in three groups: ① polymerization agents, ② fillers, and ③ additives.

6.4.2.1.3 Resins

There are two main groups of polymeric resins: thermosetting resins and thermoplastic resins. Polymers in solid form can be distinguished by the way that the polymer chains are connected.

Thermosetting resins such as polyester, vinyl ester, phenolic resin, and epoxy resin are the products of polymerization chemical reaction. When heated, this reaction results in the formation of amorphous three-dimensional cross-linked molecular structure. Thermosetting resins become immiscible after curing because of their own irreversibility, so they cannot be reprocessed. Thermosetting resins have good fiber impregnation capacity and very good adhesive properties. In addition, they exhibit low viscosity, allowing for high processing speeds.

Thermoplastic resins (including polypropylene, polyamide, polyethylene, and polybutene) are not cross-linked and will not undergo any chemical changes during processing. Their molecular chains are held together by hydrogen bonds or by weak van der Waals forces. Because the thermoplastic resin will not form irreversible structure after heating, it can be recycled and reprocessed. Compared with thermosetting resin, thermoplastic resin has poor impregnation performance and adhesion to absorbance, and its higher viscosity leads to greater processing difficulty, thus increasing the manufacturing cost.

Due to the above characteristics, thermosetting resins have higher potential in civil engineering applications, so they have been used in most commercial FRP products. Typical physical and mechanical properties of the most common thermosetting resins (polyester, vinyl ester, epoxy, and phenolic) are listed in Table 6.3. These resins are isotropic viscoelastic materials.

Polyester accounts for about 75% of the resin currently used in FRP products. This figure is due to a good balance between their properties

Table 6.3 Physical and mechanical properties of thermoset resins.

Property	Test standard	Polyester	Epoxy	Vinylester	Phenolic
Tensile strength [MPa]	ISO 527, ASTM D 638	20–70	60–80	68–82	30–50
Elasticity modulus in tension [GPa]		2.0–3.0	2.0–4.0	3.5	3.6
Maximum strain in tension [%]		1.0–5.0	1.0–8.0	3.0–4.0	1.8–2.5
Density [g/cm^3]	ISO 1183, ASTM D 1505	1.20–1.30	1.20–1.30	1.12–1.16	1.00–1.25
Glass transition temperature [°C]	ISO 11357-2, ISO11359-2, ASTM E 1356, ASTM E 1640	70–120	100–270	102–150	260

(mechanical, chemical, electrical), dimensional stability, ease of processing (viscosity reduction, reasonable pot life, possible changes in matrix composition during polymerization reactions), and relatively cheap prices.

Epoxy resins are commonly used in structural strengthening applications with special requirements (in terms of stiffness, strength, service temperature, and durability) or in combination with carbon fibers (strips, sheets, etc.) or as structural adhesives alone. Although epoxy resins are difficult to process due to their high viscosity and long curing time, they have lower shrinkage (1.2%–4.0% and 8.0%, respectively) compared with polyester. These properties indicate that they have excellent adhesion and environmental degradation resistance.

Compared with polyester and epoxy resin, vinyl ester resin has medium performance and cost. In particular, vinyl esters combine the improved properties of epoxy resins and more easily processed polyesters. Vinyl ester has better durability than polyester, which explains why most commercially available FRP bars used to strengthen concrete members are made of vinyl ester matrix. There are also several types of vinyl ester used for FRP profiles with special durability requirements.

Phenolic resin is less flammable and produces less smoke than other thermosetting resins at high temperature, which is the fire reaction characteristics of phenolic resin. In addition, phenolic resin has similar cost to polyester, and has good dimensional stability, which can maintain its adhesion and mechanical properties at relatively high temperature. The main disadvantage is that it is difficult to strengthen and cure the phenolic resin. In addition, the phenolic resin is brown and not easy to be colored.

6.4.2.1.4 Polymerization agents

The polymerization is triggered by the combination of polymerization agent and base resin. Curing reactions of polyester and vinyl ester resins are usually initiated by organic peroxides (by thermal activation) in amounts ranging from 0.25% to 1.50% (by weight of resin). The polymerization of epoxy resins is initiated with curing agents (or hardeners) of the amine type, which are typically added at ratios of 25%–50% (also by weight of the resin).

6.4.2.1.5 Fillers

The use of fillers in polymer matrix can not only reduce the final cost of FRP products but also improve some properties, which cannot be achieved when only resin and fiber are used. For example, fire resistance can be improved by using fillers to reduce the organic content in the matrix. In addition, the filler can improve the dimensional stability of the matrix by reducing its shrinkage, so as to avoid cracks in discontinuous areas or parts with high resin content. Cracking prevention may also increase the resistance to environmental degradation agents. The hardness, creep property, fatigue, and chemical corrosion resistance of the matrix can be improved by adding fillers to the matrix.

At present, the fillers used in FRP products include aluminum silicate, calcium carbonate, calcium sulfate, and alumina trihydrate. In particular, the last two compounds are usually used to improve the fire resistance of composite materials and reduce their flammability and smoke generation.

If the content of filler material in FRP material reaches 40%–65% of the total weight of matrix, then the FRP material can be used as nonstructural material. However, in typical FRP bars and pultruded FRP profiles, the filler content is 10%–30% of the weight of the resin matrix. For small pultrusion parts reinforced mainly by unidirectional roving, the filler content is usually lower, generally less than 5%. FRP strips are generally not filled.

Although some performance improvements can be achieved by using fillers, it should be emphasized that the addition of fillers to the resin system usually reduces the critical mechanical properties of FRP materials.

6.4.2.1.6 Additives

Adding a variety of additives into the resin system can improve the processability of the material, and improve the performance of the final product or simply change some properties. Additives are generally used to fulfill one of the following objectives:

① Reduction of the tendency to attract electrical charge (antistatic agents)
② Prevention of cracking, disintegration, discoloration, or loss of gloss due to ultraviolet radiation (UV stabilizer)
③ Promotion of cellular structure (foaming agents), which reduces density, materials costs and shrinkage, improving, in addition, the electrical and thermal insulation
④ Obtaining a certain color (pigments or colorants)
⑤ Facilitate removal from molds (release agents)
⑥ Reduction of shrinkage (as low profile or shrink additives)
⑦ Reduce smoke and flammability on fire (as flame retardant additive)
⑧ Inhibition oxidation of polymers (as antioxidants)
⑨ Reduction of voids content

⑩ Increase of toughness (rubber or other elastomeric materials)
⑪ Increase of electromagnetic interference (adding conductive materials) and electrical conductivity (by adding metal and/or carbon particles)

The amount of additives is much smaller than that of fillers, generally less than 1% of the weight of the matrix. However, although the amount of additive is relatively small, it may have a significant effect on the physical and mechanical properties of FRP products (similar to fillers).

6.4.2.2 Philosophy in the development of FRP composites

When developing FRP materials for civil engineering, it should be noted that their properties are affected not only by the composition and arrangement of the reinforcing fibers (type and orientation) and the constituent materials (mechanical properties of fibers and matrix) but also by the interaction between fibers and matrix. The interaction depends on the mechanical compatibility and adhesion between the fiber and the matrix, and also on the angle between the mechanical loading direction and the fiber.

The intrinsic concept of FRP product development involves the careful selection of materials based on specific requirements for the design of properties such as chemical resistance, strength, stiffness, and bulk density. Each different civil engineering structural application has specific performance requirements, and FRP products with different functions are designed and manufactured using this design approach.

Exercises

1. What are the requirements from the structure of a polymer to ensure that it behaves as an elastomer?
2. A given polymer is characterized by the following transition temperatures: $T_s = -10°C$ $T_m = 180°C$. The polymer can be obtained in two forms: (a) with a high degree of crystallinity of 80% and (b) with a low degree of crystallinity of 20%.
   What are the differences or similarities between these two polymers when the service temperatures are (i) 280°C (ii) +20°C?
3. Polyethylene can be obtained in the form of a flexible and soft material which can be used for packaging or as a rigid solid which can be applied to produce pipes. What are the differences in the structure of two forms of polyethylene which lead to these properties?

4. What are the possible applications of the three different polymers in a service temperature in the range of 10–30°C:

    polymer (a): $Tg = 100°C$
    polymer (b): $Tg = 20°C$
    polymer (c): $Tg = -50°C$

5. Discuss the limitations of characterization of the performance of the adhesion characteristics of a given polymer adhesive by the bond strength value obtained by one particular test method.

CHAPTER 7
Asphalt

Asphalt is one of the oldest materials used in architectures. Asphalt mixtures were used in 3000 B.C., continuing the use of the wheel by 1000 years. Asphalt was exploited from natural pools in different places throughout the world before the mid-1850s, such as the Trinidad Lake asphalt, which is still mined. However, the application of asphalt cement has been widely promoted with the finding and refining of petroleum in Pennsylvania. More asphalt cement came from natural deposits by 1907. Today, nearly all asphalt cement is from refined petroleum.

Bituminous materials are divided into asphalts and tars. Several asphalt products are used; asphalt is mainly used in pavement construction, and also as sealing and waterproof agent. Tars are produced by the destructive distillation of bituminous coal or cracking petroleum vapors. Tar is used mostly for waterproofing membranes in China and United States, such as roofs. Tar may also be used for treatments of pavement, especially where fuel spills may dissolve asphalt cement, such as on parking spaces and airport aprons.

This chapter discusses the types, uses, and chemical and physical properties of asphalt. The asphalt concrete is also presented which is used in road and airport pavements, composed of asphalt cement and aggregates (Fig. 7.1).

7.1 Asphalt cement

7.1.1 Introduction

Asphalt cement, asphalt diluent, and asphalt emulsion are three forms which are used in pavement asphalt. Asphalt cement is a mixture of hydrocarbons of different molecular weights. The properties of asphalt depend on its chemical composition and molecular weight distribution. The asphalt becomes harder and more viscous as the distributions shift toward heavier molecular weights. Asphalt cement is a semisolid material at room temperatures that cannot be applied easily as a binder without being heated.

Civil Engineering Materials
ISBN 978-0-12-822865-4
https://doi.org/10.1016/B978-0-12-822865-4.00007-6

Figure 7.1 Photo showing asphalt cement concrete. *(After Ye Ting, Study on performance of permeable asphalt concrete with modified recycled coarse aggregate, Nan Chang University, Jiangxi Province, 2020.)*

Liquid asphalt products have been developed and can be used without heating such as cutbacks and emulsions.

Although the liquid asphalts are convenient, they cannot produce a quality of asphalt concrete equivalent to what can be produced by heating neat asphalt cement and mixing it with selective aggregates. Asphalt cement has brilliant adhesive characteristics so it can become a superior binder for pavement applications. In fact, it is the most common adhesive material for pavement.

A diluent is produced by dissolving asphalt cement in a lighter molecular weight hydrocarbon solvent. The solvent evaporates when the diluent is sprayed on a pavement or mixed with aggregates, leaving the asphalt residue as the binder. In the past, cutbacks were universally used for highway construction. They were efficient and could be applied easily in situ. However, the use of cutbacks has been severely limited by three disadvantages. First, the use of these costly solvents as a carrying agent for the asphalt cement is no longer cost-effective as petroleum has become more expensive. Second, due to the volatility of the solvents, cutbacks are hazardous materials. Finally, application of the cutback releases environmentally polluted hydrocarbons into the air. In fact, cutback materials are absolutely prohibited in many districts with air pollution issues.

Scattering the asphalt in water as emulsion could be an alternative to dissolving the asphalt in a solvent, as shown in Fig. 7.2. In this process, the asphalt cement is physically cracked into micron-sized globules that are

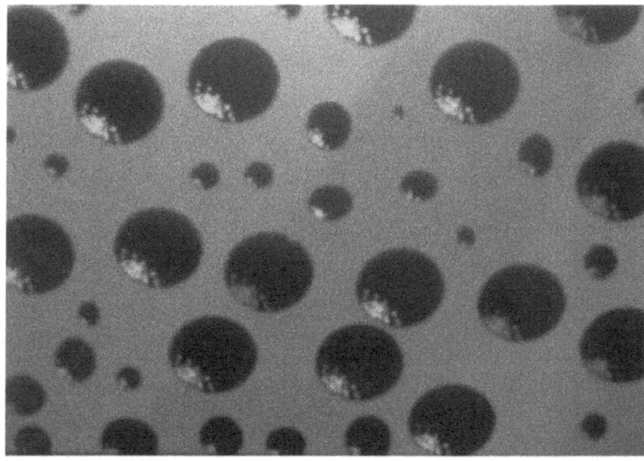

Figure 7.2 Photo of magnified asphalt emulsion showing minute droplets of asphalt cement dispersed in a water medium. *(After Ou Yangjian, Study on rheological properties of fresh cement emulsified asphalt mortar, Harbin Institute of Technology, Heilongjiang Province, 2015.)*

mixed into water containing an emulsifier. Emulsified asphalts typically compose of about 60%–70% asphalt cement, 30%–40% water, and a fraction of a percent of emulsifying agent. There are many kinds of emulsifying agents; mainly they are a soap material. There are two different components of emulsified molecules, the head of which has electrostatic charge, and the tail of which has a high affinity for asphalt. Positive charges produce cations in the emulsion, while negative charges produce anions. When asphalt is introduced into the water with the emulsifier, the tail of the emulsifier is attached to the asphalt and the head is exposed. The electric charge of the emulsifier causes a repulsion between the asphalt globules, which keep them separate in the water. Asphalt will not float or sink in water since the specific gravity of asphalt is very near that of water, the globules have a neutral buoyancy. When the emulsion is mixed with aggregates or used on a pavement, the water evaporates, and the asphalt globs come together to forming the binder. The phenomenon of separation between the asphalt residue and water is called as breaking or setting. The rate of emulsion setting can be controlled by variating the type and amount of the emulsifier.

Since most aggregates bear either positive surface charges (such as limestone) or negative surface charges (such as siliceous aggregates), they tend to be consistent with anionic or cationic emulsions, respectively. However, emulsions that bond well to aggregate-specific types can be produced by some emulsion manufacturers, regardless of the surface charges.

Although emulsifiers and cutbacks can be used for the same purposes, the use of emulsions is rising because they do not include hazardous and costly solvents.

Asphalt concrete can also be produced by a process which does not require healing of the composition materials. An asphalt cement is combined with a solvent firstly to form a so-called liquid asphalt which is fluid at environment temperatures. This material is then mixed with aggregates to form a "cold-mixed" asphalt concrete, which is placed and consolidated similarly to hot-mixed, hot-laid asphalt concrete mixtures. When the solvent evaporates and secondary bonds are reformed in the asphalt cement, hardening occurs.

7.1.2 Composition and structure

7.1.2.1 Chemical composition

Asphalt cements have a very intricate and fickle chemical composition, which lie on the origin of the raw petroleum or petroleum from which they are derived and the processing to which they have been endured. They are composed of high-molecular-weight hydrocarbons with the general formula $C_nH_{2n+b}X_d$, where X represents some elements as sulfur, nitrogen, oxygen, or trace metals; d is usually tiny; and b may be negative. The elemental composition commonly lies within the following limits (Table 7.1).

The number of carbon atoms (n) ranges from about 25 to 150 in asphalt cement molecules, giving molecular weights from about 300 to 2000. These molecules can be divided into three main types: ① aliphatic or paraffinic, in which the carbon atoms are linked in straight or branched chains; ② napthenic, in which the carbon atoms are linked in simple or intricate (condensed) saturated rings (saturated means that the highest possible hydrogen to carbon ratio is probably existed); and ③ aromatic, in which the carbon atoms are linked in particularly steady benzene rings. Asphalt cements compose of an intricate combination of all three types of molecules.

Table 7.1 Elemental composition of asphalt cements.

Elements	Percent/%	Elements	Percent/%
Carbon	80–87	Nitrogen	0–1
Hydrogen	9–11	Sulfur	0.5–7
Oxygen	2–8	Trace metals*	0–0.5

Note:*Iron, nickel, vanadium, and calcium.

Some elements like oxygen, nitrogen, sulfur, and/or trace metals in asphalt cement still have a substantial effect on the physical and rheological properties of the material in spite of their relatively small content. The heteroatoms nitrogen, oxygen, and sulfur form functional or polar groups and hydrocarbon molecules to which such groups are attached are capable of strong intermolecular associations through secondary bonding mechanisms. As a result, some properties of this material such as boiling point, solubility, and viscosity will be affected as if it has a much higher molecular weight.

Because asphalt cements contain such a large number of molecules with differing chemical composition and structure, complete chemical analysis is obviously impractical, if not impossible. Commonly, therefore, an asphalt cement is separated into a small number of component fractions firstly on the base of the size and the reactivity and/or polarity of the various kinds of molecules present, and then it can be analyzed. However, the separated fractions only can be defined in terms of the particular procedure used since they are still complex mixtures and not distinct chemical species, because the fractionation is not necessarily dependent on chemical composition. From one asphalt cement source to another, the same generic fraction can vary considerably in composition and properties. In addition, the relevance between compositional features based on such separation techniques and the characteristics or in-service performance is indeterminate at this time. Nevertheless, component fractionation can be used to assess changes which occur in a specific asphalt cement during its manufacture and use.

7.1.2.2 Physical structure

To understand the performance and characteristics of asphalt cements, it is necessary to have some understanding of their structure (i.e., the physical arrangement of their components). First, asphalt cement can be regarded as a colloidal suspension composed of dispersed (continuous) phase and dispersed (discontinuous) phase, and its interface material plays a role in preventing the dispersion and agglomeration. Table 7.2 summarizes the

Table 7.2 Constituents of asphalt cement.

Phase	Component	C/H ratio[a]	Contribution
Dispersion	Oils	<0.4	Viscosity and fluidity
Dispersed	Asphaltenes[b]	>0.8	Strength and stiffness
Interfacial	Resins	−0.6	Adhesion and ductility

Note: [a]Number of carbon atoms/number of hydrogen atoms; [b]Typical formula: $C_{84}H_{97}S_{3.2}O_{2.5} = C_nH_{2n-71}X_{5.7}$, $C/H = 0.87$.

composition of these phases, their approximate chemical composition, and their general contribution to the performance of asphalt cement. Their general molecular structure is shown in the three-component (ternary) phase diagram in Fig. 7.2, which represents the possible components of asphalt cement.

This phase separation is essentially based on the molecular size, so it depends on the solubility of different components. But due to molecular structure and polarity, there are some differences. The oil phase is mainly composed of nonpolar, uncondensed hydrocarbon ring molecules saturated with long-chain molecules, while the asphaltene phase is composed of polar, condensed aromatic ring molecules joined with chain molecules. These resins have intermediate components, with some condensed ring molecules and some side chains; without them, the suspension will decompose because the other two components are insoluble.

It should be emphasized that these phases exist as a continuum, and there is no obvious boundary between them. The most polar and largest molecules that make up the asphaltene component contribute to interaction and association, so they are concentrated in colloidal droplets or micelles. The degree of association, and therefore the size of the micelles, depends on the concentration of asphaltene, the dispersing ability of the other components, and the temperature. These micelles, each layer surrounded by a layer of resin, are randomly distributed in the oil phase, and the polarity and size of molecules in each phase usually decrease with the increase of the distance from the center core. In fact, these resins act as solvents for asphaltenes, while oils act as solvents for resins, thus forming chemically stable structures.

According to its structure, asphalt cement can be divided into three different types: sol, gel, and sol—gel (Fig. 7.3). In the sol—gel type, the asphaltene micelles are separated and widely dispersed in the oil phase, and the asphalt cement shows the basic viscous (Newtonian) fluid behavior, almost no elasticity. Micelles are still discrete in gel type, but micelles are strongly bound to a complex three-dimensional network through intermolecular attraction, so asphalt cement exhibits elastic, inelastic, and permanent deformation (non-Newton) behavior. The sol—gel structure is between the other two structures with the micelles bound, but not as closely as in the gel type. When load is applied, the sol—gel asphalt cement exhibits elastic behavior first, followed by viscous behavior, and hence they are viscoelastic in nature. This type of asphalt cement accounts for a large proportion of the asphalt cement used in road construction.

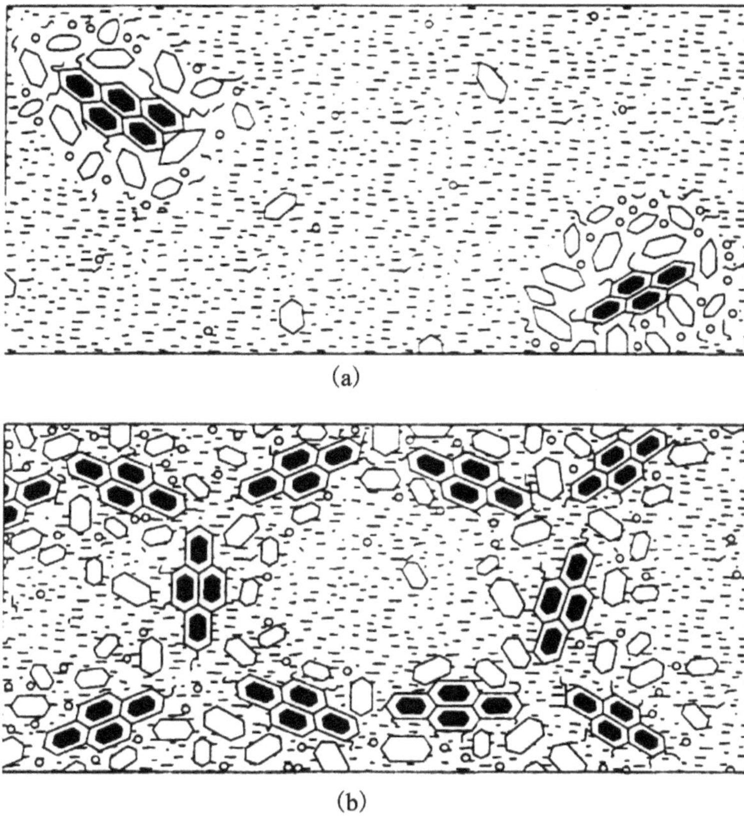

Figure 7.3 The internal structure of asphalt cement: (A) sol, (B) gel. *(After the Shell Bitumen Handbook, 1990, taken from J. M. Illston, ed. Construction Materials, E&FN SPON, 1994.)*

The structure of an asphalt cement and therefore its performance, as noted previously, are a function of temperature as well as its chemical nature and the relative volumes of its constituents. With the increase of temperature, the solubility of asphaltene in resin increases, while the solubility of resin in oil increases, and the viscosity of material decreases. With the decrease of temperature, asphaltenes become difficult to dissolve, micelles combine in an orderly structure, and the material becomes more viscous. Finally, when the temperature drops below the glass transition temperature (T_g), the structure is effectively frozen and the material becomes hard and brittle, which shows as viscoelastic solid. *TG* depends to a certain extent on the composition of asphalt cement, especially the content.

7.1.3 Properties

Due to the difficulty of complete chemical characterization, specifications for asphalt cements are currently based only on the physical and rheological properties of these materials. However, in most cases, the correlation between physical and rheological properties and engineering performances has been proven to be reliable, although all tests used to measure these properties have been limited to some extent. Many of these tests are empirical, most of which do not provide information for the entire range of typical in-service operating temperatures. However, physical and rheological properties will continue to provide a reasonable method for evaluating the quality of asphalt cement without the need for chemical analysis of these materials.

7.1.3.1 Aging

It must be noticed that the performance of an asphalt cement in an asphalt concrete mixture will be significantly different from that before the production of the mixture. Asphalt cements are submitted to heating for variable time periods and at a wide range of temperatures during the mixing, curing, and in-service life of an asphaltic concrete. During these processes, substantial changes occur in the structure and composition of the asphalt molecules, which is mainly due to the volatilization of light hydrocarbon components and the reaction with oxygen in the environment. These changes cause the asphalt cement to harden or become less ductile in turn, a phenomenon known as age hardening or aging. In addition, asphalt cements exist in asphaltic concretes in the form of thin films as shown schematically in Fig. 7.4. Their behavior and properties in this form are quite different from those of "bulk" materials, and the oxidation or age-hardening reaction takes place much faster.

Standardized equipment and procedures have been developed to simulate the mix production, construction, and in-service aging processes, so that changes in the performance of asphalt cement can be evaluated. Two frequently used artificial aging methods are the rolling thin-film oven (RTFO) procedure (ASTM D-2872 or AASHTO2 T-240) and the pressure-aging vessel (PAV) procedure (AASHTO Provisional Standard PP1). The RTFO procedure, in which a thin film of asphalt cement is exposed to heat and airflow in an oven without a break, is supposed to imitate the effects of asphalt concrete mixing and construction procedures on the performance of the cement. It can be used for two purposes, to

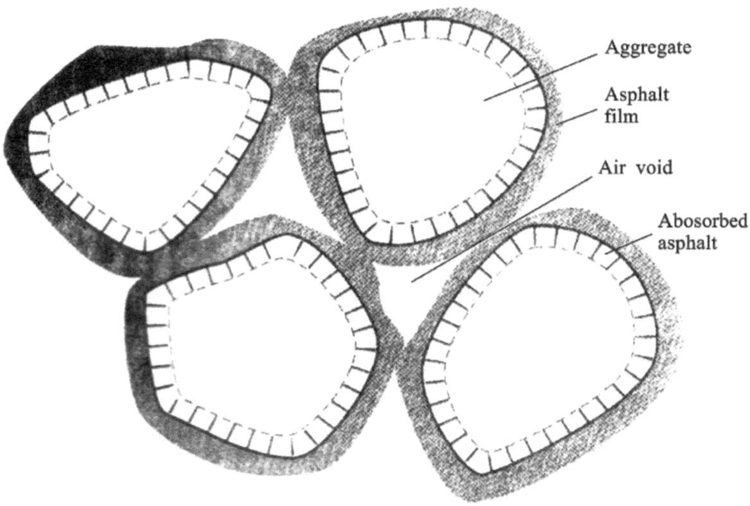

Figure 7.4 Structure of asphalt concrete paving mixture showing aggregates, air voids, and asphalt cement binder. *(After Wang Lusheng, Study on the influence of different coarse aggregates on the adhesion and road performance of cement emulsified asphalt mixture, Changsha University of Technology, Hunan Province, 2019.)*

provide an aged asphalt cement sample for additional tests, and to provide an indication of the mass quantity of the volatiles lost during the heating and mixing process.

The PAV procedure, in which a thin-film sample of asphalt cement aged in the RTFO in advance is exposed to high pressure and temperature, is supposed to imitate the effects of long-term in-service conditions. Therefore, these procedures can be used to prepare and test samples under conditions that represent the critical stages of the service life of asphalt cement.

7.1.3.2 Viscosity and consistency

7.1.3.2.1 Rheological behavior

Viscosity is the basic material property that relates the rate of shear strain in a fluid to the applied shear stress: in fact it is the resistance to flow. In Newtonian fluids it is mathematically described by the coefficient of viscosity, η, which is expressed in units called poise (*P*), where 1 poise is 1×10^{-1} Pascal seconds (Pa s). The engineering term consistency is an empirical measurement of the resistance of a fluid to continuous deformation when it is subjected to a shearing stress. Generally speaking, sol-type asphalt cements are Newtonian fluids, whereas gel-type and sol–gel-type are non-Newtonian under the typical asphalt concrete mixing temperatures (to 140°C). Since the flow curve of non-Newtonian fluids can be expressed by a power relationship, their log(σ)−log(γ) plots arc linear and, therefore, often used by paving engineers (Fig. 7.5).

The rheological properties of asphalt cements are closely related to temperature, and therefore the temperature of a material must be stated when the rheological properties are reported. As shown in Fig. 7.6, the viscosity−temperature relationship is the opposite: viscosity decreases with the increase of temperature. This relation changes with the different asphalt compositions Viscosity−temperature plots for all asphalt cements appear to approach a limiting value of 109 P as temperature is decreased. Prolonged aging at high temperatures, as well as the film thickness, would affect such relations.

As mentioned before, the viscosity/consistency of asphalt cements increases with time due to aging or age hardening. The relationship between viscosity/consistency and aging time are nonlinear and of the form

$$\eta = b \cdot t^m \tag{7.1}$$

or

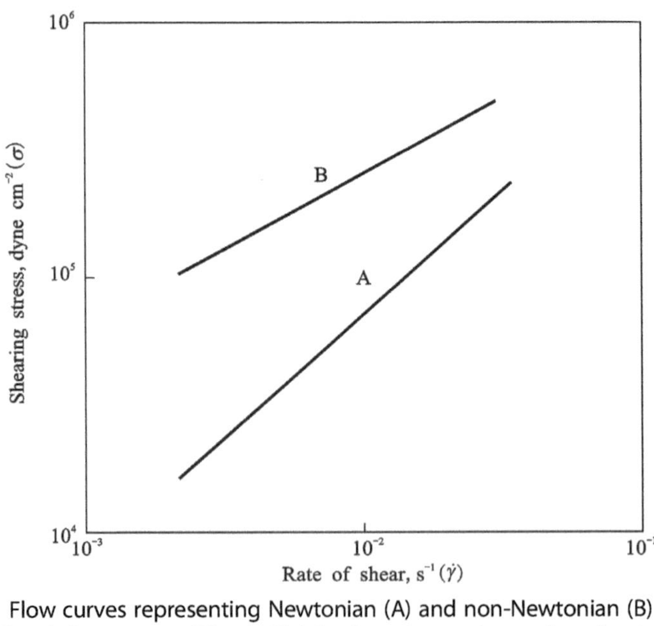

Figure 7.5 Flow curves representing Newtonian (A) and non-Newtonian (B) rheological behavior of asphalt cements. *(After Li Wei, Study on interaction and mechanism of cement emulsified asphalt, Southeast University, 2018.)*

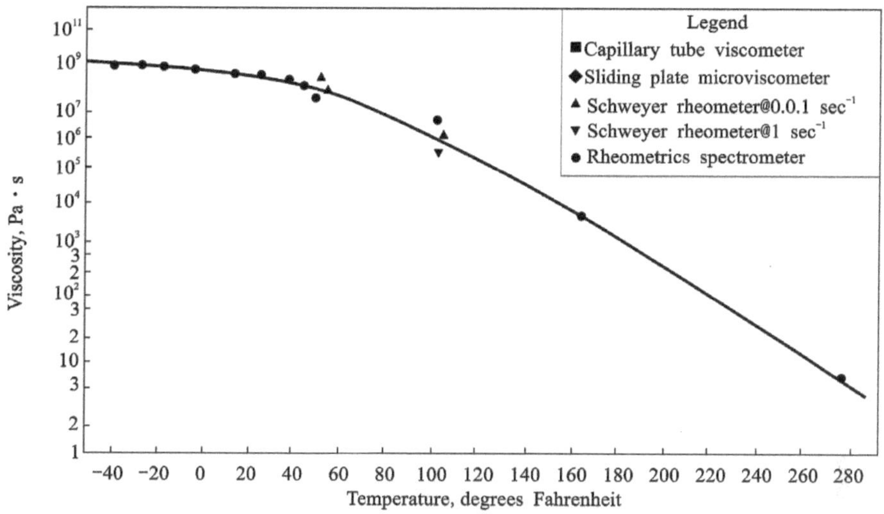

Figure 7.6 Effect of temperature on the viscosity of asphalt cement. *(After J. W. Button, J. A. Epps, D. N. Little, and B. M. Gallway, Asphalt Temperature Susceptability and Its Effect on Pavements, Transportation Research Record 843: Asphalts, Asphalt Mixtures and Additives, Transportation Research Board, National Research Council, Washington, D.C., 1982, p. 123.)*

$$\log\eta = \log b + m\,\log t \tag{7.2}$$

where η is viscosity (Poise); t is time of aging (hours); b is constant and is intercept of log–log plot; m is constant and is slope of log–log plot.

The slope, m, of the log n–log t plot is called the aging degree, or asphalt aging index: there is reasonable correlation between this parameter and the degree of complex or non-Newtonian flow of the material.

The maximum degree of asphalt cement aging usually occurs in the surface of an asphalt concrete pavement, where the binder b is most easily exposed to oxygen and ultraviolet light.

7.1.3.3 Rheological TESTS

The consistency of asphalt cements is usually measured by penetration test (ASTM D-5), as shown in Fig. 7.7. Consistency is measured according to the depth (in units of 0.1 mm) of a standard needle penetrates a sample of the material under standard loading, time, and temperature. Penetration does not properly represent a basic material and is only suitable for comparative purposes, but it was the principal test method for grading asphalt cements for paving purposes, and still continues to be used in some

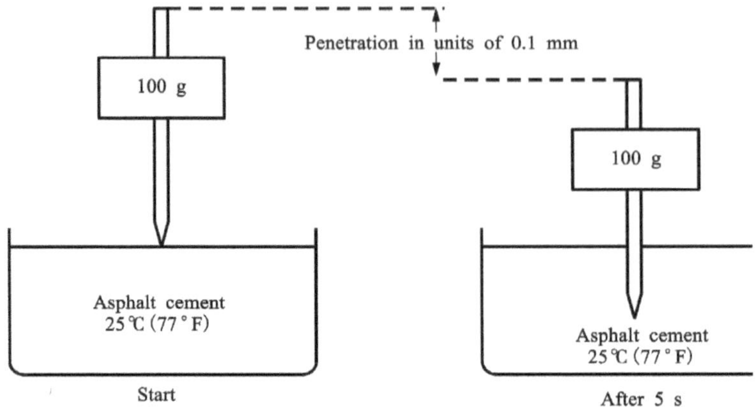

Figure 7.7 Penetration test (ASTM D-5). *(After The Asphalt Handbook, Manual Series No. 4[MS-4], The Asphalt Institute, 1989, p. 39.)*

areas of North America. There is no unique relationship between viscosity and penetration that will be applicable to all asphalt cements.

There are several ways to measure the viscosity of asphalt cements. Capillary flow under either gravity (ASTM D-2170) or vacuum (ASTM D-2171) is the most common. However, with the emergence of more modern asphalt cements classification system, it is necessary to determine the flow characteristics through tests, which provide more physical significance. Superwave requires two kinds of tests: rotating coaxial cylinder viscometer (Chapter 4) and dynamic shear rheometer Asphalt cement classification system (Section 7.1.4). The rotational viscometer (RV) test (ASTM D-4402) is used to ensure that original or "tank" asphalt cement has sufficient fluidity to mix easily with the aggregate to produce asphalt concrete. The more complex dynamic shear rheometer (DSR) test (AASHTO Provisional Standard TPS) measures the rheological behavior of both original and aged (RTFO and/or PAV) materials at medium and high temperatures. With this apparatus, the elastic and the viscous components of the behavior of asphalt cements can be measured, and a more complete picture obtained of their behavior under in-service conditions.

7.1.3.4 Stiffness

Like other viscoelastic material, the deformational behavior of asphalt cements also depends on the rate or duration of loading or load application. At low temperatures and/or short durations of loading, the behavior is mainly elastic: at high temperatures and/or long durations of loading, the viscous component becomes the dominant one. Within the actual conditions encountered in the pavement structure, the performance of asphalt cement is between these limits, which can be described by stiffness or stiffness modulus (which is the creep modulus defined in Chapter 4), defined as

$$S_{t,T} = (\sigma/\varepsilon)_{t,T} \tag{7.3}$$

where $S_{t,\mathrm{T}}$ is the stiffness or stiffness modulus of the material for a particular lime of loading (*t*) and temperature (*T*); σ is stress at time t and temperature (*T*); ε is strain at time/and temperature *T*.

The time-of-loading and temperature dependence of the stiffness of an asphalt material is shown in Fig. 7.8. In a short loading period, the curves become horizontal, indicating the equality of essentially elastic behavior with the stiffness approximately and the elastic modulus value. At intermediate loading times, the material shows viscoelastic behavior, with both elastic and viscous deformations contributing significantly: the stiffness decreases nonlinearly with an increase of loading lime. At very long loading times, the property approaches that of a purely viscous Newtonian fluid, with the stiffness continuing to decrease, but it is nearly linear, and approaching $3\eta/t$, where η is the viscosity indicating purely viscous behavior.

The similarity of the shape of the curves in Fig. 7.9 shows that they could be made to coincide by horizontal movement. In other words, there is a value of temperature that has the same effect on the response of the

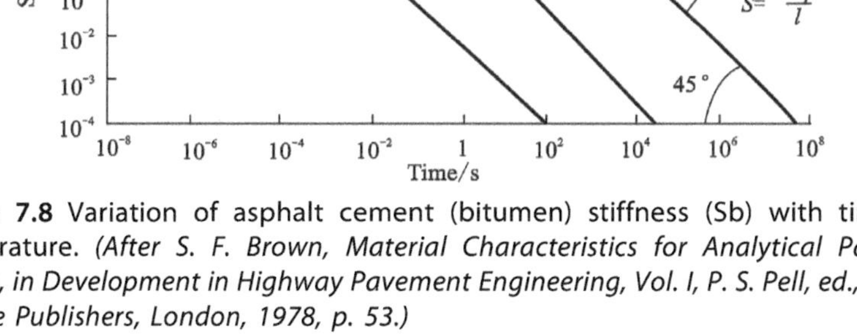

Figure 7.8 Variation of asphalt cement (bitumen) stiffness (Sb) with time and temperature. *(After S. F. Brown, Material Characteristics for Analytical Pavement Design, in Development in Highway Pavement Engineering, Vol. I, P. S. Pell, ed., Applied Science Publishers, London, 1978, p. 53.)*

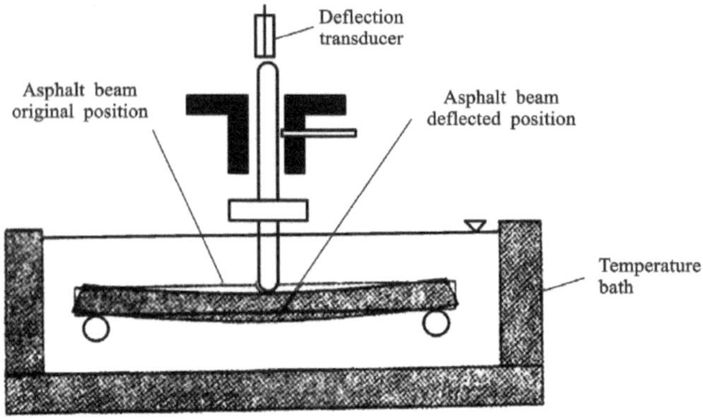

Figure 7.9 Bending beam rheometer test. *(After Superpave-Performance Graded Asphalt Binder Specifications and Testing, Asphalt Institute Superpave Series No. 1[SP-1], Asphalt Institute, Lexington, Ky., 1994, p. 33.)*

material as a certain loading time. This interchangeability of time and temperature is an important aspect of viscoelastic behavior. Therefore, by carrying out such tests at different temperatures, the laboratory test results determined in a limited time range can be extended to a very large time range.

At low temperatures, the stiffness of asphalt cement is too high to permit determination of rheological properties by using the rheological methods described previously. Thus, a method based on a bending beam rheometer (BBR) is used in the Superpave classification system. In this method (AASHTO Provisional Standard TPI), both RTFO and PAV aged samples are subjected to a constant bending load at constant temperature. The resulting deflection over time (creep) is used to determine the stiffness properties of the material.

7.1.3.5 Temperature susceptibility

The change in stiffness of an asphalt cement with temperature is basically related to a change in viscosity, so it is related to the proportion of elastic and viscous components in the deformation behavior of viscoelastic materials. The greater the rate of change of viscosity/consistency with temperature, the higher the sensitivity of asphalt cement to temperature, (i.e., the more liable it is to fracture, or crack, at low temperatures or to deform excessively, rut, at high temperatures; see Fig. 7.10). There does not appear to be any correlation between the parameters used to describe temperature sensitivity and the structure and/or composition of an asphalt cement.

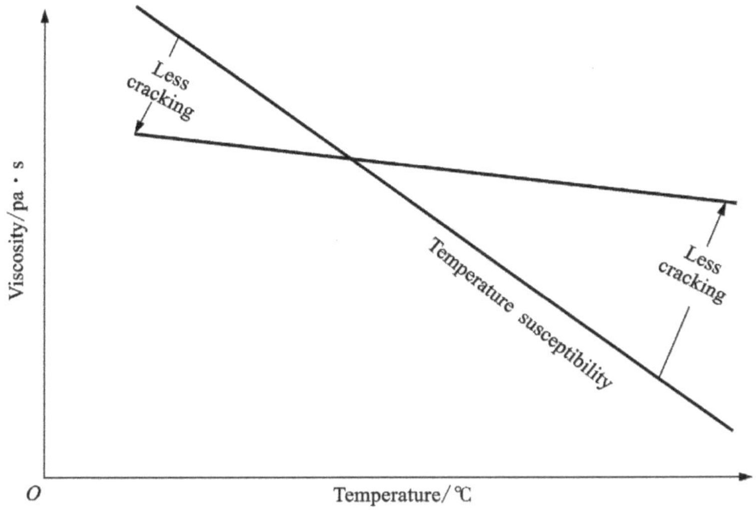

Figure 7.10 Temperature susceptibility of asphalt cements. *(After Ma Cunfei, Synthesis and properties of amphoteric amide asphalt emulsifier, Shandong University, Shandong Province, 2018.)*

Temperature sensitivity can be evaluated using the viscosity–temperature susceptibility (VTS), defined as

$$VTS = \frac{\log \eta_2 - \log \eta_1}{\log T_2 - \log T_1} \quad (7.4)$$

where η_1 is the viscosity at temperature T_1 and η_2 is the viscosity at temperature T_2. This relationship has been used to develop a standard viscosity–temperature chart for asphalt cements (ASTM D-2493), similar to Fig. 7.6. In most VTC calculations, viscosities measured at 60°C (140°F) and 135°C (275°F) are used, but other temperature ranges can be chosen.

7.1.3.6 Tensile properties

The tensile strength of an asphalt cement is significant under conditions of promoting its elastic behavior and brittle fracture. The strength of the material depends on its viscoelastic and physical properties as well as the loading rate and temperature conditions. It has been generally observed that the tensile strength reaches the maximum value of 2–4 MPa, and decreases with the increase of load duration and temperature, as shown in Fig. 7.11. With the decrease of oil film thickness, the strength of asphalt binder

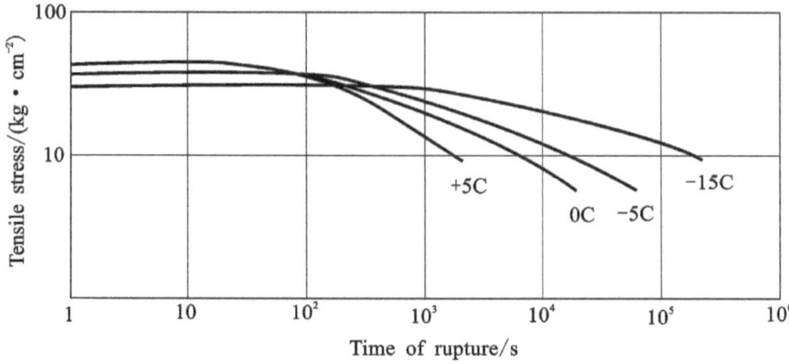

Figure 7.11 Effect of duration of load and temperature on the tensile strength of sol-type asphalt cement. *(After Fang Bowen, Study on adsorption and degradation of tunnel vehicle exhaust based on asphalt pavement carrier, Nanjing University of Forestry, Jiangsu Province, 2016.)*

increases, which is usually attributed to ① a decrease in the ability of the material to flow, ② a decrease in the probability of flaws and/or cavities in the material, and ③ an increase in the effect of molecular orientation at the surface of the material. In addition to strength, it is necessary to evaluate the ductility or elongation of asphalt cement before fracture, especially its sensitivity to service temperature and loading conditions. Therefore, a direct tension test (AASHTO Provisional Standard TP3) is also used in the Superpave classification system.

7.1.4 Characterization of asphalt cement

Many tests can be used to characterize asphalt cement. Some tests are commonly used by highway agencies, while others are used for research. Because the characteristics of asphalt are highly sensitive to temperature, all asphalt tests must be carried out at a specific temperature within very tight tolerances.

7.1.4.1 Performance Grade characterization approach

Prior to SHRP research, the asphalt cement specifications were generally based on measurements of viscosity, penetration, ductility, and softening point temperature. These measurements are not enough to properly describe the viscoelastic and failure characteristics of asphalt cement, which are necessary to link asphalt binder properties to properties of mixture and pavement. The new Performance Grade binder specifications were designed to provide performance-related properties that can be reasonably related to pavement performance in a rational manner.

The characterization of asphalt binder by the Performance Grade tests are carried out at pavement temperatures to indicate the upper, middle, and lower range of service temperatures. The measurements are obtained at temperatures consistent with the distress mechanisms. Therefore, unlike previous specifications, which require testing at a fixed temperature and changing the requirements of different grades of asphalt, the Performance Grade specifications require testing at the critical pavement temperature and determining the criteria for all asphalt grades. Therefore, the Performance Grade philosophy ensures that the asphalt properties meet the specification criteria at the critical pavement temperature.

The binder specifications require three pavement design temperatures: a maximum, an intermediate, and a minimum temperature. Based on the weather information from 7500 weather stations, the algorithm included in SHRP software can be used to generate the maximum and minimum pavement temperatures for a given geographic location in the United States. The maximum pavement design temperature is selected as the average temperature of the highest pavement for seven consecutive days. The minimum pavement design temperature is the minimum pavement temperature expected over the life of the pavement. The maximum and minimum pavement design temperatures plus an average of 4°C are defined as the intermediate pavement design temperature.

Laboratory tests to assess rutting potential use the maximum pavement design temperature, while tests that evaluate fatigue potential use the intermediate pavement design temperature. Thermal-cracking tests use the minimum pavement design temperature plus 10°C (18°F). Increase the minimum pavement design temperature by 10°C to reduce the testing time. Based on the principal of the time—temperature superposition, test results obtained at a higher temperature and shorter load duration are equivalent to tests performed at a lower temperature and longer load duration.

7.1.4.2 Performance Grade binder characterization

The asphalt binder is characterized by several tests in the Performance Grade method. Some of these tests have been used for asphalt testing before, while others are new. The following discussion summarizes the main steps and the necessity of SHRP tests. The test temperature are selected based on the temperature at the design location except the rotational (Brookfield) viscometer, solubility, and flash point tests. The specific test temperatures used for various binders for each test are indicated in the binder specification. There are four tests performed on the neat or tank asphalt: flash point, solubility, rotational viscosity, and dynamic shear rheometer. In order to simulate the effect of aging on the properties of the binder, the short-term and long-term pretreatment of binder was carried out in RTFO and PAV. Before determining their characteristics with respect to fatigue and low-temperature cracking, samples are conditioned with both the RTFO and PAV.

7.1.4.3 Rolling thin-film oven

The RTFO program is used to simulate the short-term aging of asphalt during the production of asphalt concrete. In the RTFO method (ASTM D2872), the asphalt binder is poured into special bottles. The bottles are placed on a shelf in a forced-draft oven, at a temperature of 163°C (325°F) for 75 min. The rack rotates vertically, exposing fresh asphalt continuously. The binder in the rotating bottles is also affected by air jet to accelerate the aging process. The Performance Grade specifications limit the amount of mass loss during RTFO conditioning. Before rutting potential with the dynamic shear rheometer and conditioning with the PAV, RTFO conditioning is used to prepare samples. Under the penetration and viscosity grading methods, the penetration or viscosity of the aged binder is usually tested and the results are compared with those of new asphalt.

7.1.4.4 Pressure-aging vessel

The PAV is composed of a temperature-controlled chamber, and pressure- and temperature-controlling and measuring devices (ASTM D6521). The asphalt binder is first aged, using the RTFO (ASTM D2872). A specified thickness of residue from the RTFO is placed in the PAV pans. The asphalt is then aged for 20 h at the specified aging temperature in a vessel under 2.10 MPa (305 psi) of air pressure. Aging temperature, which is selected according to the grade of the asphalt binder, ranges between 90 and 110°C (194 and 230°F). It is necessary to use a vacuum furnace to remove any bubbles from the sample before testing, since this procedure forces oxygen into the sample.

The PAV is designed to simulate the oxidative aging of asphalt binders during pavement service. After 5–10 years in the field, residue from this process may be used to estimate the physical or chemical properties of an asphalt binder.

7.1.4.5 Flash point

At high temperatures, asphalt may flash or ignite in the presence of open flame or spark. The flash point test is a safety test, which measures the temperature at which the asphalt flashes; asphalt cement can be heated to a temperature below this, without fire hazard. The Cleveland open cup method (ASTM D92) requires the standard brass cup to be partially filled with asphalt cement. The asphalt is then heated at a prescribed rate with a small flame periodically passed over the surface of the cup. The flash point is the temperature of the asphalt when the volatile fumes coming off the sample will sustain a flame for a short period of time. The lowest temperature volatile fumes can maintain the flame for an extended period of time is the ignition point.

7.1.4.6 Rotational viscometer test

The rotational (Brookfield) viscometer test (ASTM D4402) consists of a rotational coaxial cylinder viscometer and a temperature control unit. The test is carried out on unaged binders. The asphalt binder sample is placed in the sample chamber at a temperature of 135°C (275°F); then both are placed in the thermocell. There is a spindle placed in the asphalt sample and it is then rotated at a specified speed. The viscosity is determined by the amount of torque required to rotate the spindle at the specified speed. The spindle size used is based on the measured viscosity. The Performance Grade specification limit is expressed in Pascal seconds (Pa·s) equal to cP multiplied by 1000. The viscosity is recorded as the average of three readings, once per minute, to the nearest 0.1 Pa s. In addition to testing at the temperature required by the specification, additional tests are carried out at higher temperatures to determine the temperature sensitivity relationship between compaction and mixing temperatures required for the mix design process.

7.1.4.7 Dynamic shear rheometer test

The dynamic shear rheometer test system consists of an environmental chamber, two parallel metal plates, a loading device, and a control and data acquisition system (AASHTO T315). The dynamic shear rheometer is used to measure three specifications in the Performance Grading system. For the neat binder and rutting potential tests, the test temperature is equal to the upper temperature of the asphalt binder grade (e.g., a PG 64−22 is tested at 64°C). The sample size is 25 mm in diameter and 1 mm thick for these tests. Before the rutting potential is tested, the sample is placed in the

RTFO for processing. When evaluating fatigue potential, the intermediate temperature is used; 25°C for PG 64–22. The sample size is 8 mm in diameter by 2 mm thick. Before the test, the sample is placed in the RTFO for adjustment, followed by the PAV.

7.2 Liquid asphalts

The viscosity of an asphalt cement can be reduced by combining with low viscosity (i.e., a solvent), thus forming liquid asphalt. After laying the material, the viscosity and hence the viscoelastic nature of the material increase gradually as the solvent evaporates away from the asphalt cement. There are liquid asphalts of two mainly types: cutbacks and emulsions.

Cutback asphalt is a mixture of asphalt cement and hydrocarbon solvent. There are three common types of cutbacks, which vary according to the rate at which the material cures or hardens: slow-curing, medium-curing, and rapid-curing. The rate of cure depends on the rate of solvent evaporation, and therefore on the type of solvent used. The composition of typical cutbacks is shown in Table 7.3.

Emulsified asphalts are produced by breaking asphalt cement into very fine droplets or particles, and dispersing these in a mixture of water and a surface-active emulsifying agent. According to the type of ionic charge which is induced on the dispersed asphalt cement droplets, these liquid asphalts are classified as anionic, cationic, or nonionic. As shown in Fig. 7.12, the charged asphalt droplets are attracted to oppositely charged aggregate particles. Cationic emulsions are most commonly used.

Table 7.3 Composition of cutback asphalts.

Type	Base asphalt cement	Solvent	Solvent concentration (% volume)
Slow-curing (SC)	Low viscosity/high penetration	Diesel fuel	0–50
Medium-curing (MC)	Medium viscosity/medium pene ration	Kerosene	15–45
Rapid-curing (RC)	High viscosity' low penetration	Naptha/ gasoline	15–45

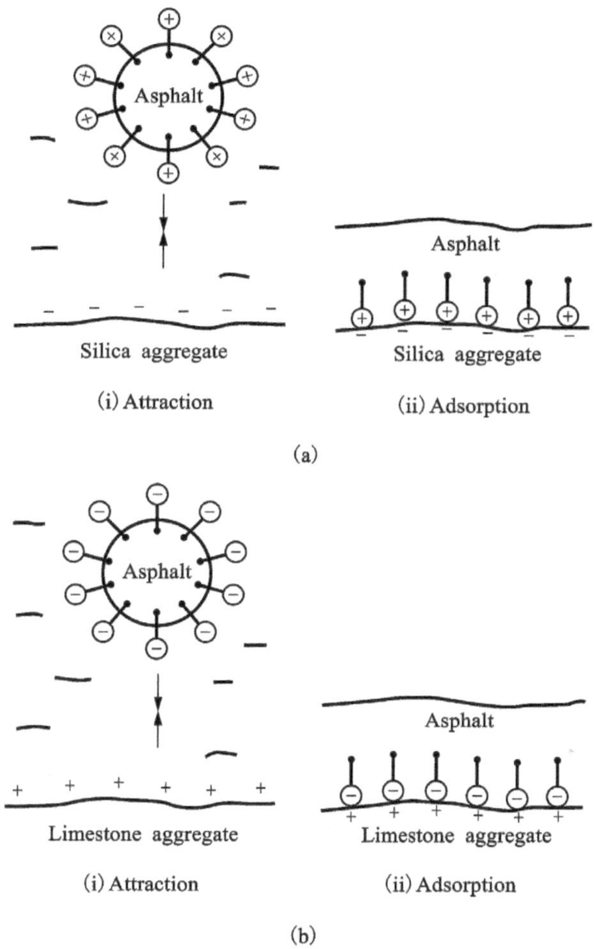

Figure 7.12 The mechanism of asphalt emulsion attraction to aggregate surface: (A) cationic emulsion attracted to negatively charged silica aggregate, (B) anionic emulsion attracted to positively charged limestone aggregate. *(After Chen Huan, Study on the stability of SBS modified asphalt emulsion under long time strong stirring, Central South University, Hunan Province, 2012.)*

An ordinary emulsion generally contains 55%–70% asphalt cement and 0.5%–3.0% emulsifiers. As the asphalt cement content increases, it becomes unstable because the dispersed asphalt cement droplets begin to coalesce at a content of about 80%; the emulsion is then said to be broken. The breaking process is caused by the evaporation of the water phase from the emulsion, with the increase of viscosity.

7.3 Asphalt concrete

7.3.1 Introduction

Asphalt concretes can be classified according to the following characteristics:

① Type of binder. Asphalt cements are the primary type of binder used, but liquid asphalts (cutbacks and emulsions) are usually used in some certain applications.

② When the aggregate is uniformly graded from the maximum size down to the filler to obtain a minimum void content and a high stability, mixtures are classified as dense-graded (well-graded), or if there is little or no fine material so that the void spaces in the compacted aggregate are relatively large, called open-graded.

③ Production methods. There are two mainly types of asphalt concrete: hot-mixed, hot-laid mixtures, which are mixed, placed, and compacted at various high temperatures determined by the viscosity—temperature relationship of the binder; and cold-mixed, cold-laid mixtures, which are mixed, placed, and compacted at, or slightly higher than the ambient temperatures, with liquid asphalt as the binder.

The discussion in the remainder of this chapter, unless noted otherwise, will apply to well-compacted, hot-mixed and hot-laid asphalt concrete mixtures, which contain asphalt cement and a dense-graded aggregate.

7.3.2 Composition and structure

An asphalt concrete is essentially a kind of complex or composite material, which is composed of two components or phases, namely asphalt cement and aggregate, which are physically combined. In order to characterize the structure and hence, the behavior of such a material fully, it is necessary to determine its geometrical and compositional variables. As shown in Fig. 7.13, either the asphalt cement or the aggregate acts as the continuous phase because of the proportionability of an asphalt concrete mixture. In the first case, Fig. 7.13A, since the asphalt cement is the matrix and the aggregate is the filler, the deformational behavior is more viscous or viscoelastic. In the second case, Fig. 7.13B, since the aggregate particles are in intimate point-to-point contact acting as a continuous phase, the behavior will be essentially that of a solid, with the asphalt cement filling the void spaces between the particles and helping to restrain their relative motion. To a considerable extent, the optimum structure is dependent on the desired properties of the mixture. In general, however, it more closely

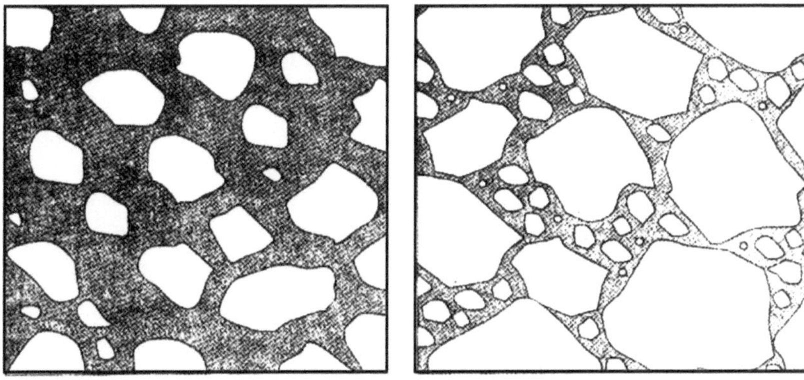

(a) (b)

Figure 7.13 Asphalt concrete mixture with (A) open-graded aggregate and excess asphalt cement, (B) dense-graded aggregate and sufficient asphalt cement. *(After Lun Jubin, Influence of aggregate gradation on performance of asphalt concrete for core wall, Xinjiang Agricultural University, Xinjiang Province, 2017.)*

approximates the second case, which is composed of a dense and interlock structure of aggregate particles with sufficient asphalt cement between the particles and in the interparticle void spaces to bind them together to protect them from harmful environmental impact.

As suggested by the previous discussion, aggregates account for 90% or more, by weight, of asphalt concretes with asphalt cement (and, in practice, air voids) making up the remainder of the mixture. From the point of view of their use in these mixtures, the important characteristics of aggregates were described in detail before.

It is of paramount importance on achieving the desired behavior and properties to densify the asphalt concrete mixtures by compaction. In fact, asphalt concretes are compacted to a certain percentage (typically greater than 90%) of their theoretical maximum density (i.e., the density of a voidless mixture of the asphalt cement and aggregate). In the field, this is usually achieved by the use of self-propelled steel-wheeled or pneumatic-tired rollers, and must be carried out while the asphalt cement flows sufficiently to act as a lubricant and, if necessary, assists the repositioning of the aggregate particles into a denser and more stable configuration. As mentioned earlier, the discussion in this chapter will apply to sufficiently well-compacted asphalt concrete mixtures.

7.3.3 Response to applied loads

The response of an asphalt concrete to external loads is a function of several properties of the material, which are discussed in this section.

7.3.3.1 *Stiffness*

As defined earlier in this chapter (Eq. 7.3), stiffness is expected to depend on both the asphalt cement and the concentration and geometric properties of the aggregate. From an engineering point of view, the stiffness characteristics of these mixtures can be determined by using simple axial loading conditions.

Data obtained under both creep (long time of loading) and dynamic (short time of loading) tests show that the stiffness of an asphalt concrete depends on the volume concentration of the aggregate in the mixture and stiffness of the asphalt cement (Fig. 7.14):

$$S_{con} = S_{ac}\left(1 + \frac{2.5}{n}\frac{C_1}{1 - C}\right)^n \tag{7.5}$$

where S_{con} = the stiffness of the mixture; S_{ac} = the stiffness of the asphalt cement; $n = 0.83 \log[(4 \times 10^5)/S_{ac}]$ and

$$C_v = \frac{\text{volume of compacted aggregate}}{\text{volume of compacted mixture}}$$

This formula is suitable for well-compacted mixtures with voids of about 3% and C_v values between 0.7 and 0.9. Mature stiffness is also affected by aggregate characteristics, even if not implied in Eq. (7.5), such as particle shape, surface texture, and gradation: A more rigid mixture than round, smooth, and open graded aggregates is usually produced by more angular, coarse, and dense graded aggregates.

7.3.3.2 *Stability*

The characteristic of asphalt concrete resisting permanent deformation under traffic loading, is called stability. Marshall test is an unconfined compression test using the special configuration, which can determine the stability of asphalt concrete, shown in Fig. 7.15A. The maximum load is defined as stability in this test, in load units (Fig. 7.15B).

Because stability is mainly due to mechanical or frictional interlock between the aggregate particles, the maximum value is just enough to fill the interparticle voids in a mixture with dense graded aggregate and asphalt

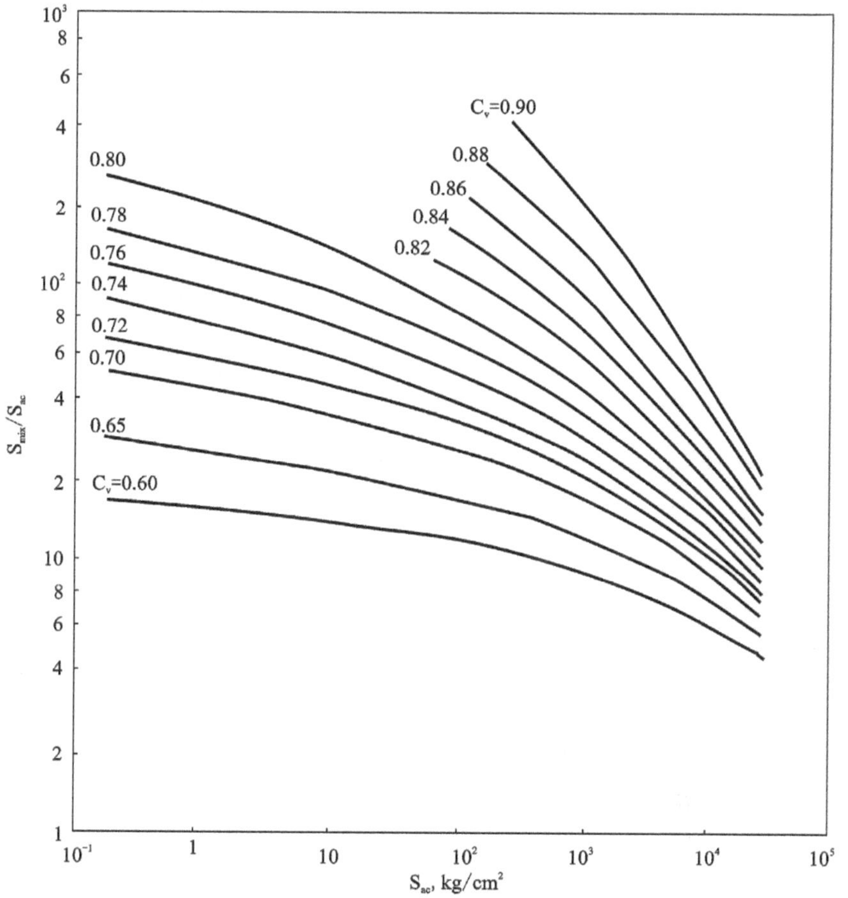

Figure 7.14 Effect of asphalt cement stiffness and aggregate content on the stiffness of asphalt concrete. *(After W. Heukelom and A. J. G. Klomp, Road Design and Dynamic Loading, Proc. Association of Asphalt Paving Technologies, 33, 1964, pp. 92–125.)*

cement content (Fig. 7.16B). If the content of asphalt cement is too high, the asphalt cement will act as a lubricant and reduce the degree of aggregate particle contact, and the stability of the mixture will be reduced (Fig. 7.16A). On the other hand, if the asphalt cement content is less than that required to fill the voids, the stability of the mixture is low because aggregate particles can move with each other under the applied load. Therefore, when the mixture is in a given degree of compaction and contains a given aggregate and asphalt cement, there is an optimal asphalt cement content with the maximum stability, as shown in Fig. 7.16C.

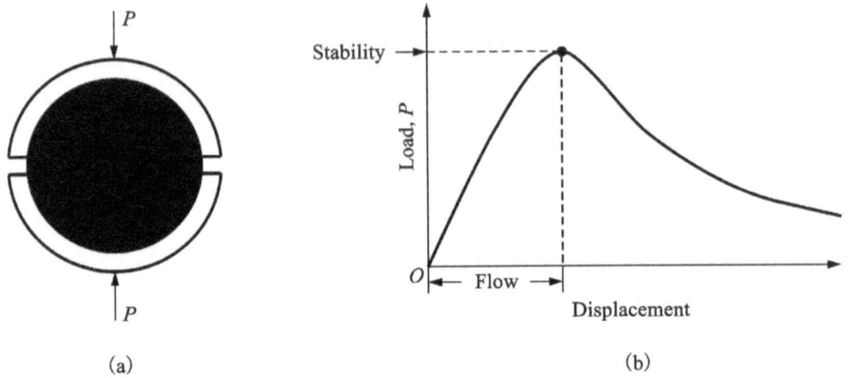

Figure 7.15 Marshall stability test configuration (A) and resulting load-displacement curve (B) showing stability and flow values. *(After Li Hai, Study on performance of modified recycled aggregate asphalt concrete, Nanchang University, Jiangxi Province, 2018.)*

7.3.3.3 Flexibility

Flexibility refers to the fact that asphalt concrete can adapt to the minor and long-term settlement of the base or subgrade layers under the pavement structure without fracture. Generally, the relatively open graded aggregate mixture and high asphalt cement content can improve the flexibility, and these factors lead to the decrease of mixture stability. Therefore, a compromise must be reached between good stability and sufficient flexibility.

7.3.3.4 Fatigue resistance

Fatigue resistance refers to the ability of the mixture to resist fracture or cracking failure under repeated loading conditions related to vehicle traffic. Fatigue resistance of mixtures with higher asphalt cement contents and densely graded aggregates is generally greater than those with low cement contents or open-graded aggregates.

7.3.3.5 Tensile (fracture) strength

The adhesion between the binder and the aggregate, the tensile strength of the binder, the amount of binder in the mixture, and the voids content of the mixture determine the maximum tensile strength or fracture strength of the mixture. As we have seen, the tensile strength of the binder itself is directly related to its viscoelastic and physical properties, and at low temperatures and high loading rates, the tensile strength is approximately constant and decreases with the increase of temperature or the decrease of loading rate.

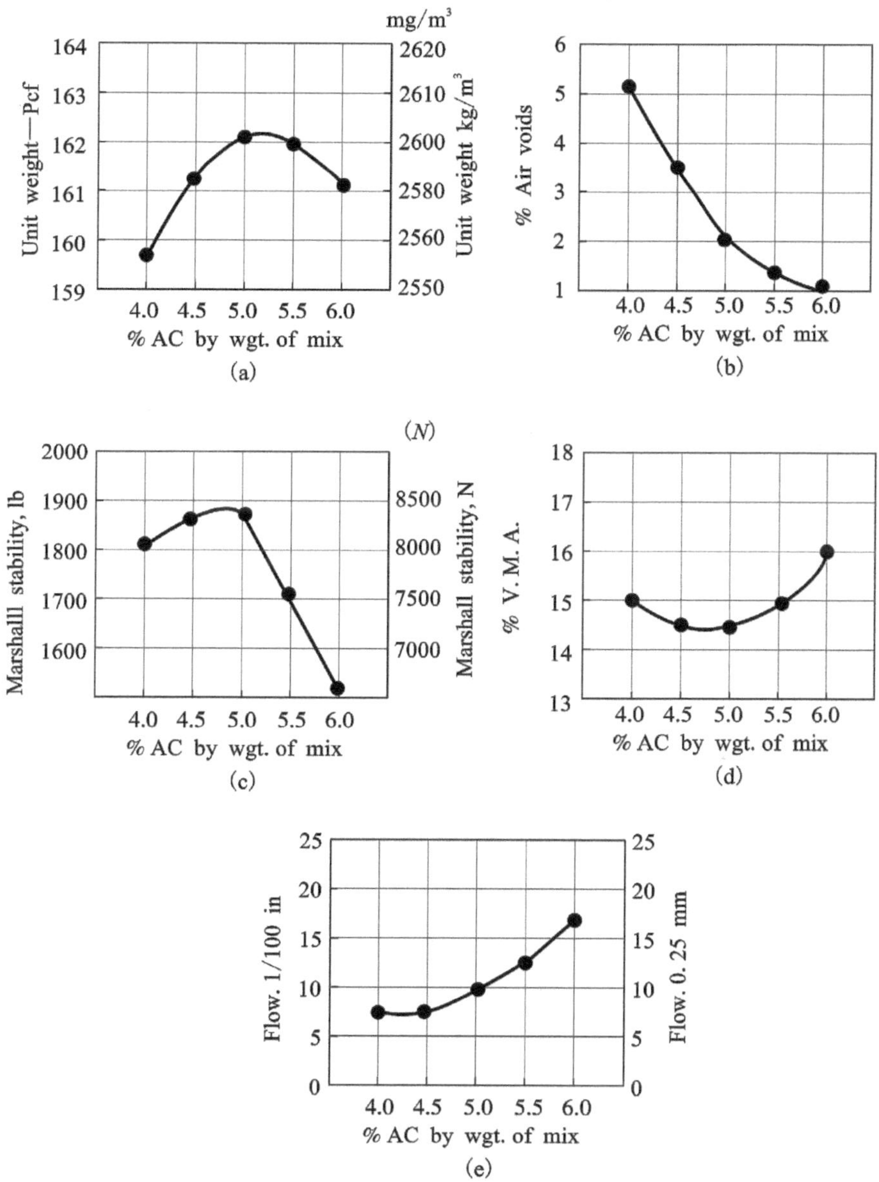

Figure 7.16 Typical Marshall test properly versus asphalt cement (AC) content curves for asphalt concrete unit weight (A), air voids content (B), stability (C), Voids-in-Mineral-Aggregate (V.M.A.) (D), and flow (E). *(After The Asphalt Handbook, Manual Series No. 4 [MS-4], The Asphalt Institute, 1989.)*

The maximum tensile strength of asphalt concretes is 4–10 MPa under rapid loading rates and/or low temperatures, and the strain at failure is $1000*10^{-6}$ to $1200*10^{-6}$. In addition, the limit values of tensile strength and failure strain can be correlated with a corresponding limiting value of stiffness modulus when the stress–strain behavior under these conditions is nearly linear (i.e., elastic behavior).

Under the following conditions, the tensile strength of the mixture is important: (1) when an asphalt concrete pavement layer is subjected to heavy loads at low temperatures, (2) when the subgrade under the layer is relatively weak, and (3) when volume changes in the mixture itself, or high tensile stresses are induced in the layer by temperature changes, volume changes in the subgrade.

7.3.3.6 Permanent deformation

Normal and/or shear stresses caused by the applied load will lead to excessive permanent or plastic deformation in an asphalt concrete. Under traffic loading, this deformation can take several forms like shoving (or pushing), rutting, slippage, or corrugation, known in practice. Rutting, as the most common form, are some channelized depressions in the wheelpath formed by densification (consolidation) and/or lateral movement of the mixture.

The rut tendency of an asphalt concrete is influenced by the following factors; those related to the mixture itself include filler content, shape of the aggregate particles, and asphalt cement content, with higher cement and/or filler contents and rounded coarse and/or fine aggregate particles, which increase rutting tendency. Laboratory studies have shown that these factors also contribute to increase the creep tendency of the mixture, and there is a correlation between rutting test data and creep. The mechanisms of deformation are considered to be similar, so a relatively simple laboratory creep test can be used to assess the rutting potential of the mixture.

7.3.4 Response to moisture

7.3.4.1 Permeability

The difficulty of air, water, and/or water vapor entering or passing through the asphalt concrete depends on the permeability of the asphalt concrete, or the content of voids and the degree of interconnection in the mixture. Low permeability and hence good durability are caused by high asphalt cement contents, good compaction, and dense aggregate gradations.

7.3.4.2 Durability

The durability of an asphalt concrete is the ability to resist disintegration due to weathering and the abrasive action of traffic. Durability is maximized when the aggregate particles are completely covered with asphalt cement and the mixture contains no air voids to facilitate the entry of air (which increase the rate of age hardening of the binder) and water (and the resulting possibility of stripping and/or freezing) or light. Therefore, durability considerations limit the acceptable air content in the compacted paving and placed mixture to less than about 5%.

The durability characteristic of the aggregate is another important ability: It must have sufficient strength and toughness to resist the forces associated with the traffic loading and, if necessary, to resist damage due to freeze-thaw action.

7.3.4.3 Stripping (moisture-induced damage)

The phenomenon, the bond between aggregate particles and asphalt cement binder is broken by the action of water, is called stripping. Some mechanisms cause stripping, the most important being (1) when the water droplets migrate to the asphalt aggregate interface through the asphalt membrane, the emulsion spontaneously forms, and (2) rupture of the asphalt film by interfacial tension at the air—water—asphalt interface. There are two basic categories for adhesion of asphalt to aggregate tests, loss of strength tests (ASTM D-1057) and stripping tests (ASTM D-3625).

In practice, by paying particular attention to construction details, many of the factors contributing to the potential for stopping in asphalt concrete pavement mixtures can be avoided. Examples of such details include the following:

① Not using known hydrophilic aggregates;
② Reducing the void content of the asphalt concrete;
③ Using higher temperatures in the mixing phase to drive off water and to reduce asphalt cement viscosity, thereby facilitating aggregate coating;
④ Precoating aggregates with bitumen or diluents prior to mixing with asphalt cement; and
⑤ Washing aggregates to remove any coatings.

7.3.5 Response to temperature

The proportional influence of the coefficient for each of its components is reflected by the coefficient of thermal expansion and contraction of the mixture reflects. For asphalt cements the coefficient is essentially constant at

6×10^{-4}/°C for temperatures above the glass transition temperature, and at 2×10^{-4}/°C to 4×10^{-4}/°C for temperatures below. For aggregates, the coefficient is much smaller, generally between approximately 5×10^{-6} and 13×10^{-6}/°C. Therefore, the expansion or contraction of an asphalt concrete mixture at temperature reflects the reaction of asphalt cement binder to a great extent.

Thermal cracking (or low-temperature cracking, as it is more commonly called) is a result of the shrinkage or thermal contraction of an asphalt concrete under freezing conditions. In practice, this occurs when such thermal shrinkage is resisted (for example, by friction with an underlying layer in a pavement structure), and the resulting tensile stresses exceed the tensile (fracture) strength of the mixture. It usually starts from the exposed surface of an asphalt concrete layer and extends downward over time. Under low temperatures, asphalt concrete mixtures with high stiffness modulus are most prone to low-temperature shrinkage cracking.

7.3.6 Response to chemicals

Asphalt cements are highly resistant to the action of most alkalis, acids, and salts, as are most aggregates used to produce asphalt concrete mixtures. Therefore, when exposed to such chemicals, we would expect these mixtures to react little. However, asphalt cements readily dissolve in volatile petroleum solvents, which may bring difficulties to asphalt concrete pavements of parking lot for vehicles using petroleum-based fuels or lubricants (overflow of this kind of fluids can lead to disintegration and destruction of the asphalt concrete).

7.3.7 Additives and fillers

It is a common way to add fillers and additives to the asphalt concrete or asphalt cement to improve their performance.

7.3.7.1 Antistripping agents

Stripping is the main durability problem of asphalt concrete paving mixtures in most parts of North America, and antistripping agents are widely used. Many of these agents are inorganic chemicals used alone or in combination with organic acids, or organic compounds (either cationic or anionic in nature or a combination of both). By forming water-resistant complexes, they migrate to the aggregate—asphalt interface and enhance the bonding and its stability between these two components.

Hydrated lime, or calcium hydroxide, is the most commonly used adhesion promoter. It function is to stabilize any line material or clays that may have contaminated the aggregate surface, to provide calcium binding sites for asphalt cement, and act as a filler.

Before mixing with aggregates, it is a common but inefficient method to add antistripping agents which function at the binder—aggregate interface. Only when the viscosity of the asphalt cement is low enough can the agents migrate to the interface. Although it is cost initially and not convenient to use the agent directly on the surface of the aggregate as the pretreatment of aggregate, it may be a more effective method to introduce these agents into the mixture.

7.3.7.2 Asphalt cement modifiers

To retard the oxidative aging or hardening process, antioxidants, like lead diethyldithiocarbamate, are added to asphalt cements. As the temperatures encountered in the mixing process, these materials must be nonvolatile and stable.

Natural and synthetic rubbers have been widely used in asphalt concrete mixtures: styrene-butadiene rubber (SBR) is the material most extensively used. In the form of an emulsion or latex or as finely divided solids, these materials can be added to asphalt cement. In general, small quantities of rubber (i.e., less than 5% by weight) will increase the viscosity of an asphalt cement, reduce its temperature susceptibility, retard its rate of oxidation, and improve its adhesion to the aggregate particles.

7.3.7.3 Recycling agents

Recycling of asphalt materials is becoming a common practice due to the decreasing availability and increasing cost of new aggregate, the increasing cost of asphalt cements, and the reduction of available funds for transportation facilities. To meet desired specification requirements, recycling agents are hydrocarbon products with physical and chemical characteristics selected to restore aged asphalt cements. There are basically two common types: softening agents and rejuvenating agents. Softening agents are usually crude oil fractions with proper viscosity which only decrease the viscosity of the aged asphalt cement. The composition of the rejuvenating agent is similar to that of the asphalt cement which changes during the aging process. It plays a role in restoring the chemical and physical properties of the aged materials.

7.3.7.4 Extenders

As an additive, filler can replace a part of asphalt cement in asphalt concrete, so its usage is reduced. It can also lead to better economics or performance improvements. Elemental sulfur is the most popular and commonly used extender so far. There are two ways to disperse sulfur into asphalt concrete: (1) premixing, that is, adding aggregate after mixing liquid sulfur and hot asphalt cement; or (2) mixing mixer repair, that is, adding liquid sulfur and hot asphalt cement to the mixture respectively and mixing them during the mixing process. Obviously, it is necessary to add or modify the asphalt concrete mixing plant to complete the two processes.

The addition of sulfur can reduce the viscosity of asphalt cement in the normal temperature range, so that it can be used for mixing and laying asphalt concrete mixture. In addition, sulfur acts to increase the stiffness of an asphalt cement, and hence that of an asphalt concrete mixture, at higher service temperatures while having little or no effect at lower service temperatures. Therefore, the temperature sensitivity of binder and concrete mixture will be improved.

7.3.7.5 Fillers

In order to achieve a grading closer to the maximum density and minimum void ratio, filler as a very fine material will be added to the aggregate of asphalt concrete. Limestone powder is the most common filler, and other fine minerals such as stone dust, hydrated lime, and Portland cement are also used.

The size of filler cone makes it play a dual role in asphalt concrete mixture. They fill the voids and provide many points of contact between the larger edges and corners of the aggregate. In addition, the addition of filler usually leads to "hardening" of asphalt cement, such as the increase of viscosity or consistency, increase of tensile strength, and decrease of ductility. The magnitude of these effects depends on the type of filler used and its concentration.

7.3.8 Superpave mix design

The Superpave mix design process consists of (AASHTO R35-09).

① Selection of aggregates

② Selection of binder

③ Evaluation of moisture susceptibility

④ Determination of the design binder content
⑤ Determination of the design aggregate structure

The use of Superpave mixture as aggregate must meet the requirements of source and consistency. Source requirements are determined by the owner/purchasing agency and specified for each stockpile. Source properties may include soundness, Los Angeles abrasion, and deleterious materials. Consistent requirements are part of the national Superpave specification and relate to aggregate mixing. Consistencies similar to the following are necessary properties of the aggregate:

① coarse aggregate angularity measured by the percentage of fractured faces
② sand equivalency (ASTM D2419)
③ flat and elongated particles (ASTM D4791)
④ fine aggregate angularity (AASHTO TP33)

As shown in Table 7.4, the level of traffic and the depth of material used under the pavement determine the specification limits for these characteristics.

Only the well graded aggregate can be used in asphalt concrete. It is suggested that 0.45 power char should be used in Superpave mix design. The grading curve must be between control points. The grading requirements for a 12.5 mm (1/2 in.) nominal-sized Superpave mix are shown in Fig. 7.17.

In order to prevent segregation of the slurry, the aggregates used should be stacked into stockpiles according to their sizes. In the mix design of asphalt concrete mixture, the designer must select the materials that meet the requirements of source, consensus, and gradation.

Table 7.4 Superpave consensus aggregate properties.

Design level	Course aggregate angularity (% min)	Fine aggregate angularity (% min)	Flat and elongated (% max)	Sand equivalency (% min)
Light traffic	55/–	—	—	40
Medium traffic	75/–	40	10	40
Heavy traffic	85/80*	45	10	45

*85% of coarse aggregate has one fracture surface and 80% has two or more fracture surfaces.

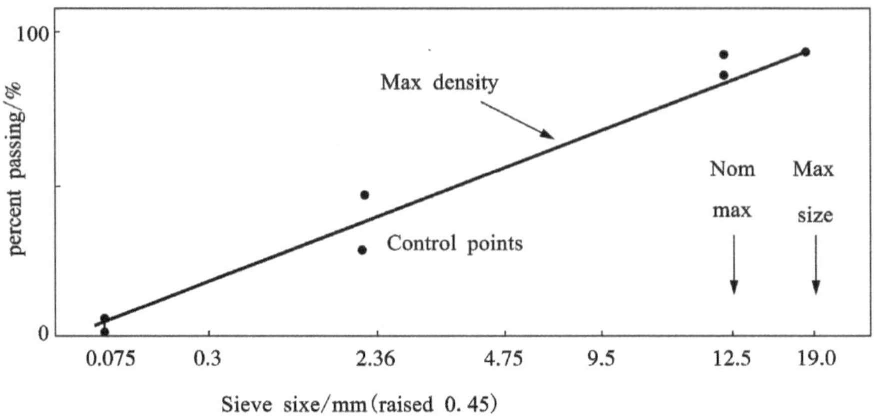

Figure 7.17 Superpave gradation limits for 12.5 mm nominal maximum size. *(After Li Hai, Study on performance of modified recycled aggregate asphalt concrete, Nanchang University, Jiangxi Province, 2018.)*

7.3.8.1 Binder selection

As mentioned earlier, the binder is selected based on the highest and lowest pavement temperatures. In addition to the specification tests, the specific gravity and the viscosity versus temperature relationship for the selected asphalt binder must be measured. The specific gravity is required for the volumetric analysis.

The relationship between viscosity and temperature needs to be determined before selecting the mixing and compacting temperature required for the mixture, which named viscosity—temperature relationship. The Superpave method requires mixing the asphalt and aggregates at a temperature at which the compacting temperature corresponds to a viscosity of 0.280 ± 0.030 Pa s and the viscosity of the asphalt binder is 0.170 ± 0.020 Pa s.

7.3.8.2 Design aggregate structure

The trial specimens are prepared with three different asphalt contents and different aggregate gradations after selecting the appropriate aggregate, binder, and modifiers (if any). The equation for estimating the optimum asphalt content is generally used to determine the specimen prepared for the design of aggregate structure. However, these equations are empirical and the designer is free to estimate the asphalt content. Specimens are compacted using the Superpave gyratory compactor with a constant vertical pressure of 600 kPa (87 psi) and a gyration angle of 1.16° (internal mold

Table 7.5 Number of gyrations at specific design traffic levels (AASHTO R35-09).

	Traffic level (10* ESAL*)			
	<0.3	**0.3–3**	**3–30**	**>30**
N_{ini}	6	7	8	9
N_{des}	50	75	100	125
$N_{\max}$	75	115	160	205
Typical application	Very light traffic volumes where truck traffic is prohibited or very limited. Limited to local traffic or special purpose roads, e.g., access to recreational sites.	Collector roads and access streets and county with medium traffic.	Many two-lane, multilane, divided, and partially or completely controlled access roadways may include medium to highly trafficked city streets, many state routes. US highways and some rural interstates.	Majority of interstate system and special applications such as truck weight stations or truck climbing lanes on two lane roads.

Note:*ESAL is the 18.000-lb equivalent single axle load. It is a design factor used in the design of pavement that considers both traffic volume and loads.

measurement). The number of gyrations used for compaction is determined based on the traffic level, as shown in Table 7.5.

As shown in Table 7.6, the Superpave method identifies three key stages of compaction: the initial, the design, and the maximum. The design compaction level Ndes corresponds to the expected degree of compaction at the completion of the construction process. The maximum compaction nmax corresponds to the ultimate density level of the pavement after several years of traffic.

The initial compaction level N_{ini} was implemented to assist with identifying "tender" mixes. In the construction process, the tender mixture lacks stability, so it will displace under the roller instead of densification.

Samples are compacted with N_{des} gyrations for the initial stage of determining the design aggregate structure. The volumetric properties are

Table 7.6 Superpave mix design criteria.

Design air voids	**4%**					
Dust-to-effective Asphalt1	0.6–1.2					
Tensile strength ratio	80% min					
Minimum VMA (%)	Nominal maximum size (mm)					
	37.5	25	19	12.5	9.5	4.75
	11	12	13	14	15	16
Gmm, and VFA requirements						
Design EASL in millions	Percent maximum theoretical specific gravity			Percent voids filled with Asphalt2,3,4		
	N_{init}	N_{des}	$N_{\max}$			
<0.3	≤ 91.5	96	≤ 98.0	70–80		
0.3–3	≤ 90.5	96	≤ 98.0	65–78		
3–10	≤ 89.0	96	≤ 98.0	65–75		
10–30	≤ 89.0	96	≤ 98.0	65–75		
>30	≤ 89.0	96	≤ 98.0	65–75		

Notes: ① Dust-to-binder ratio range is 0.9–2.0 for 4.75 mm mixes; ② For 9.5 mm nominal maximum aggregate size mixes and design ESAL and three million, VFA range is 73%–76% and for 4.75 mm mixes the range is 75%–78%; ③ For 25 mm nominal maximum aggregate size mixes, the lower limit of the VFA range shall be 67% for design traffic levels <0.3 million ESALs; ④ For 37.5 mm nominal maximum aggregate size mixes, the lower limit of the VFA range shall be 64% for all design traffic levels.

determined by measuring the bulk specific gravity (G_{mb}) of compacted mixtures and the maximum theoretical specific gravity (G_{mm}) of loose mixtures with the same asphalt content and aggregate composition. The volumetric parameters, VTM, VMA, and VFA, are checked and determined according to the criteria in Table 7.6. In the Superpave method, two additional parameters were evaluated: the dust-to-effective binder content ratio and percent of G_{mm} at N_{ini}. The percent of G_{mm} at N_{ini} is determined as

$$\text{Percent } G_{mm,Nini} = \text{Percent } G_{mm,Ndes} \frac{h_{des}}{h_{ini}} \tag{7.6}$$

where h_{des} and h_{ini} are the heights of the specimen at the design and initial number of gyrations, respectively. Note that the percent $G_{mm,Ndes}$ is equal to (100-VTM).

The dust-to-effective binder ratio is the percent of aggregate passing the 0.075 mm (#200) sieve divided by the effective asphalt content, which is calculated as follows:

$$P_{ba} = 100\left(\frac{G_{se} - G_{sb}}{G_{sb}G_{se}}\right)G_b \tag{7.7}$$

$$P_{be} = P_b - \left(\frac{P_{ba}}{100}\right)P_s \tag{7.8}$$

$$D/B = \frac{P_D}{P_{be}} \tag{7.9}$$

where D/B is dust-to-binder ratio; P_{ba} is percent absorbed binder based on the mass of aggregates; G_{sb} is bulk specific gravity of aggregate; P_D is percent dust, or % of aggregate passing the 0.075 mm sieve; P_{be} is percent effective binder content.

The Superpave method requires the determination of the volumetric properties at 4% VTM. Samples prepared to evaluate the design aggregate content do not necessarily meet the binder content required to achieve this void level. Therefore, the results of the volumetric evaluation are "corrected" to 4% air voids as

$$P_{b,est} = P_{bt} - 0.4(4 - VTM_t) \tag{7.10}$$

$$VMA_{est} = VMA_t + C(4 - VTM_t) \tag{7.11}$$

$$C = 0.1 \text{ for } VTM_t < 4.0\%$$

$$C = 0.2 \text{ for } VTM_t \geq 4.0\%$$

$$VTM_{est} = 100\frac{VMA_{est} - 4.0}{VMA_{est}} \tag{7.12}$$

$$\%G_{mmNini.est} = \%G_{mmNini} - (4.0 - VTM) \tag{7.13}$$

$$P_{be,est} = P_{b,est} - \frac{P_s G_b (G_{se} - G_{sb})}{G_{se}G_{sb}} \tag{7.14}$$

$$D/B_{est} = \frac{P_D}{P_{be,est}} \tag{7.15}$$

where $P_{b,est}$ is adjusted estimated binder content; VMA_{est} is adjusted VMA; VMA_t is VMA determined from volumetric analysis; VFA_{est} is adjusted VFA; VTM_t is VTM determined from the volumetric analysis; $P_{be,est}$ is adjusted percent effective binder; D/B_{est} is adjusted dust-to-binder ratio; P_D is percent aggregate passing 0.075 mm sieve.

In Eq. (7.12), the void fraction is assumed to be 4.0%, which is the target void fraction for mix design. The adjusted results of the design aggregate structure evaluation are compared to the Superpave mix—design criteria, as shown in Table 7.6.

The design aggregate blend refers to the mixture whose volume parameters after adjustment meet all standards. If more than one mixture meets all the criteria, the designer can use discretion for the selection of the blend. A pair of samples is compacted at the adjusted binder content for the selected aggregate blend. The average percent of maximum theoretical specific gravity is determined and compared to the design criteria. If successful, the program slurry continues to determine the design binder content, otherwise the design process is restarted and another mixed aggregate is selected.

Exercises

1. What is "aging" (age-hardening) of an asphalt cement? Briefly describe two methods of artificially aging an asphalt cement.
2. Describe the variation in the stiffness or stiffness modulus of an asphalt cement with time-of-loading.
3. What is the relationship between the "time of aging" of an asphalt cement and the viscosity? What is the "asphalt aging index"?
4. What is the effect of a decrease in the thickness of an asphalt cement film on the apparent tensile strength of the material? What are the probable causes of this effect?
5. What is the temperature susceptibility of an asphalt cement? What are the practical consequences of using an asphalt cement with a high temperature susceptibility?
6. What is a "liquid" asphalt? How is an emulsified liquid asphalt produced? What is "breaking" of an emulsified liquid asphalt?
7. What are the differences between traditional viscosity grading systems and the Superpave grading system for asphalt cements, in terms of the relationship between properties and the temperatures at which they are measured?
8. Identify and describe three characteristics that can be used to classify asphalt concrete mixtures.

9. What are the principal factors affecting the bonding of an asphalt cement to an aggregate particle surface?
10. In practice, how is adequate densification of asphalt concrete mixtures achieved?
11. In terms of aggregate gradation and asphalt cement content, what is the optimum structure for an asphalt concrete mixture?
12. What properties of the aggregate used in an asphalt concrete mixture will affect the stiffness of the mixture?
13. What is the principal factor affecting the stability of an asphalt concrete mixture?
14. Under what practical circumstances is the tensile strength of an asphalt concrete mixture important?
15. What is "Cutting" of an asphalt concrete mixture? What factors affect the tendency of an asphalt concrete to rut?
16. What is "Stripping" of an asphalt concrete mixture? Identify three measures that can be taken to reduce the potential for stripping to occur.

CHAPTER 8

Cement-based composites

8.1 Cement asphalt composite

8.1.1 Introduction

The high-speed railway plays a significant role in the transportation system, and greatly promotes the capacity of transportation and economic development. The track system is the most vital structure of high-speed railways because it ensures the safety and stability of the operation of a train. Compared with ballast tracks, ballastless tracks have a more extensive application in high-speed railways due to multifold advantages including a reduction in the structural height and lower maintenance requirements, and, hence, higher availability and increased service life. There are currently five types of ballastless tracks in China: China Railway Track System (CRTS)-I type slab track (Fig. 8.1), CRTS-II type slab track (Fig. 8.2), CRTS-III type slab track, CRTS-I type double-block track, and CRTS-II type double-block track. In CRTS-I and CRTS-II type tracks, cement asphalt mortar (hereinafter abbreviated CA mortar) is adopted as the key material of the layer, whereas in CRTS-III type track, self-compacting concrete is used.

In the structure of the ballastless slab track, CA mortar injected between the concrete roadbed and the track plate is one of the most important parts, which functions as a cushion layer to play important roles such as supporting, adjusting, load transferring, vibration absorption, shock insulation, etc., and improve the durability of the track structure as well as the ride safety and comfort level of the high-speed trains. CA mortar consists of cement, asphalt and/or polymer emulsion, fine aggregate, and chemical admixtures, such as superplasticizer (SP), expansive, air-entrained, defoaming agent, etc. It is an organic-inorganic composite with mechanical properties that are highly sensitive to time and temperature. Two kinds of CA mortar are often used according to the structure of the ballastless track, one with the high elastic modulus (7000−10,000 MPa) and high compressive strength (more than 15.0 MPa at the curing age of 28 days is required) used in Bögl slab track

Civil Engineering Materials
ISBN 978-0-12-822865-4
https://doi.org/10.1016/B978-0-12-822865-4.00008-8

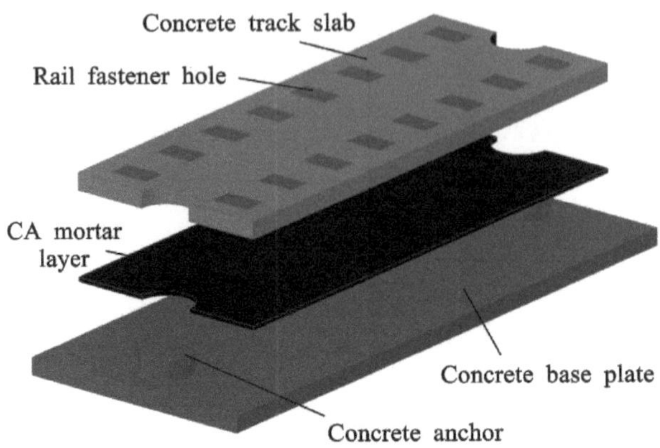

Figure 8.1 Structure of CRTS I type slab track.

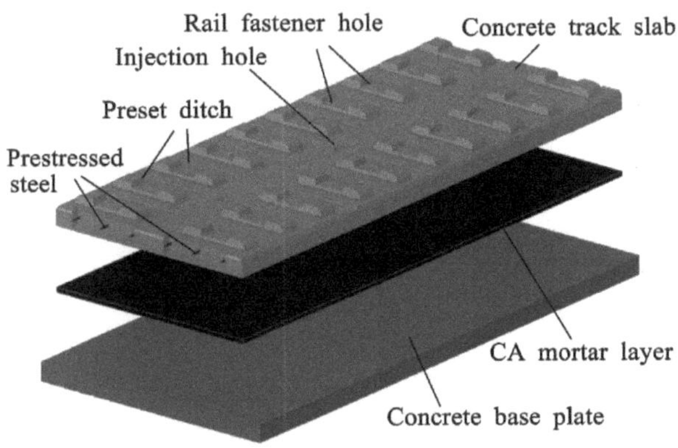

Figure 8.2 Structure of CRTS II type slab track.

firstly developed in Germany, in which anionic asphalt emulsion is utilized, and another one with the low elastic modulus (100–300 MPa) and low compressive strength (only require a compressive strength of over 1.8 MPa at 28 days) previously employed in the shinkansen slab track developed in Japan, in which cationic asphalt emulsion is used as the main raw material. Both CA mortars are now introduced into China and employed in the construction of the high-speed railway, The CRTS I ballastless slab track adopts CA mortar with low elastic modulus, whereas CRTS II ballastless slab track employs CA mortar with high elastic modulus. In China's high-speed railway system, more than 6000 km of lines implemented with CRTS I or II ballastless slab tracks have been built, such as Beijing–Shanghai High-Speed Railway, Beijing–Guangzhou High-Speed Railway, Harbin–Dalian High-speed Railway, and the Shanghai–Kunming High-Speed Railway.

8.1.2 Raw materials

Hardened CA mortar can be treated as a three-phase composite, including asphalt agglomerates, cement hydrates, and sand. Asphalt is a viscoelastic material that is characterized by high viscosity and low elasticity. Besides, the mechanical properties of asphalt are sensitive to service temperature and loading frequency. In contrast, hardened cement paste and sand are generally considered elastic materials that are characterized by high elasticity and low viscosity with mechanical properties that are not sensitive to service temperature and loading frequency. Hence, the mechanical properties of each component and their volume fractions collectively affect the mechanical properties of the composite.

8.1.2.1 Asphalt emulsion

Asphalt emulsion is water-continuous dispersions of fine asphalt droplets, typically 1−10 μm in diameter, manufactured with the aid of a colloid mill. The asphalt emulsion plays a significant role in the properties of CA mortar, the CA mortar shows a different rheological property, mechanical property, and durability compared with ordinary cement mortar. As above mentioned, asphalt is a typical temperature-dependent and time-dependent material. As a result, the mechanical properties of CA mortar are also temperature-dependent and time-dependent. Cationic asphalt emulsion is used to make CRTS I CA mortar with low elastic modulus and anionic asphalt emulsion is utilized to make CRTS II CA mortar with high elastic modulus. The properties of cationic asphalt emulsion and anionic asphalt emulsion are listed in Tables 8.1 and 8.2.

8.1.2.2 Cement

After hardening, the cement paste is mainly responsible for the strength of CA mortar. The Ordinary Portland cement (strengthen grade $\geq$ 42.5) complying with GB175 and sulfate aluminum cement (strengthen grade $\geq$ 42.5) complying with GB20472 can be used.

8.1.2.3 Sand

Sand (fine aggregate) shall be normal, hard, and durable. It shall not contain a harmful quantity of dirt, mud, or other organic substances. Natural sand (from river or mountain) or artificial sand (silica sand, etc.), can be used. The properties of sands for CRTS I CA mortar and CRTS II CA mortar are shown in Tables 8.3 and 8.4, respectively.

Table 8.1 Properties of cationic asphalt emulsion.

Item			Value
1	Appearance		Light brown liquid, uniform, and no mechanical impurities
2	Particle ion change		Cationic
3	Ngualla viscosity (25°C)		5–15
4	Residue on the sieve (1.18 mm)/%		<0.1
5	Storage stability (1d,25°C)/%		<1.0
6	Storage stability (5d,25°C)/%		<5.0
7	Low temperature storage stability (−5°C) (1)		No coarse particles or lumps
8	Mixability/%		<1.0
9	Evaporation residue	Residue content/%	58–63
		Penetration (25°C, 100 g)/0.1 mm	60–120
		Solubility (trichloroethylene)/%	>97
		Ductility (5°C) (2)/cm	≥20
		Ductility (15°C)/cm	≥50

Table 8.2 Properties of anionic asphalt emulsion.

1	Residue on the sieve/%		<0.1
2	Polarity of particles		Anionic
3	Particle size/μm		Particle mean size ≤ 7 Modulus particle size ≤ 5
4	Adaptability		At least 70 mL sample flow out in 20 s
5	Storage stability (1d,25°C)/%		<1.0
6	Storage stability (5d,25°C)/%		<5.0
7	Low temperature storage stability (1) (−5°C)		No coarse particles or lumps
8	Evaporation residue	Residue content/%	≥60
9		Penetration (25°C, 100 g, 5 s)/0.1 mm	40–120
10		Softening point (globe law)/°C	≥42
11		Solubility (trichloroethylene)/%	≥99
12		Ductility (25°C)/cm	≥100
13		Ductility (5°C) (2)/cm	≥20

Table 8.3 Properties of fine aggregate used in CRTS I CA mortar.

	Item	Value
1	Maximum size/mm	2.36
2	Fineness modulus	1.4–1.8
3	Apparent density/g·cm^{-3}	≥2.55
4	Saturated-surface-dried moisture/%	<3.0
5	Clay lump content/%	<1.0
6	Mud content/%	<1.0
7	Organic content	Lighter than standard
8	Chloride content/%	<0.01

Table 8.4 Performance indicators of fine aggregate used in CRTS II CA mortar.

	Item	Index	
		River sand	Machine-made sand
1	Maximum size/mm	1.18	
1	Apparent density/g·cm^{-3}	≥2.55	
2	Moisture content/%	<0.1	
3	Water absorption/%	<3.0	
4	Block content/%	<1.0	—
5	Mud content/%	<2.0	—
6	Stone power content/%	—	<5.0
7	Robustness/%	≤	
8	Organism (colorimetry)	Qualified	
9	Chloride content/%	<0.02	
10	Sulfide and sulfate content (converted into SO_3)/%	≤0.5	
11	Alkali activity (fast mortar stick method)/%	≤0.10	

8.1.2.4 Aluminum powder and expansive agent

After the fresh CA mortar is grouted into the sealed space of the filling layer, a gap between the CA mortar layer and concrete slab may appear due to the combined effects of the loss of free water and chemical shrinkage of Portland cement. Therefore, the aluminum powder complying with GB/T 2085.2 and expansive agent complying with GB 23439 or JCJ 476 are used to compensate for the volume shrinkage of CA mortar. Aluminum powders react with OH- to form H_2 and the formation of H_2, causing the volume expansion of CA mortar. An expansive agent can form expansive products during cement hydration. Aluminum powder and expansive agent can significantly compensate for the early age (0-7d) shrinkage of CA mortar.

8.1.2.5 Other materials

In CRTS I CA mortar, the polymer is also used to further modify the framework then to improve the properties of CA mortar. The properties of the polymer are shown in Table 8.5.

In CRTS I CA mortar, the silicon-based antifoaming agent and the colophony-based air entrainment agent are both used. The role of the antifoaming agent is to remove the big bubbles that emerged from the process of fast mixing CA mortar, and the air entrainment can bring entrained air (closed-cell microbubble approximately less than 200 μm) into concrete. These cell microbubbles can significantly improve the properties of CA mortar, such as the rheological property, the dynamic mechanic's property, and the resistance to frost damage.

In CRTS II CA mortar, the water-reduce admixture complying with GB8076 or JG/T is used, and the silicon-based antifoaming agent is also used.

8.1.3 Mix proportion and mixing

As to concrete, the design method for the mix proportion of concrete is well known, the dosage per 1 m^3 of different raw materials, such as cement, fine aggregate, coarse aggregate, mineral admixtures, and water-reducing agent, so on, can be accurately calculated based on the combined requirements of the strength grade, the workability, and durability of concrete. However, there is not a special method to calculate the mix proportion of CA mortar based on the workability, mechanical properties, and durability, because it is a kind of organic-inorganic material and its compositions and properties are much complicated than concrete. There are several basic regulations for the mix proportion design of CRTS I CA mortar and II type CA mortar.

Table 8.5 Properties of polymer.

	Item	Value
1	Density/$g \cdot cm^{-3}$	1.0 ∓ 0.1
2	Solid content/%	45 ∓ 3
3	Mix ability with cement/%	<1.0

8.1.3.1 CRTS I CA

① The content of cement is limited in the range from 320 to 390 kg/m^3;
② The water/cement ratio (*W/C*) should be controlled less than 0.9;
③ The asphalt (+polymer)/cement ratio *(A/C)* cannot be less than 1.40.

8.1.3.2 CRTS II CA

① The content of cement should be less than 400 kg/m3;
② The water/cement ratio (*W/C*) should be controlled less than 0.58;
③ The asphalt/cement ratio (*A/C*) cannot be less than 0.58.

Based on the above basic regulations, the mix proportions of CA mortar will be calculated, tested, and adjusted to meet the standard values of different properties shown in Tables 8.7 and 8.8. Table 8.6 shows a type mix proportion of CRTS I CA mortar and a type mix proportion of CRTS II CA mortar.

It can be found that two parameters of *W/C* and *A/C* play the key roles in the mix proportion design and the properties. On the other hand, the two CA mortars are both porous materials, 6%—12% air content is required in the fresh CRTS I CA mortar, and 0%—10% air content is required in the fresh CRTS I CA mortar. Therefore, the air volume in the hardened mortar is also a significant parameter to influence the properties of the mortar. The effects of *W/C*, *A/C*, and the air volume on the properties of CA mortar will be discussed in Section 8.1.5.

CA mortar is mixed by a CA mortar plant car. The picture of the CA mortar plant and organization is shown in Fig. 8.3. The mixer is the key component of the plant car, there are three types of the mixer: twin-shaft type, planetary type, and turbine type.

The mixing process of CA mortar is very important because the air content, the homogenization, and the workability of mortar are significantly influenced by the mixing process. The mixing process includes the procedure of feeding raw materials, the mixing speed, and mixing time. Normally, the solid materials (including cement, fine and coarse aggregate, mineral admixtures) are mixed for 30 s at the first step, and then the liquid materials (including water and water reducing agent) are mixed for 2—3 min for concrete. However, in the mixing process of CA mortar, the liquid materials (emulsified asphalt, water, and other liquid agents) are mixed at the first step, and then the solid materials are mixed. The parameters setting for mixing speed and mixing time are also more

Table 8.6 The type mix proportions of CRTS I and CRTS II CA mortar (kg/m^3).

	Asphalt	Polymer	Water	Sand	Cement	Expansive agent	Aluminum powder	Antifoaming agent	Air entrainment agent	Water reducing agent
CRTS I	441	63	63	630	284	32	0.041	0.158	7.9	
CRTS II	255	/	180	945	472.5	52.5	/	0.5	/	2

Table 8.7 Properties of CRTS I CA mortar.

	Item		Value
1	Mixing temperature		5–40°C
2	Apparent density		>1300 kg/m^3
3	Air content		6%–12%
4	Flow time		18–26s
5	Working time		≥30 min
6	Separation degree		<1.0%
7	Bleeding rate		0%
8	Compressive strength	1d	>0.10 MPa
		7d	>0.70 MPa
		28d	>1.80 MPa
9	Elastic modulus (28d)		100 –300 MPa
10	Expansion rate		1.0%–3.0%
11	Freezing resistance		After 300 times freeze–thaw cycle tests, the relative dynamic modulus ≥ 60%, mass loss rate ≤ 5%
12	Weather resistance		No spelling, no cracking, and the relative compressive strength > 70%
13	Fatigue failure[a](100 million times, 12Hz)		≤ 0.10 mm
14	Low temperature cracking resistance (–40°C)[a]		≥ 1.0 mm
15	Low temperature flexure/compression ratio (–40°C)[a]		≥ 0.20
16	Low temperature elastic modulus (–40°C)[a]		100 –300 MPa

[a]This test is carried out when cement asphalt mortar is used in cold areas.

complicated than concrete mixing. Fig. 8.4 shows the two typic mixing processes of CRTS I CA mortar and CRTS II CA mortar. The effects of the mixing speed and mixing time on the properties of fresh mortar will be discussed in Section 8.1.5.

Table 8.8 Properties of CRTS II CA mortar.

	Item		Value
1	Mixing temperature		5–35°C
2	Apparent density		$\geq$ 1800 kg/m^3
3	Air content		4.0% – $\leq$ 10.0%
4	Spread*		D5 $\geq$ 280 mm t280 $\leq$ 16s D30 $\geq$ 280 mm t280 $\leq$ 22s
5	Flow time		80–160s
6	Separation degree (1d)		$\leq$ 2.0%
7	Bleeding ratio		0%
8	Expansion ratio		0%–2.0%
9	Flexural strength	1d 28d	$\geq$0.5 MPa $\geq$3.0 MPa
10	Compressive strength	1d 28d	$\geq$1.0 MPa $\geq$15.0–25 MPa
11	Elastic modulus (28d)		7000–15,000 MPa
12	Freezing resistance (28d)		Mass loss $\leq$ 1000 g/m^2 Relative dynamic modulus $\geq$ 60%
13	Fatigue failure (water saturation, 10000times, 28d)		No crack
14	Weather resistance (500h, 28d)		No crack, no spelling Relative compressive strength $\geq$ 80%
15	Shrinkage ratio		$\leq 800 \times 10^{-6}$

* D5 is the diameter of spread when the mortar is just poured from the mixer; D30 is the diameter of spread after the mortar is poured from the mixer for 30 min; t280 is the time when the diameter of the fresh CA mortar reaches 280 mm.

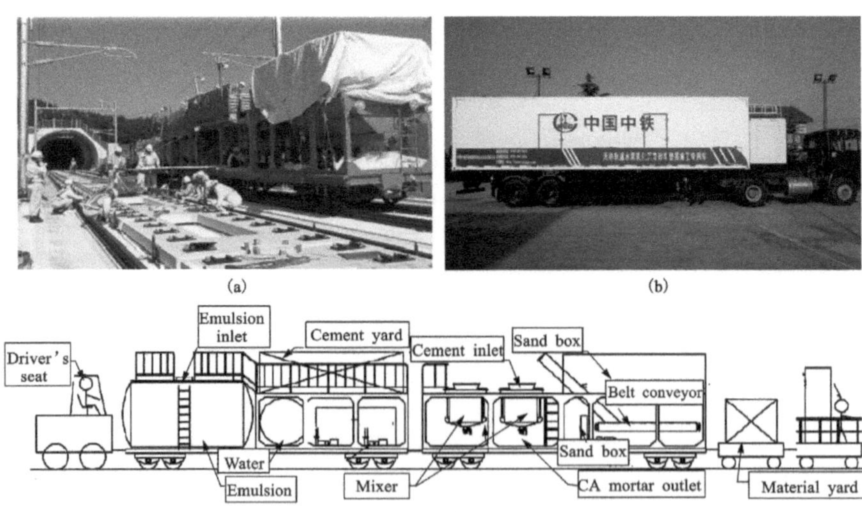

Figure 8.3 CA mortar plant car: (A) car used in Japan; (B) car used in China; (C) the organization of car.

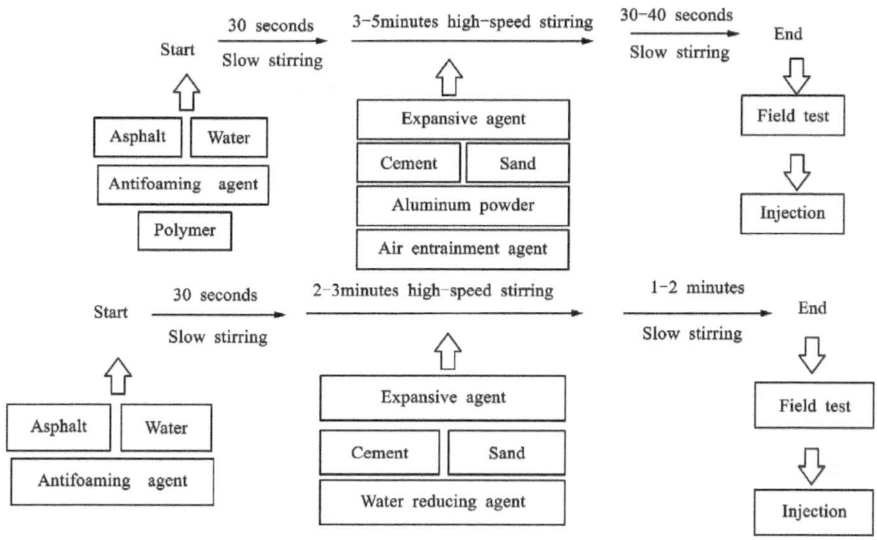

Figure 8.4 The mixing process of CA mortar (Upper: CRTS I CA mortar, bottom: CRTS II CA mortar).

8.1.4 Hardening and structure

When the inorganic cement is mixed with organic emulsified asphalt, the interaction mechanism is complicated. Normally, the emulsifier in the emulsified asphalt becomes unstable because the cement paste is alkaline. This emulsified asphalt undergoes an de-emulsification reaction in the alkaline solution due to Ca(OH)2 formation, damaging the asphalt micelles in emulsified asphalt, which then congregate into asphalt balls and float. On the other hand, the anionic emulsified asphalt has a beneficial effect on cement hydration and possessed a higher pH value; however, acidic cationic emulsified asphalt can neutralize alkaline cement slurry, resulting in a relatively small increase in pH value of CA mortar.

During the hardening process, the further cement hydration absorbs the water in emulsified asphalt, then the spaces between the micelles in the emulsified asphalt are reduced, the electrical attraction and probability of collision between micelles increases, causing the integration of asphalt micelles. Water loss in emulsified asphalt is accompanied by the combined process of cement hydration, progressive de-emulsification reaction of emulsified asphalt, and water evaporation. With the progress of the water loss, the workability of CA mortar is further reduced, the asphalt film is concentrated, the strength is developed and the physical structure of

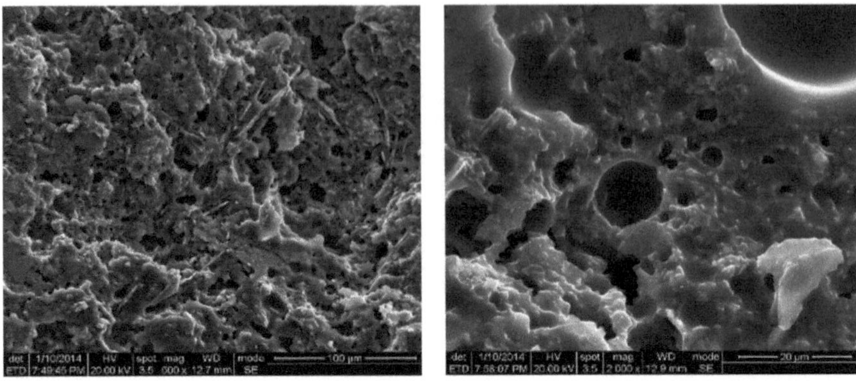

(a) CRTS I CA mortar

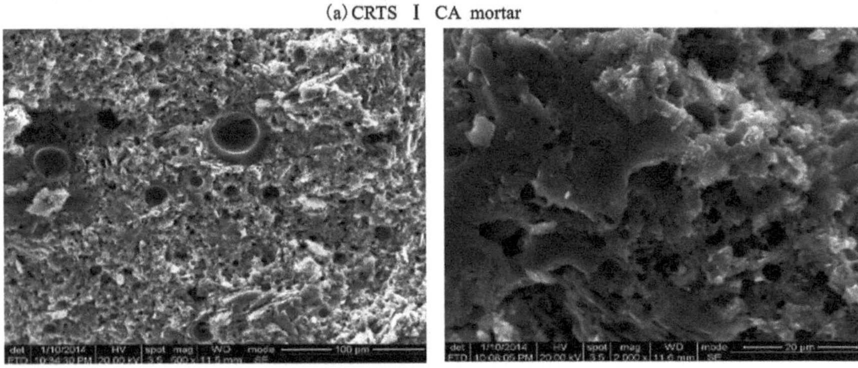

(b) CRTS II CA mortar

Figure 8.5 Microstructure of two types CA mortars.

hardened CA mortar is formed at last. The microstructures of hardened CRTS I CA mortar and CRTS II CA mortar can be explained by the scanning electron microscope image. It can be found in Fig. 8.5:

① The two CA mortars are both porous materials with different frameworks;

② In CRTS I CA mortar, the organic asphalt phase forms the matrix, the inorganic cement hydration products and unhydrated cement particles are dispersed in the matrix as the reinforcing phase, and a large number of pores, where the pore wall is rich in organic—inorganic gel, are observed. There is 6%—12% air content in the fresh mortar;

③ In CRTS II CA mortar, the inorganic cement phase (including the hydration products and unhydrated cement particles) becomes the matrix, the asphalt phase films are embedded in the inorganic cement phase or enriched in the pore wall. The air content of fresh CRTS II CA mortar A is required to be 0%—10%.

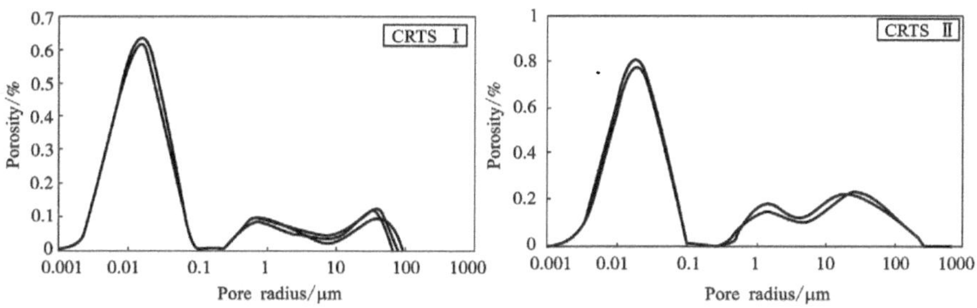

Figure 8.6 Pore size distributions of hardened CA mortars.

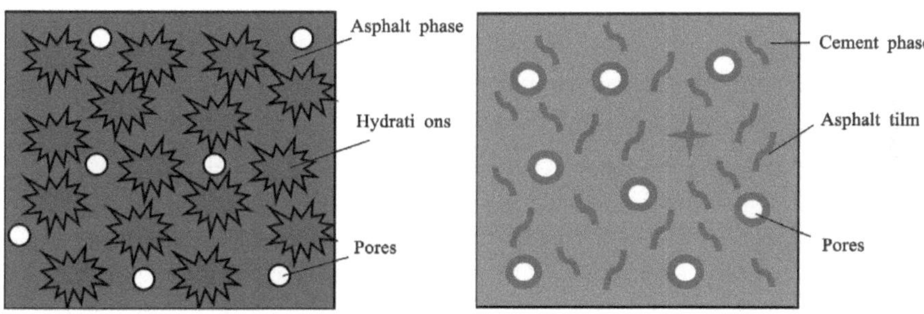

Figure 8.7 Physical structure of CA mortar (Left: CRTS I; Right: CRTS II).

④ Fig. 8.6 shows the pore size distribution curves in the typic CRTS I mortar and CRTS II mortar used in situ construction by nuclear magnetic resonance. It can be seen that the pore size distribution curves in both CRTS I and CRTS II mortars exhibited three main peaks, centered at 20 nm, 1 μm, and 40 μm for CRTS II and 25 nm, 16 μm, and 50 μm for CRTS I mortar, respectively. CRTS I CA mortar has a coarser pore structure than that of CRTS II CA mortar. It is illustrated that the increase in the *W/C* ratio does not change the pore size distribution but enhances the porosity of the most probable pores.

According to the structure characteristics shown in Fig. 8.5, the physical structures of these two types of CA mortars can be sketched as shown in Fig. 8.7.

8.1.5 Properties

To control the site construction quality of the hardened CA mortar layer, the required values of the different properties of the two types of CA mortars are listed in Tables 8.7 and 8.8 complying with TB 10754-2018.

8.1.5.1 Density and air content

The density of CRTS I CA mortar is less than that of CRTS II CA mortar because much more asphalt is used. The densities of these two mortars are both measured by a 1000 mL conical flask. The apparent density (ρ_{app}) is calculated according to Eq. (8.1):

$$\rho_{app} = \frac{M_C}{V} \tag{8.1}$$

where M_C is the mass of mortar in the conical flack; V is the volume of the conical flask.

As to the porous materials, a certain amount of air content is necessary because it can significantly improve the segregation resistance of fresh mortar and the freeze–thawing resistance, the fracture energy and permeability of the hardened mortar, etc., about 10% of air content is required in the two fresh CA mortars. The content and structure of air bubbles are influenced by the asphalt, the antifoaming agent, the air entrainment agent, and the aluminate powder.

The air content of CRTS I CA mortar is calculated by Eq. (8.2):

$$P_{air} = \frac{\rho_{theor} - \rho_{app}}{\rho_{theor}} \tag{8.2}$$

where P_{air} is the air content; ρ_{app} is the apparent density; ρ_{theor} is the theoretical density, which is calculated by

$$\rho_{theor} = \frac{C + A + S + W}{C/\rho_C + A/\rho_A + S/\rho_S + W/\rho_W} \tag{8.3}$$

where C, A, S, and W are the mass (g) of cement, asphalt, sand, and water in a fresh CA mortar, respectively; ρ_C, ρ_A, ρ_S, *and* ρ_W are the density (g/cm^3) of cement, asphalt, sand, and water, respectively.

The air content of CRTS II CA mortar is measured by the mortar air content meter complying with JGJT70 shown in Fig. 8.8.

8.1.5.2 The properties of the fresh CA mortar

In in situ construction, CA mortar is grouted into a narrow space between the concrete roadbed and precast track slabs (shown in Fig. 8.9). Thus, fresh CA mortar must possess adequate flowability and flowability retention for a certain period to ensure construction in succession. Furthermore, a good segregation resistance is required for fresh CA mortar to prevent itself from segregation in and after construction.

Figure 8.8 The mortar press air meter.

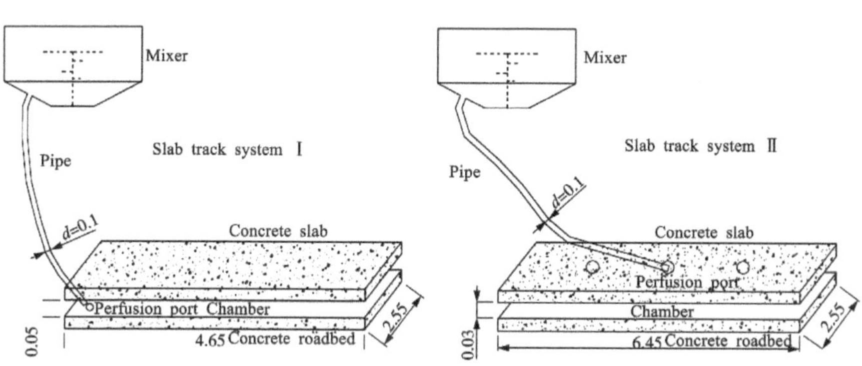

Figure 8.9 Geometry and dimensions (in meters) of the slab track systems and the grouting procedure of CRTS I and CRTS II CA mortars.

The properties of the flow time, working time/spread test, and separation degree are shown in Tables 8.7 and 8.8 that are used to comprehensively evaluate the flowability and uniformity of the two fresh CA mortars.

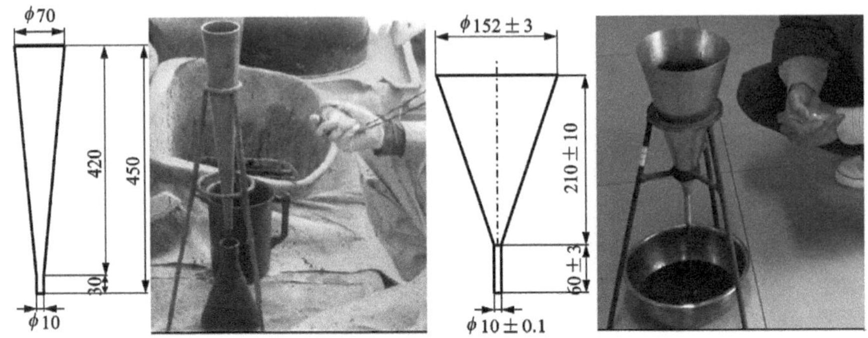

Figure 8.10 Funnels for CRTS I CA mortar (A) and CRTS II CA mortar (B).

8.1.5.2.1 Flow times

The flow times of the two fresh CA mortars are measured by two funnels of different sizes shown in Fig. 8.10.

The test steps are listed as follows:

① The funnel is vertically erected on the bracket;

② The nozzle of the funnel is pressed by a finger, and the funnel is filled with the fresh mortar to level the upper edge;

③ Release the finger and the mortar continuously flows out from the funnel, and the flow time (in seconds, accurate to 0.1s) for the mortar flowing from the beginning to the end is measured by a stopwatch. The flow time is measured for three times, and the mean value is taken.

8.1.5.2.2 Working time

The working time is defined as a period that the fresh mortar is able to retain acceptable flowability similarly during the initial stage. After this period, the mixture will lose flowability and cannot be appropriately cast. The working time of the fresh CRTS I CA mortar is required larger than 30 min s, this means that the flow time can be kept at the level of 18–26s for 30 min after the mortar is poured from the mixer.

The working time of the fresh CRTS II CA mortar is evaluated by the spread test shown in Fig. 8.11.

The test steps are listed as follows:

① The glass pane is horizontally placed on the ground, and a steel cylinder, with inner diameter 50 ± 1 mm and height 190 mm ± 2 mm, is put on the center of the glass pane and filled with fresh CA mortar.

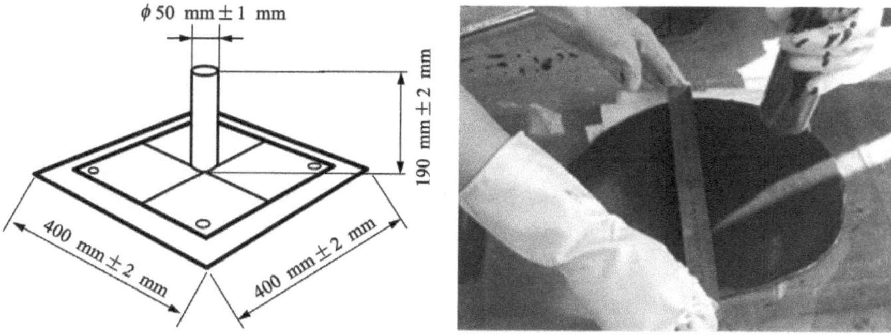

Figure 8.11 Spread test for CRTS II CA mortar.

② The cylinder is lifted rapidly to about 150 mm high.
③ The flow time (T280) when the diameter of the CA mortar reaches 280 mm is recorded. After the tested CA mortar becomes static, the orthogonal diameters (D) of CA mortar in the glass pane are measured, the mean value is taken.
④ The working time is evaluated by the diameters and the flow times of CA mortar measured after the fresh mortar is poured from the mixer for 5 and 30 min, respectively.

8.1.5.2.3 Separation rate and bleeding ratio

The separation rate tests of CRTS I and II CA mortar are similar, the test procedure of separation rate is as follows:
① The fresh CA mortars are poured into three cylinders with a size of 50 mm inner diameter and 50 mm height.
② The CRTS I CA mortar specimens are removed from the molds after 23 h, and then cut into halves.
③ The CRTS II CA mortar specimens are demoulded after 24 h and moved into an environmental box at $20 \pm 2°C$ and $\geq 95\%$ RH for 7 days, then the specimens are placed in the environmental box at $20 \pm 2°C$ and $\geq 60\% \pm 5\%$ RH for 21 days and cut into halves.
④ The density of each half is measured. According to Eq. (8.4), the separation rate was calculated by the difference in density between the top and bottom halves.

$$R_{sep} = \frac{\rho_{bott} - \rho_{top}}{2(\rho_{bott} + \rho_{top})} \times 100\% \tag{8.4}$$

The bleeding ratio is just required for the fresh CRTS I CA mortar, the fresh mortar is poured into a polythene cylinder bag, with diameter 50 mm and height 500 mm at least. Then the bag is hanged for 24 h to observe if there is bleeding on the surface of hardened mortar.

8.1.5.3 The factors influencing the properties of fresh CA mortar

The water to cement ratio (W/C), asphalt to cement ratio (A/C), and sand to cement ratio (S/C) are the main factors influencing the properties of fresh mortar.

8.1.5.3.1 Working time

The working time is important for the fresh CA mortar casting in succession, the factors influencing the working time are shown as follows:

① The cement hydration. For example, compared with Portland cement, the sulfate–aluminum cement in CA mortar can significantly decrease the setting time of the CA mortar.

② The compatibility between asphalt, cement, and sand. If the stability of asphalt in cement paste is poor, it will greatly shorten the working time.

③ The content of asphalt. Normally, the CA mortar with higher A/C has a longer working time.

④ A larger W/C ratio also leads to a longer working time of CA mortar.

8.1.5.3.2 Flowability and uniformity

Concerning CRTS I CA mortar, it can be reasonably imaged that mixtures with higher W/C have a faster time to flow out of the funnel nozzle (shown in Fig. 8.11). The reduction in W/C remarkably prolonged the flow time with the agglomeration phenomenon of CA mortar particles at the nozzle of the steel funnel (Fig. 8.12). However, the sand sinking will

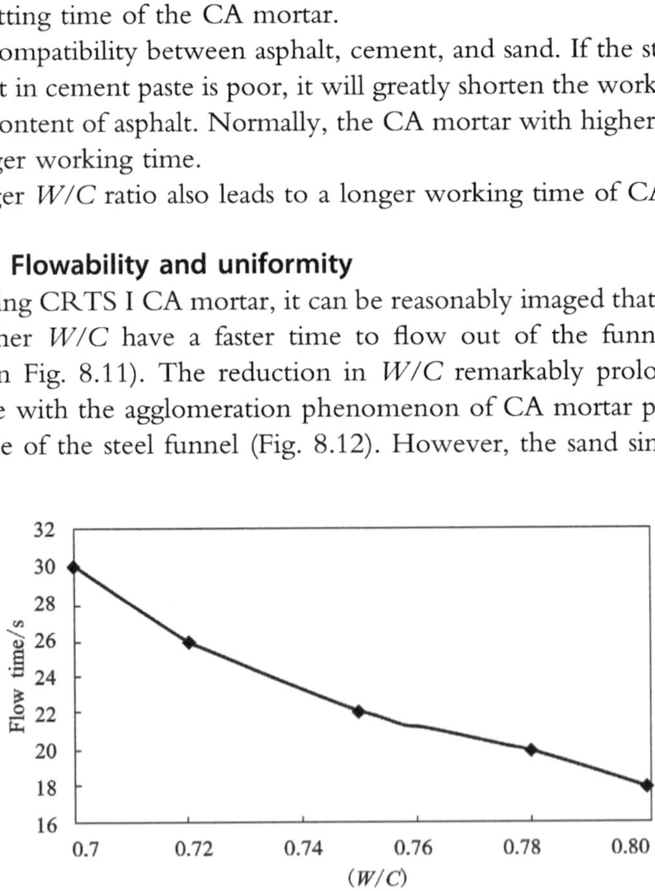

Figure 8.12 Effect of W/C on the flow time of CRTS I CA mortar.

occur in the fresh mortar if W/C is too large, this will lead to a larger separation ratio and larger bleeding ratio.

The workability of fresh CA mortar is strongly affected by A/C and emulsion type. The increase in A/C results in a decrease in workability. The increase in A/C causes congregation of cement grains and asphalt emulsion particles, entrapping water and thus decreasing the flow of CA mortar (Fig. 8.13).

With the increase of the S/C ratio, the flow time of CA mortar increases (shown in Fig. 8.14) due to the decrease of cement paste content. On the other hand, the sand particles size distribution and fineness also influence the flowability of mortar. The mortar with continuous-graded sand has a shorter flow time than the mortar with open-graded sand.

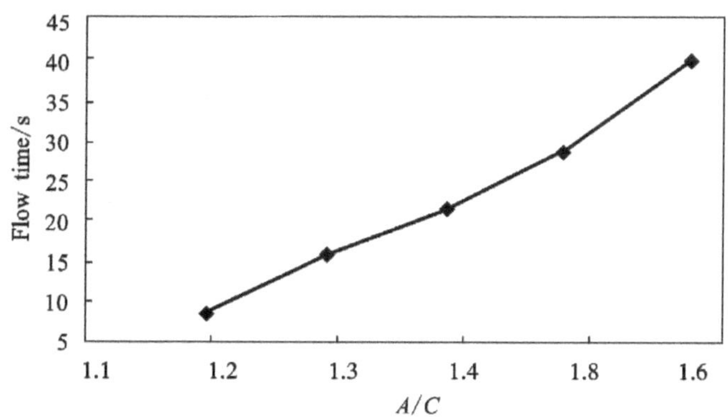

Figure 8.13 Effect of A/C on the flow time of CRTS I CA mortar.

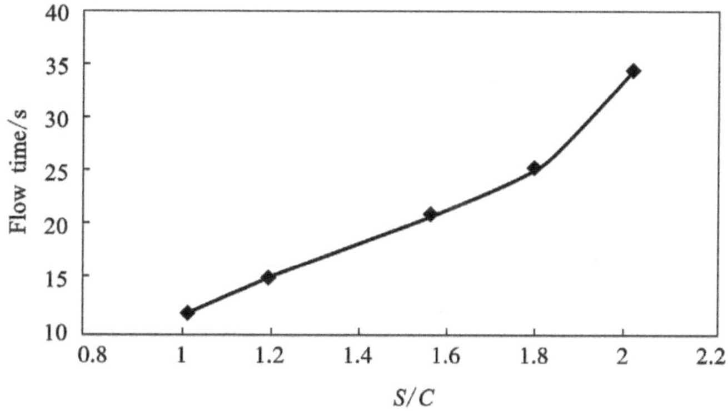

Figure 8.14 Effect of S/C on the flow time of CRTS I CA mortar.

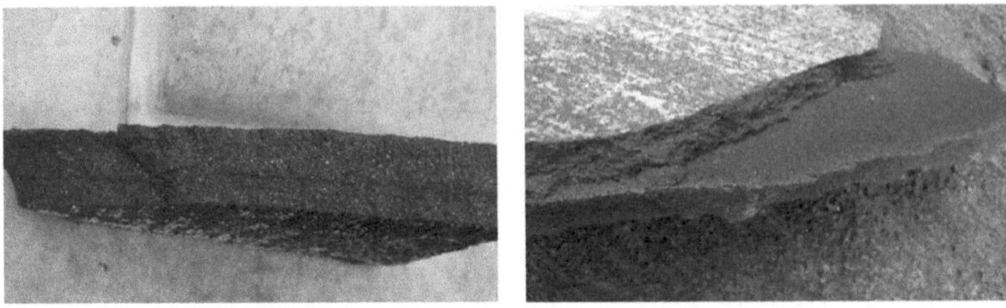

(a) CRTS I CA mortar (a) CRTS II CA mortar

Figure 8.15 the separation of hardened CA mortar due to the bubble gathering.

The flow time decreases with the increasing fineness modulus of sand due to the decrease of the specific surface area. However, the separation ratio may become larger with the increase of S/C and the particle size of sand.

Concerning CRTS II CA mortar, the effects of W/C and A/C on the workability show similar results to CRTS I CA mortar, but CRTS II CA mortar prepared with the anionic asphalt emulsion shows better workability than that prepared with the cationic asphalt emulsion (CRTS I) because the anionic asphalt emulsion disassembles flocculated cement grains more effectively than the cationic emulsion. The S/C does not significantly influence the workability of CRTS II CA mortar.

As shown in Tables 8.7 and 8.8, about 10% of air content is necessary for both CRTS I CA mortar and CRTS II CA mortar. The air content can affect the flowability of mortar, and the uniformity of CA mortar is significantly affected by the bubble stability. Normally, the flow time increases with the increase of the air content. The stability of bubbles is very important, especially the big bubbles, which affects not only the separation ratio but also the construction quality of the hardened CA mortar. As shown in Fig. 8.15, the hardened mortars are separated into several layers due to the bubbles gathering in the surface or middle of the CA mortar layers.

8.1.5.4 Early-age deformation of CA mortar

As shown in Tables 8.7 and 8.8, a 1.0%–3.0% expansion ratio is required for CRTS I CA mortar and 0%–2.0% expansion ratio for CRTS II CA mortar. The expansion ratio of CA mortar is measured by a 250 mL graduated cylinder and a vernier caliper with a precision of 0.02 mm (shown in Fig. 8.16).

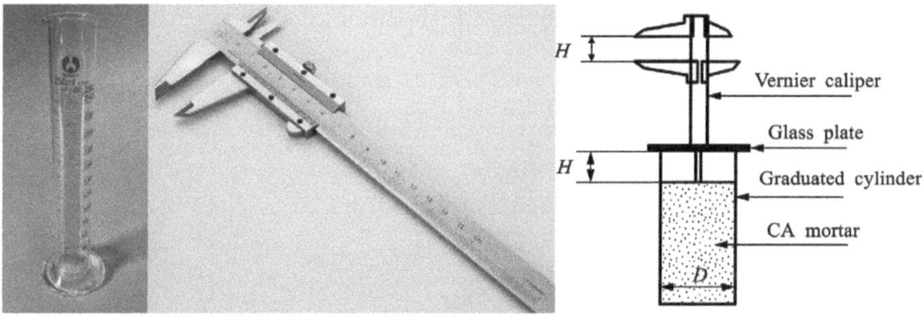

Figure 8.16 CA mortar expansion ratio measurement.

The fresh CA mortar is poured into the graduated cylinder, and the cylinder is covered by a glass plate to prevent the moist exchange. The depth (H) between the glass plate and the surface of the mortar is measured by the vernier caliper as illustrated in Fig. 8.16. The inner diameter (D) of the graduated cylinder, the initial depth after casting (H_0), and the depth after 24 h (H_{24}) are measured, respectively. The expansion rate (R_{ex}) is calculated following Eq. (8.5).

$$R_{exp} = 0.000314 \times (H_0 - H_{24}) \times D^2 \quad (8.5)$$

The aluminum powder (AP) and expansive agent are the main two factors affecting the expansion ratio of CA mortar.

The effect of aluminum powder (AP) on the expansion ratio is shown in Fig. 8.17. The volume of the plain CA mortar (AP content = 0%) sharply decreases and then tends to be stable. With AP addition, CA mortar

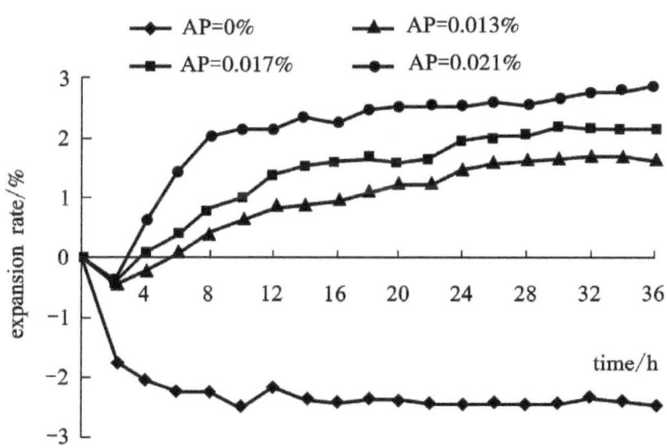

Figure 8.17 Effect of aluminum powder on the expansion ratio of CA mortar at 20°C.

exhibits expected expansive patterns with the increase of AP dosage, but a slight shrinkage is noted in the first 4–7 h. Volume expansion starts slowing down gradually after about 24 h.

On the other hand, the temperature and A/C ratio also play important role in the expansion ratio: higher temperature can lead to a larger evaporation ratio and the expansion ratio of CA is restrained with the increase of A/C ratio.

As abovementioned, the aluminum powders react with OH^- to form H_2, and the formation of H_2 causes the volume expansion of CA mortar. Therefore, H_2 content and formation rate control the development of expansion ratio, and H_2 content and formation rate are controlled by the AP dosage and OH^- concentration in the cement–asphalt paste. So, it is obvious that CA mortar presents larger expansion ratio with the increase of AP dosage, high temperature can accelerate the cement hydration to lead to form high OH^- concentration, and more H_2 formation to generate higher expansion ratio (shown in Fig. 8.18), but larger A/C ratio means less cement is used in CA mortar and H_2 formation is slowed down to cause lower expansion ratio (shown in Fig. 8.19). The shrinkage occurs in the first 4–7 h because little H_2 is formed at the early age of cement hydration.

The addition of expansive agent results in similar expansive to CA mortar with AP as shown in Fig. 8.17.

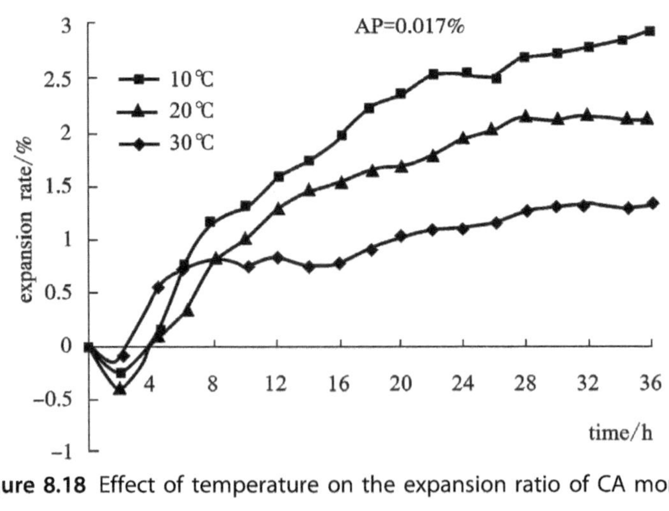

Figure 8.18 Effect of temperature on the expansion ratio of CA mortar.

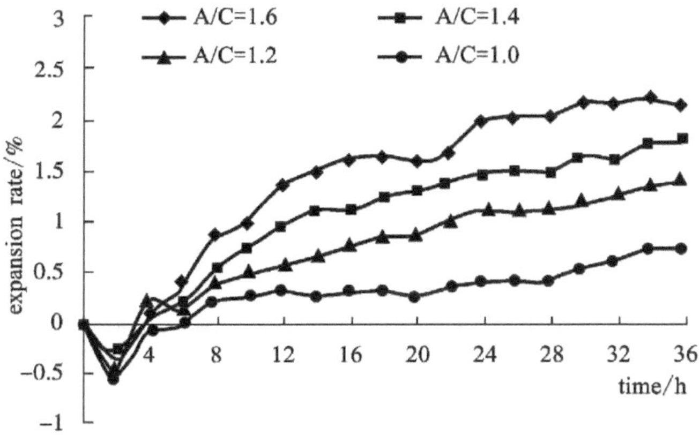

Figure 8.19 Effect of A/C on the expansion ratio of CA mortar.

8.1.5.5 The mechanical properties of the hardened CA mortar

8.1.5.5.1 Compressive strength and flexural strength

As shown in Tables 8.7 and 8.8, the compressive strength is only required for the CRTS I CA mortar, and the compressive strength and flexural strength are both required for the CRTS II CA mortar.

For CRTS I CA mortar, cylinders of Φ50 mm × 50 mm are immediately cast in the molds (shown in Fig. 8.20) after mixing, and specimens (three specimens for one group) are covered with plastic sheets and placed in an environmental room at (20 ± 3) °C and (65 ± 5)% RH. After 23 h curing, the specimens are demoulded and the casting surfaces of the specimens are smoothed by gypsum. Then, the specimens are cured in the

Figure 8.20 Cylinder molds for CRTS I CA mortar.

room for the remaining time. At the age of 1, 3, and 7d, the diameters of the bottom face are measured three times by a vernier caliper with a precision of 0.02 mm, and the averaged value (D_{avg}, 0.02 mm) is calculated, then the maximum force (P, 1N) of the specimen is measured by the uniaxial compression test. The compressive strength (σ, 0.01 MPa) is calculated according to Eq. (8.6),

$$\sigma = \frac{P\,(N)}{D_{avg}(mm^2)} \tag{8.6}$$

And the averaged value of three specimens is calculated.

For CRTS II CA mortar, the fresh mortar is cast in the molds of 40 mm × 40 mm × 160 mm (as the same as the cement mortar according to Chinese National Standards GB/T 17671), and specimens (three specimens for one group) are covered with plastic sheets and placed in a standard curing room at 20 ± 3°C and ≥ 95% RH. The specimens demoulded after 24h are cured first at 20 ± 1°C and 95% relative humidity to 7d and then cured at (20 ± 2)°C and (65 ± 5)% relative humidity for the remaining time. At the ages of 1 and 28d, the compressive strength and flexural strength are tested according to Chinese National Standards GB/T 17671.

The main factors influencing the compressive strength of CRTS I CA mortar include the S/C ratio, A/C ratio, and W/C ratio. The effect of S/C on the compressive strength is shown in Fig. 8.21A, it can be found that a larger S/C ratio causes higher compressive strength. The effect of the A/C ratio on the compressive strength is given in Fig. 8.21B, the results indicate that the compressive strength remarkably decreases with the increase of the A/C ratio.

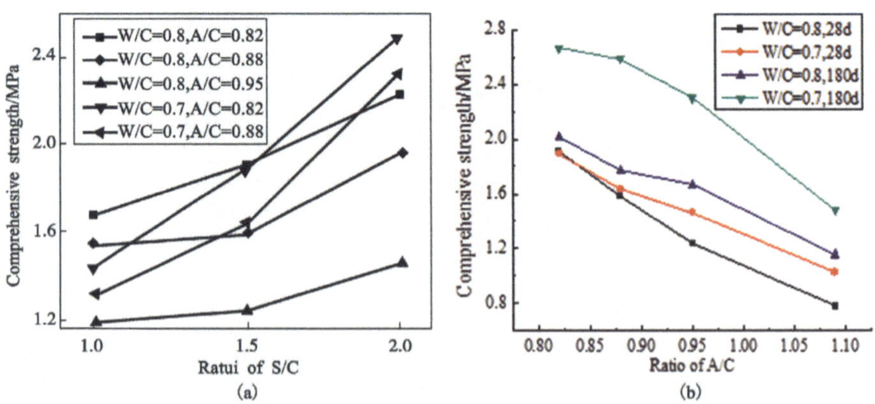

Figure 8.21 Effects of *S/C*, *A/Cs*, and *W/C* ratios on the compressive strength of CRTS I CA mortar (A: Effect of S/C on different CA mortars; B: Effect of A/C on different CA mortars).

It also can be observed in Fig. 8.21B that the compressive strength increases with the decrease of W/C and the increase of curing age.

For CRTS II CA mortar, *A/C* ratio and W/C ratio are the two main factors affecting the compressive strength and flexural strength. Similar to CRTS I CA mortar the compressive strength increases with the decrease of *W/C* ratio and A/C ratio. However, the flexural strength increases with the increase of the A/C ratio and the decrease of *W/C*.

As abovementioned, the CA mortar is an organic–inorganic composite and shows significant viscoelastic characteristics. During the service period of a high-speed railway, the temperature caused by climate change and the loading rate induced by the wheel–rail interaction of a high-speed train are two significant factors in the performance of a CA mortar. It is generally accepted that the compressive behavior of CA mortar is degraded by increasing temperature, and the temperature dependence of the compressive strength is more significant when the A/C ratio increases. The compressive strength increases with a higher loading rate, and the compressive strength of CA mortar with a higher *A/C* ratio increases more with the increasing loading rate.

8.1.5.5.2 Elastic modulus

For CRTs I CA mortar, cylinder samples as same as the compressive strength test are used to measure the elastic modulus. After 28d curing, the diameter of the bottom face and the height of every cylinder are measured three times by a vernier caliper with a precision of 0.02 mm, respectively, and the averaged diameter (D_{avg}, 0.02 mm) and average height (H_{avg}, 0.02 mm) *are* calculated. Then, the stress–strain curves of specimens are measured by an electronic universal testing machine system. Firstly, the specimen is preloaded four times at an unloading rate of 1.0 mm/min to reach 0.1 MPa. Then, the experimental data (σ_{max} and ε_{max}, $\sigma_{3/4}$ and $\varepsilon 3/4$) are collected at the fifth time shown in Fig. 8.22, and the elastic modulus (E_c) is calculated according to Eq. (8.7)

$$Ec = \frac{\left(\sigma_{max} - \sigma_{3/4}\right) H_{avg}}{\varepsilon_{max} - \varepsilon_{3/4}} \tag{8.7}$$

For CRTS II CA mortar, specimens with a size of 100 mm × 100 mm × 300 mm are made to measure the elastic modulus. The same casting and curing process as those in the compressive strength test is carried out. The elastic modulus is tested according to Chinese National Standards GB/T 50081.

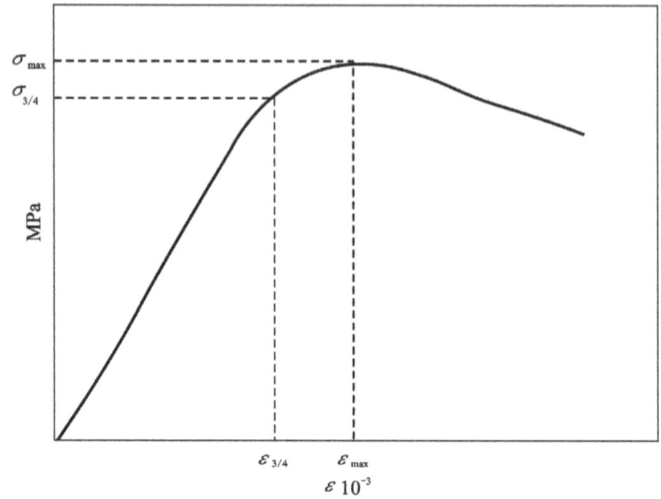

Figure 8.22 Stress–strain curve of CRTS I CA mortar.

Similar to the compressive strength, the elastic modulus is also significantly influenced by the *S/C* ratio, *A/C* ratio, *W/C* ratio, curing age, temperature, and loading rate and shows a similar trend to the compressive strength affected by these factors. For example, the compressive strength and elastic modulus are both degraded by increasing temperature, and the compressive strength and elastic modulus are both found to increase with increasing loading rate, despite the asphalt to cement ratio.

8.1.5.5.3 Freezing and thawing resistance and low-temperature properties at −40°C

When CA mortar is applied in the frigid zone where the practical temperature of the service environment of CA mortar is in the range of −40~−60°C (such as in northeast regions of China), the freezing and thawing resistance of CA mortar is significantly important. The freeze–thaw cycles lead to performance deterioration of CA mortar, particularly, its dynamic mechanical properties due to the strong temperature sensitivity. Meanwhile, the pore structure and fracture shape of CA mortar are changed due to the freeze–thaw cycles, resulting in the damage of microstructure.

The freezing and thawing test of CRTS I CA mortar is referenced as the rapid freezing and thawing method of concrete shown in GBT 50082, the steps of the test are shown as follows:

① The fresh mortar is cast in the prism molds of 100 mm × 100 mm × 400 mm, after 24 h the specimens are demoulded and cured at (20 ± 3)°C and (65 ± 5)% relative humidity for 28d. A reference mortar with better freezing and thawing resistance is also cast.

② After 28d curing, the prism specimens (including the reference specimen) are immersed below the water level for 20 mm at (20 ± 3) °C.

③ After 4d immersion, the specimens are removed out from the water and the surface water is absorbed by a wet cloth, then the mass (*W*) and the fundamental transverse frequency (f0) of the specimens are measured.

④ The specimens are vertically placed into the rubber boxes, and the water is poured into the box to immerse the specimen for 3 mm from the top surface of the specimen to the water level. The reference specimen is placed in the central position of the freezing and thawing machine.

⑤ The quick freezing and thawing test conditions are determined as follows: the freezing temperature is (−18 ± 2) °C, the thawing temperature is (5 ± 2) °C, and a cycle time is 2−4 h.

⑥ After 25, 50, 75, 100, and 125 (or more times) cycles of freezing and thawing, the mass (*W*) and the fundamental transverse frequency (fn) are measured.

⑦ Relative dynamic modulus of elasticity (*RP*) and mass variation (Δ*W*) are measured according to Eqs. (8.8) and (8.9):

$$RP = \frac{f_n^2}{f_0^2} \times 100\% \tag{8.8}$$

$$\Delta W = \frac{W_0 - W_n}{W_0} \times 100\% \tag{8.9}$$

The freezing and thawing test of CRTS II CA mortar is referenced as the one side freezing and thawing method of concrete shown in GBT 50082, the steps of the test are shown as follows:

① Insert a PTFE gasket (Teflon material, the size of the PTFE gasket is about 150 ×150 ×2 mm) in the middle of a cube mold with a side length of 150 mm, the fresh mortar is cast in the mold into two parts.

② The specimens are covered with plastic sheets and placed in a standard curing room at 20 ± 3°C and ≥ 95% RH. The specimens demoulded after 24 h are cured first at 20 ± 1°C and 95% relative humidity to 7d and then cured at (20 ± 2) °C and (65 ± 5)% relative humidity. After 7d curing, cut the specimen to a height of 110 mm (cut off the forming surface). The number of specimens in each group is not less than 5.

③ The initial mass (W_0) of the specimen is weighed. Four sides of the specimen (besides the surface with the PTFE gasket and its corresponding

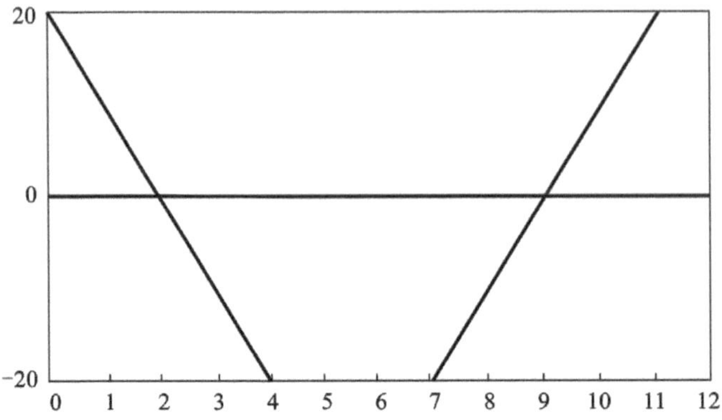

Figure 8.23 Temperature cycle process.

surface) are sealed with silicone rubber or resin at 2–4 d before the stipulated test age.

④ At the stipulated test age, the specimens are placed in the test container with an unsealed face immersed underwater level of (10 ± 1)mm height, the ambient temperature is (20 ± 2)°C. After 7d immersion, the initial value TCS ($0.1\mu_s$) of the ultrasonic transmission time of the specimen is tested.

⑤ The freeze–thaw temperature is between +20 and −20 °C, the temperature cycle process is illustrated in Fig. 8.23.

⑥ After every four cycles of freezing and thawing, the exfoliation and the ultrasonic transmission time of specimens are measured and calculated.

The exfoliation (μ_s) is calculated according to Eq. 8.10, and the per-unit exfoliation (MN 1g/m^2) amount of each specimen is calculated according to Eq. 8.11 after cycles.

$$\mu = \mu_b - \mu_f \tag{8.10}$$

where μ_s is the exfoliation of mortar (0.01g); μ/f is the mass of filter paper (0.01g); μ_b is the total mass of dried filter paper and exfoliations.

$$m_n = \left(\frac{\sum \mu_s}{A} \times 10^6\right) \tag{8.11}$$

where m_n is the total exfoliation amount (g/m^2) of the per-unit test surface of each specimen after n cycles; μ_s is the exfoliation amount of the specimen obtained in each test; A is the area of specimen test surface (mm^2).

The transmission time (t_c, $0.1\mu_s$) of ultrasonic waves in water is calculated according to Eq. (8.12):

$$t_c = l_c / v_c \tag{8.12}$$

where l_c (mm) is the length of ultrasonic transmission in water, v_c is the speed of ultrasonic transmission in the water.

After n cycles of freezing–thawing, the relative dynamic modulus ($R_{u,n}$) of the specimen is calculated according to Eqs. 8.13 and 8.14:

$$R_{u,n} = \tau_n^2 \times 100\% \tag{8.13}$$

$$\tau_n = \frac{t_{cs} - t_c}{t_n - t_c} \tag{8.14}$$

where τ_n is the relative ultrasonic transmission time of specimen; n is the number of freezing and thawing cycles; TCS ($0.1\mu_s$) is the initial value of ultrasonic transmission time; t_n is the ultrasonic transmission time after n cycles.

Take an average of five specimens as the relative dynamic modulus of each group of specimens.

S/C ratio, A/C ratio, and W/C ratio are still the main factors influencing the freezing and thawing resistance of CA mortar, especially the A/C ratio.

It can be observed in Fig. 8.24 that with the rise of A/C of CRTS I CA mortar, RP increases correspondingly at each specified cycle of freezing and thawing, which indicates that increasing asphalt content of CA mortar can mitigate its loss in dynamic modulus of elasticity. On the other hand, the CA mortar made with sulfoaluminate cement shows better freezing and thawing resistance than the mortar with Portland cement.

Concerning CRTS II CA mortar, the similar results of the effect of A/C ratio on freezing and thawing resistance to CRTS I CA mortar can be found in Table 8.9.

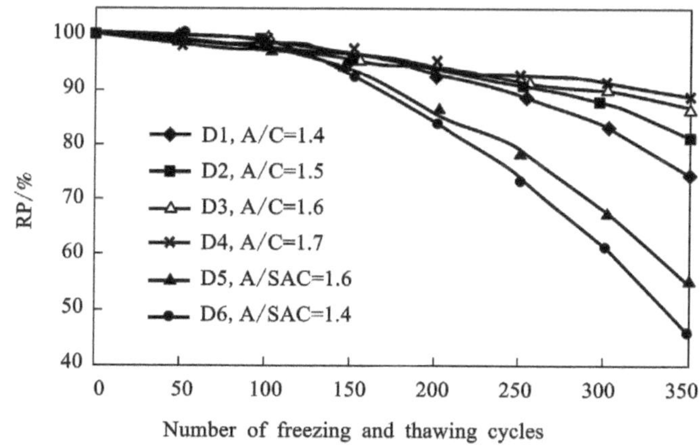

Figure 8.24 Effect of *A/C* ratio and cement on the freezing and thawing resistance of CRTS I CA mortar.

Table 8.9 Effect of *A/C* ratio on the freezing and thawing resistance of CRTS II CA mortar.

A/C ratio	$MN/g \cdot m^{-2}$	$\tau_n/\%$
0.25	856	73
0.30	630	80
0.35	541	85

The freezing and thawing cycle has a great influence on the microstructure of CA mortar. It leads to porosity increasing and obvious structural damage and the static and dynamic mechanical properties of CA mortar will be changed.

Besides the freezing and thawing resistance, the low-temperature properties at −40°C are also required for CRTS I CA mortar, including low-temperature flexure/compression ratio, low-temperature cracking resistance, and low-temperature elastic modulus. These properties are applied to evaluate the brittleness of CA mortar because the asphalt may become a glass state under low temperature, leading to the mortar become hard and brittle.

The tests of low-temperature properties are carried out at (−40 ± 3) °C. The specimens are frozen at (−40 ± 3) °C for 24 h and the clamps in the strength machine for different tests are frozen too (−40 ± 3) °C.

The test of elastic modulus at low temperature is the same as the test at (20 ± 3) °C.

The prism specimens of 40 mm × 40 mm × 100 mm are applied to measure the flexural strength and compressive strength at −40°C. The test is carried out according to Chinese National Standards GB/T 17671.

The cylinder specimens of Φ100 mm × 50 mm are used in the test of low temperature cracking resistance (shown in Fig. 8.25). A steel ball of Φ25 mm is pressed into CA mortar until a crack occurs; the depth of impression is recorded to evaluate the cracking resistance.

Normally, the mortar shows better low-temperature properties with an increase of asphalt amount in the mortar. On the other hand, using organic fiber (such as lignocellulose fiber) and polymer (such as SBR and SBS) can improve the low-temperature properties.

8.1.5.5.4 Antifatigue performance

The antifatigue performance of CA mortar is noteworthy because the high-speed train induces a strong high-frequency vibration on the track structure and CA mortar layer. The deformation of CA mortar due to the high-frequency vibration will affect the safety and comfort of the high-speed

Figure 8.25 Specimen for the low-temperature cracking resistance.

train. The fatigue deterioration of the CA mortar not only reduces its function of vibration and noise reduction but also significantly affects the dynamic performance of the ballastless slab track.

The fatigue tests of CRTS I CA mortar and CRTS II CA are similar and the steps of the test are shown as follows:

① Specimens of Φ50 mm × 50 mm are made for CRTS I CA mortar, and specimens of Φ101.6 × 63.5 mm are prepared for CRTS II CA mortar. The casting and curing process of specimens is the same as it is in the compressive strength test.

② The upper and lower surfaces of the specimen are uniformly coated with hole lubricating oil, and the surface of the pressure transfer plate is uniformly coated with silicone resin, then coated with a layer of graphite flake. The specimens are placed on the central point of the pressure transfer plate.

③ For CRTs I CA mortar, the loading frequency is 12 Hz, the dynamic load is between 19.6 and 196 N. For CRTS II CA mortar, the dynamic load is between 0.025N/mm^2 and 0.35 N/mm^2 and the pulse time is 0.2s and the span between two pulses is 1.5s.

④ After one million times loading for CRTS I CA mortar and after 10,000 times loading for CRTS II CA mortar, the height of each specimen is measured and the cumulative displacement of the specimen is calculated.

CRTS I CA mortar and CRTS II CA mortar exhibit different fatigue failure performances as shown in Fig. 8.26. With the progress of cyclic

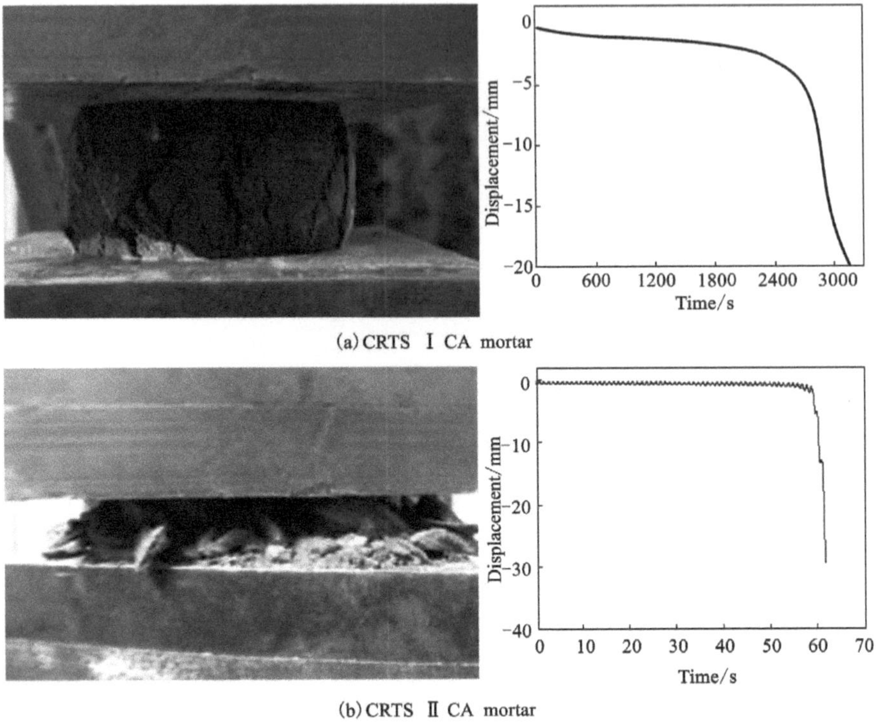

Figure 8.26 the fatigue failure of two CA mortars.

loading, it can be found that the specimen of CRTS I CA mortar has been already flattened with a lot of cracks, and the specimen of CRTS I CA mortar is still kept as a whole, but the specimen of CRTS II CA mortar has been broken into pieces (Fig. 8.26B). The fatigue performance of CRTS II CA mortar is a kind of brittle failure, and CRTS I CA mortar shows good viscoelasticity due to the high content of asphalt.

8.1.6 Construction technology

8.1.6.1 Construction steps

The mixing of CA mortar can be found in Section 8.1.3. The construction process of the fresh CA mortars includes four main steps shown in Fig. 8.27.

① The first step: the prefabricated slab is put on the top of the base plate, and the position of the slab is secured by clamps;

② The second step for CRTS I CA mortar: a polyester nonwoven cloth bag with two mouths placed between the base plate and slab;

③ The second step for CRTS II CA mortar: the space between the base plate and slab is sealed by steel formwork;

(a) CRTS Ⅰ CA mortar

(b) CRTS Ⅱ CA mortar

Figure 8.27 Construction of CA mortars.

④ The third step for CRTS I CA mortar: a funnel is tied with one of the mouths of the bag;

⑤ The third step for CRTS II CA mortar: a PVC pipe is inserted into the middle hole of the slab and a funnel is inserted into the PVC pipe;

⑥ The fourth step for CRTS I CA mortar: the fresh mortar is poured into the bag through the funnel and flow out from another mouth until the fresh mortar fills up the whole bag;
⑦ The fourth step for CRTS II CA mortar: the fresh mortar is poured into the space between slab and base plate through the funnel to fill up the whole the sealed space.

8.1.6.2 Technological test

A technological test (full-scale field test) is carried out before the in situ construction on the bridge and subgrade of CA mortar. During the technological test, the slab will be removed from the mortar layer and the mortar layer will be broken to check if the hardened mortar layer meets the quality requirements shown in Tables 8.10 and 8.11.

Table 8.10 Quality requirement of hardened CRTS I CA mortar layer.

Item	Requirements
Uniformity of top surface of mortar layer	Uniform light black-brown No accumulated pores of millimeter scale No thin mortar layer
Uniformity of mortar layer section	Uniform sections without segregation and sand settling No bubbles accumulated layer in the section No holes of diameter $\geq$1 mm

Table 8.11 Quality requirement of hardened CRTS II CA mortar layer.

Item	Requirements
Uniformity of top surface of mortar layer	Uniform light brown Clear textures corresponding with the bottom surface of slab No flowing trace of water and asphalt accumulated area No paste layer No bubbles accumulated layer The area sum of pores of diameter $\geq$10 mm is less than 0.2% of the whole surface area of mortar layer
Uniformity of mortar layer section	Uniform light brown No visual asphalt particles No bubbles accumulated layer

The bubbles accumulated layer in the middle of the CA mortar layer can be found in Fig. 8.15. Other quality defects of the CA mortar layer are listed in Fig. 8.28.

These quality defects should be resolved through technological tests. Therefore, the technological test is significantly important to control the construction quality of the hardened CA mortar layer. The mix proportion of CA mortar should be adjusted until the quality of hardened CA mortar reaches the requirement shown in Tables 8.9–8.10 and Fig. 8.29, then the

(a) Big holes in the middle of CA mortar layer
(b) Visual asphalt particles
(c) Bubbles accumulated area
(d) Paste layers on the surface of Ⅱ mortar layer
(e) Water flowing trace
(f) thin mortar layer on the surface of Ⅰ mortar layer

Figure 8.28 Quality defects of CA mortar layer.

Figure 8.29 CA mortar layers meeting the requirements in Tables 8.10 and 8.11.

working mix proportion is determined and the value range of the fresh mortar properties is targeted.

On the other hand, reliable equipment and professional constructors, mixing technology, grouting technology, and personnel organization are also important to control the construction quality of CA mortar layer and should also be practiced and determined through the technological test.

8.2 Ultrahigh-performance concrete

8.2.1 Introduction to UHPC

High-performance cement-based materials are always desirable for engineers. During the past decades, researchers have developed some cement-based with extraordinary performances, such as hot-pressed cement, macro-defect-free cement (MDF), densified with small particles (DSPs) and slurry infiltrated fiber concrete (SIFC). Although these materials have excellent performance, they can hardly be used in practice because of complicated forming process and high costs. In 1993, Richard et al. used components with increased fineness and reactivity to develop reactive powder concrete (RPC) via thermal treatment. RPC was characterized by high binder content, very low water-to-cement ratio (*w/c*), use of silica fume, fine quartz powder, and SP and fibers. It generally has high mechanical properties (compressive strength over 150 MPa) and ductility. Furthermore, it exhibited high toughness and excellent durability. One year later, De Larrard introduced the term "ultrahigh-performance concrete" (UHPC). The production of UHPC often uses thermal curing at high temperature, vacuum mixing, and pressure before and during the setting. Although these technical procedures are beneficial to mechanical properties of UHPC, they can result in low production efficiency and high energy consumption, and thus high cost. Therefore, more and more researches have been conducted on well-chosen selection of raw materials, use of common technology, and their influences on the microstructural characteristics, mechanical properties, and durability of UHPC to facilitate the production and applications of UHPC. The prestressed hybrid pedestrian bridge, which completed in 1997 at Sherbrooke in Canada, was the first engineering structure application of UHPC. In 1997 and 1998, UHPCs were cast in beams as the first industrial application. In 2001, the first UHPC road bridge was designed and constructed at Bourg-lès-Valence in France. However, there are numerous challenges before widespread implementation due to lack of commonly accepted standards for testing methods, design guides for engineers, and quality control methods in manufacturing facilities.

The compressive strength of UHPC could range from 120 to 800 MPa, approximately 3–16 times as that of conventional concrete. With incorporation of steel fiber, the ductility and energy absorption of UHPC is typical 300 times greater than that of high-performance concrete (HPC). UHPC is generally defined as concrete with a compressive strength greater than 150 MPa. UHPC typically is made with high-strength steel fibers, fine sand, cement, fly ash, a large volume of SF, and a low amount of water (a w/cm ratio less than 0.20). UHPC is nearly impermeable to carbon dioxide, chlorides, and sulfates. Its superior durability leads to long service life with reduced maintenance. The enhanced abrasion resistance provides extended life for bridge decks and industrial floors, while the enhanced corrosion resistance provides protection to areas with bad or harsh climate conditions. A significant amount of unhydrated cement in the finished product provides a self-healing potential under cracking conditions. Due to ultrahigh compressive strength, structures made with UHPC weigh only one-third or one-half of the corresponding conventional concrete structures under the same load. This weight reduction has benefit in producing more slender structures, increasing useable floor space in high-rise buildings, and reducing overall costs. Elimination of steel reinforcement bars reduces labor costs and provides greater architectural freedom, allowing nearly limitless structural member shapes and forms for architects and designers. Though UHPC possesses many outstanding properties, they have certain weaknesses. High binder content of about 800–1000 kg/m^3 affects not only the production cost but also high heat of hydration, large shrinkage. UHPC is generally costly and cannot replace the conventional concrete in most applications where the conventional mixtures can economically meet the performance criteria.

8.2.2 Raw materials

The raw materials of UHPC include cementitious component, quartz powder, quartz sand, SP, and fibers. Quartz sand usually has a particle size from 150 to 600 μm and is dimensionally the largest granular material. Powder quartz has a particle size distribution ranging from 0.1 to 100 μm and is generally considered as an inert filler. Fibers used in UHPC are often microsteel fibers with length of 13 mm and diameter of 0.2 mm.

8.2.2.1 Cementitious components

Cement content in UHPC is generally about 800–1000 kg/m^3. The high cement content affects not only the production costs but also the heat of hydration, and dimensional stability. The following sections will discuss the common cementitious components and their effects on the performance of UHPC.

① Portland cement

CEM I 52.5, ASTM Types I or II, and GB P.I cements are recommended to be used for UHPC. There are also promising tentative with other types of cement. Given the fact that the cementitious components in UHPC could not hydrate completely and act as fillers, microfine cements with particle size between that of Portland cement and silica fume are also utilized.

② Silica fume (SF)

SF is an essential constituent for UHPC. In early age, the optimum silica fume content was 20%–35% of the cement in UHPC with strength of 200 MPa. Actually, the optimum silica fume content is highly dependent on w/c, and a lower *w/c* required lower silica fume content. Under autoclave curing condition, if silica fume is absent, rapid generation of hydration products would result in formation of porous structure and low compressive strength. However, SF has limited output. Meantime, its variable carbon content can decrease the fluidity of UHPC. High unburnt coal in silica fume may result in dark surface color of concrete, which is an esthetical problem. It has been shown that pulverized granulated blast furnace slag (GGBFS) and pulverized fly ash, metakaolin, limestone powder, and rice husk ash, etc., can be used to replace silica fume in UHPC.

③ Granulated blast furnace slag (GGBFS)

GGBFS has been used as a supplementary cementitious material in concrete for many years. The compressive strength of the RPC with GGBFS of 20%, 40%, and 60% GGBFS could reach 200 MPa. When autoclaving and external pressure are applied, compressive strength could reach up to 400 MPa.

④ Fly ash (FA)

FA is a by-product from coal power station and consists of spherical particles. It is usually combined with GGBFS, SF and/or steel slag powder (SS), etc., as a binary, ternary or quaternary system. The combination of GGBFS and fly ash can enhance flexural strength, and significantly improve the toughness of concrete after all curing regimes. However, it decreases the modulus of elasticity of RPC, especially with more than 30% replacement of cement.

⑤ Metakaolin

Metakaolin is made by calcining of natural clay. Its pozzolanic reactivity is mainly controlled by calcining temperature. The use of metakaolin powder can reduce autogenous shrinkage, refine pore structure, and improve early strength and durability of concrete.

⑥ Limestone powder (LP)

Limestone powder is a nonpozzolanic mineral admixture. The use of LP can improve the fluidity of concrete, and may decrease the compressive strength of concrete. It is well accepted that LP can provide a nucleation site for CSH gel, and thus accelerate the cement hydration.

⑦ Rice husk ash (RHA)

RHA is an agricultural waste obtained by burning rice husk. When husk is completely incinerated in controlled conditions, it contains 90%–96% silica in amorphous form and can be considered as a highly active pozzolans. The average particle size of RHA usually ranges from 5 to 10 lm with a very high specific surface area (even more than 250 m^2/g). This high surface area comes from the angular and porous structure of RHA. The strength of UHPC with RHA has reached up to 210 MPa.

⑧ Nanoparticles

Nanoparticles, including nano-silica (nano-SiO_2), nano-$CaCO_3$, nano-titanium oxide (Nano-TiO_2), and nano-iron (Nano-Fe_2O_3), etc., have been added to UHPC. They have an obviously high surface area to volume ratio compared with other materials of concrete. Meantime, they can act as nuclei for cement phases, further promoting cement hydration due to their high reactivity, and as nano-reinforcement and/or as filler, densifying the microstructure and the ITZ, thereby, leading to a reduced porosity.

8.2.2.2 Aggregates

Generally speaking, aggregates are the cheapest components in concrete. However, not general aggregates but refined quartz sand with diameter less than 0.6 mm was initially used in UHPC to achieve high performance. This greatly increases its cost. In order to reduce the cost of UHPC, recycled glass cullet (RGC), natural sands, and artificial aggregates are used as replacements for refined quartz sand. It is found that the use of RGC decreases flexural strength, compressive strength, and fracture energy by approximately 15%, while natural sand results in similar mechanical properties and ductility as the refined quartz sand did. Also, it is found that iron ore tailings have the potential to be used as fine aggregate of UHPC because they are relatively inert and the particle size is significantly bigger than that of cement. In addition, crushed basalt with diameter less than 8 mm is also used to replace part or all the refined quartz sand to produce UHPC, and obtain comparable compressive strength. This indicates that the key of using aggregates is to eliminate the defects caused by coarse aggregate rather than to remove aggregates.

8.2.2.3 Superplasticizers

Typical *w/c* in UHPC lies in the range of 0.14–0.20. Polycarboxylates (PCs) are the most effective SP for UHPC, and the most important ingredient for UHPC. It was reported that the length of the side chain mainly influenced retardation time and the workability is mainly controlled by the side chain density. The addition procedure of SP also influences the workability of fresh UHPC. Stepwise addition remarkably enhances the dispersing effect of SP, and increases the flowability of UHPC, compared with the one-time direct addition. Compatibility between SP and cementitious materials is often an issue. Due to the high viscosity of UHPC, it is difficult to place UHPC on the jobsite. Thus, SPs which can decrease the viscosity of UHPC are desperately needed.

8.2.2.4 Fibers

The excellent ductability and impact resistance of UHPC are attributed to the use of proper fibers. Table 8.12 shows the typical properties of the commonly used fibers in concrete.

Steel and carbon fibers are commonly used in UHPC matrices because of their superior tensile strengths. The tensile strength of ultrahigh-performance fiber reinforced concrete linearly increases with increased fiber volume ratio from 0% to 5%. The incorporation of steel fibers in UHPC can also decrease the autogenous shrinkage. However, the increased amount of fibers increases the interaction of fibers with each other during mixing, cause balling and decreased the workability of the mixture, and hence the mechanical performances. In order to minimize the amounts of fiber and achieve desired performance, combination of two or three different fibers in UHPC matrix are often applied. The shape and geometry of fibers have significant effect on the properties of UHPC. It is found that the performance of fiber follows order of flexural performance: hooked B (HB) > twisted (T) > long smooth (LS) > hooked A (HA) fiber.

Table 8.12 typical properties of fibers in concrete.

Type	Diameter (μm)	Relative density (g cm^{-3})	Tensile strength (MPa)	Young modulus (GPa)	Elongation (%)
Asbestos	≤ 0.5	2.75	500–980	84–140	03–0.6
Acrylic	5–17	1.18	800–950	16–23	9–11
Polyester	10–80	1.38	735–1200	6–18	11–15
Polyethylene	800 –1000	0.96	200–300	5–6	3–4
Polypropylene	20–70	0.91	300–770	3.5–11	15–25
Nylon	23	1.16	900–960	42–52	18–20
PVA	1.30	1.30	600–2500	5–50	6–17
Aramid	10–12	1.44	2500–3100	60–120	2.1–4.5
Rock wool	2.7	2.7	490–770	70–119	0.6
Glass	10–16	2.74	1400–2500	70–80	2.5–3.5
Wood	25–400	1.40	50–1000	15–40	–
Carbon	7–18	1.75	1800–4000	200–480	1.2–1.6
Steel	250 –1000	7.80	280–2800	200–250	05–4.0

8.2.3 Mix design

8.2.3.1 Some theoretical principles

8.2.3.1.1 Optimize pore structure

Pore structure plays an important role in determining the strength of hardened cement-based materials. The pore size distribution, shape, and position of pores collectively affect the strength of UHPC. The lower the porosity is, the higher the compressive strength is. The goal of mix design of UHPC is to optimize its pore structure. The porosity and pore size distribution can be effectively improved by the use of SP, incorporation of very fine reactive mineral admixtures, and close packing of raw materials, and thus improve the performance of concrete.

8.2.3.1.2 Improvement in microstructure

UHPC has a very dense and uniform microstructure due to the following reasons: ① close packing of solid particles; ② hydration and pozzolanic reactions in cementitious materials; ③improvement of the interfacial transition zone between aggregates and bulk matrix. The internal microstructure of UHPC is mainly comprised of unhydrated cement clinker particles, quartz sand, and hydration products, such as CSH. Low w/b results in low porosity that restricts the space available for the growth of calcium hydroxide (CH) crystals. Use of elevated temperature curing accelerates the hydration of cement, and promotes the pozzolanic effects of mineral admixtures. It is reported that there are almost no pores observed in the nanoporous range with pore sizes up to 100 nm, and no significant CH was detected by X-ray diffraction in UHPC. The CSH gel in UHPC has high density characterized by intrinsic higher stiffness and hardness than those of low density CSH that dominates in conventional concrete. Owing to the low w/b and pozzolanic reactions between CH and reactive mineral admixture, most of the CH crystals are consumed and converts them to CSH. The ITZ in UHPC seems as dense as the matrix, since the aggregate of UHPC is very small. The homogenous structure is important for the excellence performance of UHPC.

8.2.3.1.3 Increase in toughness

Conventional concrete is a typical quasi-brittle material with low tensile strength, strain capacity, and fracture toughness. High toughness is one of the most important characteristics of UHPC. This is achieved by incorporating fibers into concrete, in this aspect, the initiation, propagation, or coalescence of cracks can be significantly delayed. When a load acts on fiber reinforced concrete, the fibers do not sustain the load directly, but the matrix does. The load is transferred to the fibers through the interface between the fibers and the matrix.

8.2.3.2 Mix design

Mix design is a selection of raw materials in optimum proportions to give concrete with required properties in fresh and hardened states for some special applications. The design of UHPC aims to achieve a densely compacted cementitious matrix with good workability and strength. Currently, there is no universally accepted design method for UHPC, and the mixture proportions of UHPC are given directly without any design procedures or explanation in most publications.

Modified Andreasen and Andersen model is often used as a target function for the optimization of the composition of mixture of granular materials.

$$P(D_i) = \frac{D_i^q - D_{\min}^q}{D_{\max}^q - D_{\min}^q} \tag{8.15}$$

where $P(D_i)$ is the fraction of the total solids being smaller than size D_i, D_i is the particle size (lm), $D_{\max}$ and $D_{\min}$ are the maximum and minimum particle sizes (μm), respectively, and q is the distribution modulus. It is recommended using a q value of 0.23 considering that a large amount of fine particles in UHPC.

In addition, software Design-Expert 8 is also used to design UHPC by optimizing the packing of particles.

8.2.4 Preparation and curing

The use of external or internal energy, such as vacuum mixing, and pressure before and during setting are effective ways to reduce the volume of pores and obtain desired mechanical property of UHPC. It is found that the compressive strength of UHPC increased with decreased mixing pressure and the lowered pressure caused a decrease in air content of concrete. Also, it is found that presetting pressure significantly improves the strength, toughness, and young's modulus of RPC. When the samples are presetting pressurized at 25 MPa, their compressive strengths and young's modulus reach 420.31 MPa and 84 GPa, respectively, while for those without presetting pressurized are 206.36 MPa and 58 GPa. However, these sophisticated technologies are too costly to be widely used in practice.

Therefore, normal preparation and curing methods are desired for the widely applicable UHPC. Normally, the ingredients of UHPC are mixed, cast, and vibrated like conventional concrete. Dry powder components are first premixed, water and SP are then added. Followed by adding fibers when mortar matrix shows appropriate flowability for good workability and adequate viscosity.

8.2.5 Properties

8.2.5.1 Workability/rheology

For UHPC with relatively low flowability, which needs to be consolidated to secure high mechanical properties, a sufficient yield stress is needed to resist fiber and aggregate segregation. However, the UHPC can be proportioned to be self-consolidating to facilitate placement. The self-consolidating property is also necessary in restrained areas where consolidation may not be practical. Given the low yield stress of self-consolidating UHPC, the plastic viscosity should be adjusted to avoid fiber segregation and enhance fiber dispersion and orientation.

Fresh UHPC can exhibit high thixotropy and viscosity due to the low water-to-binder ratio (w/b) and high binder contents. The thixotropy of cementitious suspensions originates in a change of the microstructure after removing shear stress. The change of microstructure over rest time includes physical structuration and chemical rigidification. The physical structuration is the formation of network due to flocculation of fine binder particles. The cohesiveness of the network is affected by the attractive surface interactions, including van der Waals force, electrostatic repulsion, and steric hindrance due to adsorbed polymers. The chemical rigidification corresponds to the formation of bridges among adjacent particles due to the nucleation of hydration products onto binder particle surfaces. The reversible bridge is formed by the early-age hydration products, such as metastable C—S—H gel. As the extension of rest time, the increase of stable hydration products can transform the reversible bridges to the irreversible bond. The thixotropy is generally determined by the reversible process of physical structuration and early-age chemical rigidification.

The thixotropy of UHPC is influenced by the binder type, w/b, chemical admixtures, and use of nanomaterials. In general, the use of higher binder content and addition of supplementary cementitious materials can increase thixotropy. The viscosity modifying admixture (VMA) is effective in enhancing thixotropy of cement-based materials. The propylene carbonate—based thixotropy-enhancing agent (TEA) is shown to be effective to increase thixotropy of cement-based materials.

UHPC has significantly higher viscosity than conventional concrete, which can lead to difficulties in placement operations. Low viscosity of UHPC is desired to facilitate mixing, placement, consolidation, and finishing operations. However, higher levels of viscosity of self-consolidating UHPC mixtures can lead to better fiber distribution and alignment that can enhance mechanical properties of the UHPC. Adequate knowledge of rheology is essential for proper selection of constituent materials and prediction of fresh and hardened properties of UHPC. UHPC mixtures often show shear-thickening behavior. This nonlinearity indicates that the commonly used Bingham linear model for cement-based materials is not suitable for UHPC mixtures, and that nonlinear models, such as the Modified Bingham and Herschel-Bulkley models can be applied.

High dosage rates of SPs are normally employed in UHPC to improve rheology and enhance workability retention. Polycarboxylate SP that primarily works by inducing a steric hindrance between binder particles can achieve adequate compatibility with different binder materials. It is reported that methacrylic acid ester-based and allylether-based polycarboxylate SPs are shown to be effective dispersants for UHPC. The blends of these two SPs can allow to reduce the SP demand by around 50%. This is because methacrylic acid ester-based SP disperses cement better than silica, whereas the allylether-based polycarboxylate SP fluidizes silica particularly well.

8.2.5.2 Mechanical properties

① Strength

The compressive, tensile, and flexural strengths of UHPC could vary from 150 to 800 MPa, 25–150 MPa, and 30–141 MPa, respectively. The initiation of strength gain and subsequent rate of strength gain depend on the particular UHPC constituent materials, mix proportions, and the curing conditions. The curing regimes have a significant impact on the fiber–matrix interfacial bonding strength. The strength follows the order: autoclave curing > steam curing > standard curing.

The bond strength of UHPC with fiber ranged between 4.8 and 5.5 MPa, and increased with the increasing of silica fume content to 30%. The enhancement in pullout energy due to silica fume was far more significant than that in bond strength. Compared to the matrix without silica fume, the increase in pullout energy is nearly 100% larger than that with 30% silica fume. Fibers pulled out from matrix with substantial silica fume content demonstrated a dramatically different microstructure. The pullout behavior of steel fiber in UHPC is influenced by matrix strength and fiber configuration (smooth, flat end, or hooked).

② Stress–strain relationship (Fig. 8.30)

The compressive stress–strain relationship of UHPC is the most basic constitutive relationships, and is necessary for nonlinear analysis of concrete structure. However, the study for uniaxial compressive stress–strain constitutive relationship for UHPC is still very limited. UHPC is a ductile material, its failure modes under uniaxial compression are similar to steel fiber concrete. Under compression, it behaves strain-hardening. The peak strain of UHPC significantly increases with the increasing of concrete strength. The ultimate stress coincides with peak stress, and ultimate strain coincides with peak strain.

For a given water-to-binder ratio, the peak strain increased with the increasing of steel fiber content. When the water/binder ratio is too high (up to 0.24), increasing of steel fiber content had little effect on the toughness of specimens. When the water/binder ratio is high, it is easy to form a water

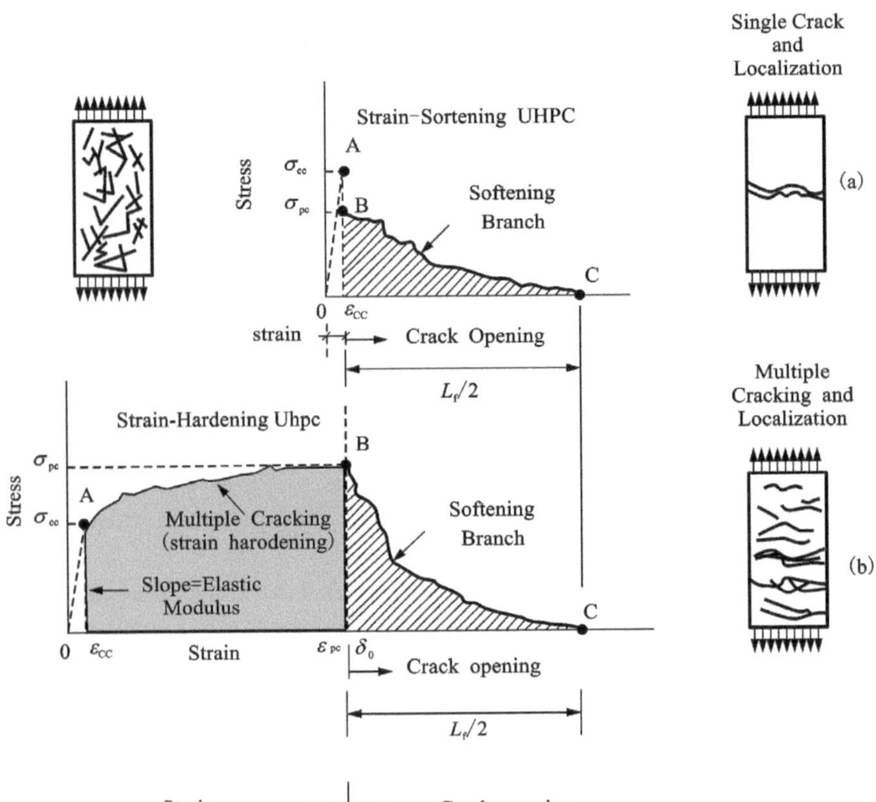

Figure 8.30 Strain–stress relationship of UHPC.

film on the surface of fiber, resulting in a more porous ITZ. Experimental results show that the elastic proportional limit of the ascending branch of stress—strain curve of UHPC is approximately 83%—95% of the peak stress attributed to the dense and homogeneous internal structure of UHPC. Also, it is found that the ultimate strain of UHPC with a length of 13 mm steel fibers are larger than that of a length of 6 mm. The failure form of the latter mainly is pulling-off instead of pulling-out.

UHPC still has much higher flexural loading carrying capacity after the peak load due to incorporation of steel fiber. The use of blast furnace slag also increased the maximum bending loads attributed to improvement of the bonding strength between the matrix and the fibers.

③ Toughness

The toughness of concrete is a synthesis of ductility and strength. The most accepted evaluation methods of toughness for conventional concrete are toughness index method, as described in ASTM C1018, JSCE SF4, and RILEM TC 162-TDF and CECS 13-2009. The content of steel fiber significantly affects the toughness of UHPC due to the cracking resistance of steel fiber. When the volume content of steel fiber was 3%, the toughness index may increase by more than 80%. Mineral admixtures could increase toughness in all curing regimes due to the improvement in bond strength between matrix and fibers, increase 18%—46% for standard room temperature curing, 24%—44% for steam curing, and 23%—39% for autoclaved curing. The other factor affects the toughness of UHPC curing regime. The toughness of UHPC after different curing regimes followed the order: standard curing > autoclave curing > steam curing.

④ Impact resistance

The incorporation of steel fibers could increase the split tensile strength and flexure strength and reduced brittleness. Incorporation of 2% and 5% steel fibers yield split tensile strengths of 21.9 and 31.6 MPa, respectively. The fracture energy of UHPC with fiber could be four times higher than that of ordinary concrete. The penetration depth and crater diameter in target specimens under impact load exhibited an overall reduction with an increase of compressive strength of UHPC. However, the trend is nonlinear. Increase in steel fiber content reduces the crater diameter and crack propagation, but does not show a significant effect on penetration depth. Attributed to hinder crack propagation, the presence of coarse granite aggregates appears to be beneficial to impact resistance. Heat-curing does not alter the impact resistance of the concrete significantly.

Compared to ordinary concrete, the impact splitting tensile strength of UHPC incorporating fibers (3% and 4%) increases by one time, and the impact axial tensile strength increases by nearly 1.5 times. It is found that the fibers of fracture surface were pulled-out rather than pulled off. Therefore, the destruction of specimen is caused by insufficient bond strength between matrix and steel fibers.

8.2.5.3 Dimensional stability

Generally, the drying shrinkage of UHPC after steam curing is very little because of the incorporation of steel fiber. Higher shrinkage rates are developed from 1 to 14 day for UHPC. The shrinkage rate increases steadily from 14 to 100 day, and then gradually reduces from 100 to 130 day. The shrinkage of concrete with silica fume is larger than that of concrete with slag. The incorporation of chemical admixtures is known to affect the porosity of the hardened cement paste, both the total porosity and the pore size distribution, which consequently affects the shrinkage behavior of the concrete.

The addition of shrinkage reducing admixture (SRA) leads to a significant reduction in the drying shrinkage due to the decrease of surface tension in the water at the pores, which results in lower capillary stresses. However, these chemical admixtures might have potential negative side effects including the loss of early strength, and a slight delay in setting time. Incorporation of fly ash could decrease the shrinkage of concrete, and it might because that unhydrated cementitious material acted as aggregate restraining shrinkage.

Autogenous shrinkage of the UHPC decreases with an increase in the water-to-binder ratio due to the more significant shrinkage restraint provided by the residual binders in the UHPC prepared with a higher water-to-binder ratio. An increase in the b/s ratio in the UHPC mix leads to a decrease in the porosity of the concrete. For UHPC prepared using a lower binder-to-sand ratio, the abundance of crystalline ettringite and calcium hydroxide presented in the pores and cracks in the concrete minimized the drying shrinkage of the concrete. The autogenous shrinkage has more significant effect than the drying shrinkage on total shrinkage of the UHPC. The porosity and the amount of unreacted binder, which are related to the degree of hydration of the concrete, are the two influential factors on the shrinkage properties of UHPC. SRA is not only able to reduce the drying shrinkage but also the autogenous shrinkage of UHPC.

8.2.6 Durability

① Permeability

Compared to ordinary concrete and HPC, UHPC is characterized by low porosity and dense microstructure, which guarantees its superior permeability resistance. It is found that the water absorption coefficient of UHPC at 90 d was approximately five times lower than that of ordinary concrete. In addition, the gas permeability coefficient of UHPC is found to be less than 1.0×10^{-19}, which is almost three orders of magnitude lower than that of the ordinary concrete. The chloride ion diffusion coefficient of UHPC is ranged from 0.2×10^{-13} to $4.1 \times 10^{-13}\ m^2/s$, depending on water-to-binder ratio, curing regime, medium solution concentration, steel fiber volume, and testing age. Due to the different applied testing conditions and mixture proportions used, it is difficult to make a quantitative comparison of the chloride permeability of those UHPCs. It is worth to mention that the chloride ion diffusion coefficient of UHPC is at least one order magnitude lower than those of HPC. The lower capillary porosity of UHPC can reduce the capillary suction and intrusion of aggressive gas or solutions, leading to improved permeability.

② Carbonation

Due to very low W/B and limit water content in UHPC, its microstructure is very dense with a very small most probable pore size. It is stated that carbonation rarely happens on UHPC, either under standard or heat-curing regime. UHPC specimens subjected to 56 and 90 d action were not found any carbonation signs. The carbonation depth of UHPC after 6 months of exposure was found to be 0.5 mm. After a 1-year exposure to 1% CO_2 atmosphere, this value was 1.5–2.0 mm. Small-scale beams of UHPC placed in a carbonation chamber that subjected to flow of 5%–100% CO_2 for 2 years showed no signs of carbonation.

③ Freeze–thaw resistance

UHPC demonstrates very high freeze–thaw resistance. It is reported that no deterioration after 300 or 600 freeze–thaw cycles and has a durability factor of 100 or greater with nearly zero mass loss. Even after 800 freeze–thaw cycles, no obvious deterioration phenomenon is observed.

It is also revealed that UHPC specimens can self-heal the microcracks when they were submerged in water after or during deterioration because of the presence of a high amount of unhydrated cement particles. This can recover the mechanical properties of UHPC from deterioration due to freeze–thaw action given sufficient immersed water.

④ Fire resistance

The absence of capillary pores with the low amount of free water in UHPC makes it more prone to explosive spalling due to increasing vapor pressure and nonuniform thermal gradient when exposed to high temperature. Three different mechanisms, including restrained thermal expansion, build-up of pore pressure, and combined thermal stresses and pore pressures, are involved in explaining the explosive spalling.

In order to alleviate the spalling of UHPC, several preventive measures can be used. The use of proper dosage of synthetic fiber with low melting point can alleviate explosive spalling. This is attributed to the formation of pores and channels from synthetic fiber melting to release vapor and reduce the build-up of pore pressure. Besides, good bond properties between the polyvinyl alcohol (PVA) fiber and the cementitious matrix, together with their higher elastic modulus than PP, can lead to better structural performance.

Exercises

1. What is the CRTS I CA mortar and CRTS II CA mortar?
2. Describe the microstructure of two type CA mortars.
3. State the factors affecting the fresh properties of CA mortars.
4. What is the effect of asphalt on the properties of hardened CA mortar?
5. What is the effect of cement on the properties of hardened CA mortar?
6. What is the effect of temperature on the properties of hardened CA mortar?
7. What is the effect of loading rate on the properties of hardened CA mortar?
8. Describe the raw materials of UHPC, and their functions in this material.
9. Describe the basic principles of making UHPC.
10. Why is the explosive spalling prone to happen to UHPC under high temperature and how to alleviate it?

Index

Note: 'Page numbers followed by "f" indicate figures and "t" indicate tables.'

A

AAM. *See* Alkali activated materials (AAM)
Absolute volume, 6
Absorption, aggregates, 86–87, 86f
Acid attack, 153
Acid corrosion, 32
Additives, 252, 283–284, 316–318
Adhesives, 252
  adhesion of
    bond force, 273
    requirements, 274–275
  composition and function
    curing agent, 272
    diluent/solvent, 272
    fillers, 273
    gel, 272
    plasticizers and toughening agents, 272
  properties of, 275–276
  types, 273, 275–276
Age hardening, 294
Aggregates
  absorption, 86–87
  characteristics of, 77–78
  classification of
    density, 70–71
    mother rock, 75–77, 78t
    origins, 71–75, 74f
    sizes, 71
  deleterious substances, 89–91
  density, 87–88
  gradation and size, 80–84, 80f–83f, 83t–84t, 85f
  maximum size of, 84–86, 85f
  mechanical properties, 88–89, 89f
  particle shape, 78–79
  significance of, 69–70
  soundness, 88
  surface texture, 78–79
  ultrahigh-performance concrete (UHPC), 366
Aging methods, 294–295, 295f
Air dry, 86
Air-entraining (AE) agents, 110–111
Air void system (AVS), 110
Alkali activated materials (AAM), 47
  activation process and products, 48–49, 49f–50f
  application, 51
  manufacture, 48
  properties, 50–51
Alkali-aggregate reaction (AAR), 147, 159–160
Alkali-silica reaction (ASR), 200
Alloying, 212–213, 234
Aluminum
  density, 236
  free alkalis, 236
  rainwater gutters and pipes, 237
American Society for Testing and Materials, 118–120
Amines, 272
Anhydrides, 272
Anisotropy, 243–244
Annealing, 212
Apparent density, 6
Apparent specific gravity, 87
Apparent volume, 6
Aramid fibers, 279
Asphalt cement, 68, 287, 288f
  aging, 294–295, 295f
  characterization
    dynamic shear rheometer (DSR) test, 305–306
    flash point, 305
    performance grade binder characterization, 303–304
    performance grade characterization approach, 302–303

Asphalt cement (*Continued*)
pressure-aging vessel (PAV), 304
rolling thin-film oven (RTFO), 304
rotational viscometer (RV), 305
chemical composition, 290–291, 290t
aliphatic, 290
aromatic, 290
napthenic, 290
diluent, 288
emulsifying agents, 288–289
liquid, 306–307, 306t, 307f
physical structure
constituents, 291–292, 291t
internal structure, 292, 293f
phase separation, 292
sol-gel type, 292
temperature, 293
properties of, 287–288
rheological tests, 297–298, 298f
scattering, 288–289, 289f
stiffness, 298–300, 299f–300f
temperature susceptibility, 300–301, 301f
tensile properties, 301–302, 302f
viscosity and consistency, 295–297, 296f–297f
Asphalt concrete, 309f
additives and fillers, 318
antistripping agents, 316–317
asphalt cement modifiers, 317
extenders, 318
recycling agents, 317
applied loads
fatigue resistance, 312
flexibility, 312
permanent deformation, 314
stability, 310–311, 312f
stiffness, 310, 311f
tensile (fracture) strength, 312–314
characteristics, 308
chemicals, 316
composition, 308–309
moisture
durability, 315
permeability, 314
stripping (moisture-induced damage), 315
structure, 308–309
Superpave mix design, 318–319, 319t, 320f
binder selection, 320
design aggregate structure, 320–324, 321t–322t
temperature, 315–316
Asphalt emulsion, 329, 330t
ASTM C 88, 88
ASTM D-5, 297–298
ASTM D-2170, 298
ASTM D-2171, 298
ASTM D-2493, 300
ASTM D2872, 304
ASTM D-4402, 298
ASTM D6521, 304
Atomic structure
crystalline SiO_2 (quartz) *vs.* amorphous SiO_2 (glass), 3, 4f
ionic bonding, 2
metallic bonding, 3
van der Waals bonding, 3
Autogenous shrinkage, 145

B

Bakelite, 262
Batching, 65, 113–114
Bending beam rheometer (BBR) test, 299–300, 300f
Bending strength, 247, 248t
Binders, 59
Bingham flow, 122–124, 124f
Biogenic (biochemical) sedimentary rocks, 75–76
Bituminous materials, 287
Blaine air-permeability method, 8, 28
Blast-furnace slag, 93t, 96–97
Bleeding, 125–127, 126f
Blended cement. *See also* Portland cement
advantages of, 37
defined, 17
Portland-limestone cement, 37
Portland-pozzolan cement, 36–37
Portland-slag cement, 36
ternary blended cement, 37
Bogue calculation, 19

Bolomy equation, 164
Bonds
primary, 2
secondary, 2
Brazil splitting test, 132–133
Brittleness, 12–13
Brownian forces, 121
Bulk density, 30, 63

C

Calcium aluminate cements (CACs), 41
application, 46–47
composition, 41
hydration
final stage, 42–43, 43f–44f
initial stage, 42
second stage, 42
manufacturing, 41
properties
durability, 45
refractory, 45–46
strength, 43
workability and setting time, 44–45
Calcium hydroxide (CH), 21, 23, 368
Calcium silicate hydrate (C-S-H), 21
inner product, 22–23
interlayer space, 26
morphology of, 22–23, 23f
outer product, 22–23
Calcium sulfoaluminate cement (CSA)
application, 40–41
composition, 37–38
hydration, 38–39
manufacture, 37–38
properties
alkalinity, 40
curing time, 40
lower carbon, 39
shrinkage, 40
strength gain, 39
Capillarity, 10
Capillary effect, 140
Capillary voids, 26–27, 27f
Carbonation, 375
Carbonic acid corrosion, 32
Carbon structural steel
designation system, 220–221
high strength low alloy, 221–223
selection of, 221
technical requirements, 221, 222t–223t
Cationic emulsions, 306
Celluloid, 261–262
Cellulose, 261
Cement asphalt (CA) composite, 327–328
construction technology
construction steps, 358–360, 359f
technological test, 360–362, 360t, 361f–362f
hardening and structure, 337–339, 338f–339f
mix proportion and mixing, 332
CRTS I CA, 333
CRTS II CA, 333, 335t, 336f, 336t, 337f
properties, 352f, 355f, 356t
antifatigue performance, 356–358, 358f
compressive strength and flexural strength, 349–351, 349f–350f
density and air content, 340
early-age deformation, 346–348, 347f–349f
elastic modulus, 351–352
flow times, 342, 342f
freezing and thawing resistance, 352–356
fresh CA mortar, 340–344, 341f
low-temperature properties, 352–356
separation rate and bleeding ratio, 343–344
working time, 342–344, 343f
raw materials, 329, 332t
aluminum powder, 331–332
cement, 329
emulsion, 329, 330t
expansive agent, 331–332
flowability and uniformity, 344–346, 345f–346f
sand, 329, 331t
Cement-based composites
cement asphalt composite, 327–362

Cement-based composites (*Continued*)
ultrahigh-performance concrete (UHPC), 362–376
Cement kilns, 69
Cement manufacturing, dry process, 18, 18f
Charpy test method, 216–217
Chemical admixtures, 102t, 103
air-entraining (AE) agents, 110–111
classification, 103, 104t
set controlling agents, 107–109
superplasticizers (SPs), 103–107, 106f
viscosity-modifying admixtures (VMAs), 111–112
Chemical composition, 1
Chemically precipitated (chemical) sedimentary rocks, 75–76
Chemical shrinkage, 144–145
China Rail Track System (CRTS), 172, 327, 328f
Chinese National Standard of Carbon Structural Steel (GB700-2006), 220–221
Chloroprene rubber, 274
Class C fly ash, 93–94
Class F fly ash, 93–94
Clastic (detrital) rocks, 75–76
Clay lumps, 91
Cleveland open cup method (ASTM D92), 305
Clinkers, 17, 19, 20f
calcium sulfoaluminate cement (CSA), 38
hydration process, 22, 22f
sintering of, 19
Coal, 91
Coarse aggregate, 71, 79
Cold bending, 215–216
Cold-rolled ribbed reinforced bars, 224
Colloidal particle interaction forces, 121
Compression strength, 129–132, 131f, 245–246, 245f
Concrete
aggregates
absorption, 86–87
characteristics of, 77–78
classification of, 70–77
deleterious substances, 89–91
density, 87–88
gradation and size, 80–84
maximum size of, 84–86
mechanical properties, 88–89
particle shape, 78–79
significance of, 69–70
soundness, 88
surface texture, 78–79
cement, 68–69
deformation, 139f
autogenous shrinkage, 145
chemical shrinkage, 144–145
creep, 141–143
creep strain, 137–138
drying shrinkage, 138–141
factors, 137–138
thermal expansion, 145–146
durability, 59–60, 147
acid attack, 153
alkali-aggregate reaction (AAR), 159–160
fire resistance, 157–159
freezing-thawing (FT) cycle, 153–157
permeability, 147–150
steel bar corrosion, 149f, 151f, 160–162
sulfate attack, 150–152, 152t
economy, 60–62, 61f
fresh state
batching, 113–114
bleeding, 125–127
cracking, 127–129
curing, 114–118
finishing, 114–118
mixing, 113–114, 113f
placing, 114–118
plastic shrinkage, 127–129
segregation, 125–127
transporting, 113–114
workability, 118–125, 119f
green aggregate, 91–92
mechanical properties
compressive strength, 129–132, 131f

elastic modulus, 133–135, 134f–136f
factors, 135–137, 137f–138f
tensile strength, 132–133, 133f
mix design, 163–164, 167t
aggregate moisture, 170
binder determination, 166, 169t
fine and coarse aggregates, 169
fly ash and slag, 165, 165t
principles, 164
sample computations, 170
sand ratio, 166, 169t
trial batch adjustments, 170
water choice, 166, 168t
w/b ratio, 165
w/c ratio, 165
prestressed steel wire, 225
steam curing, 190–202
steel, 224
supplementary cementitious materials (SCMs), 92
blast-furnace slag, 96–97
fly ash, 92–96
metakaolin, 99
natural pozzolans, 100–101
silica fume, 98–99
sustainability, 60
types of, 59, 63f
application, 63
bulk density, 63
construction method, 63–64
versatility, 59
Concrete truck mixer, 114, 114f
Condensed silica fume, 98
Consistency, 28–29
Consolidating concrete, 115–116
Constant water law, 164
Contact Metamorphism, 76–77
Copper
grades of, 234–235
unalloyed, 236
Corrosion, 15
hardened cement
acid, 32
alkali, 32
prevention of, 33–34
soft water, 31–32
sulfate attack, 32–33
steel
chemical, 232, 233f
electrochemical, 232
prevention, 233–234
steel bar, 149f, 151f, 160–162
Cracks, 5, 128
thermal stress, 37
Creep, 141, 142f
delayed elastic strain, 143
microcracking, 143
moisture movement, 143
structural adjustment, 143
Creep rupture strength, 250, 250f
CRTS III slab ballastless track, 172–173
Crushed rocks, 79
CSA. *See* Calcium sulfoaluminate cement (CSA)
Curing agents, 272
Cutback asphalt, 306, 306t

D

Deformation, 139f
autogenous shrinkage, 145
chemical shrinkage, 144–145
creep, 141–143
creep strain, 137–138
drying shrinkage, 138–141
factors, 137–138
thermal expansion, 145–146
Delayed ettringite formation (DEF), 200
Deleterious substances, 89–91, 90t
Dense slags, 79
Densified with small particles (DSPs), 362–363
Density, 6–8, 7f, 70–71
Deoxidation of steel, 219
Deterioration mechanisms, 15
Dioctyl phthalate, 272
Dissolving corrosion, 31–32
Drying shrinkage, 139
capillary effect, 140
disjoining pressure, 140–141
interlayer water movement, 141
Dry recycled aggregate, 73
Ductility, 13
Durability, 15–16

Durability (*Continued*)
asphalt concrete, 315
calcium aluminate cements (CACs), 45
ultrahigh-performance concrete (UHPC), 362–363
carbonation, 375
fire resistance, 376
freeze-thaw resistance, 375
permeability, 375
Dynamic mechanical properties, 14–15, 14f
Dynamic shear rheometer (DSR) test, 298, 305–306

E
Economy, 60–62, 61f
Efflorescence, 51
Elasticity, 11–12, 214–215, 214f
Elastic modulus, 12, 133–135, 134f–136f
Elongated, 79
Elongate minerals, 76–77
Elongation of steel, 215
Emulsified asphalts, 288–289
Envelope density, 6
Envelope volume, 6
Epoxy resin, 274–275
Equilibrium moisture content (EMC), 242
Ettringite, 24
Expansive cement, 36
Extenders, 318
External vibrators, 117

F
Fatigue resistance, 312
Ferrous metals, 205
Fiberboard, 256
Fiber reinforced polymer (FRP), 276–277
constituent materials
additives, 283–284
fibers, 277–280, 278t, 279f–280f
fillers, 282–283
polymeric matrices, 280
polymerization, 282
resins, 280–282, 281t
development, 284
Fiber saturation, 242
Fillers, 273–274, 316–318
Fine aggregate (sand), 71
Fineness, 8, 28
Fineness modulus (FM), sand, 82–83
Fire resistance, 376
aggregate, 158
cement paste, 158
microcracks, 158
Flint, 79
Flow curve, 121–122
Fly ash (FA), 68
carbon, 94
chemical effect, 95
Class C, 93–94
Class F, 93–94
factors, 92–93
factors of, 165, 165t
physical effects, 94–95, 94f
surface chemistry effect, 95–96
ultrahigh-performance concrete (UHPC), 365
Free alkalis, 236
Freezing-thawing (FT) cycle, 74
cracks, 157
ice expansion, air bubbles, 156–157
porosity and refining pores, 157
water absorption, 157
FRP. *See* Fiber reinforced polymer (FRP)

G
Geopolymers, 50
Glass fiber, 277–278
Glass reinforced plastics, 91–92
Grade 550 steel bars, 224
Grain flow, 5
Granulated blast furnace slag (GGBFS), 365
Gravel, 71, 76
Graywacke, 76
Green aggregate, 91–92
Gypsum, 19

H
Hardened cement, 25–27, 73

corrosion
acid, 32
alkali, 32
prevention of, 33–34
soft water, 31–32
sulfate attack, 32–33
Hardness, 13
Hardwood plywood, 253–254
Heat capacity, 8–9
Heavyweight aggregate, 71
High-performance concrete (HPC), 363–364
High-speed rail (HSR), 171–190, 327. *See also* Self-consolidating concrete (SCC)
High strength low alloy structural steels, 221–223
Horizontal shear, 246
Hot-pressed cement, 362–363
Hot-rolled reinforced bars, 224, 224t–226t
Hydration process, 53t, 68
AFm and AFt phases, 23
calcium aluminate cements (CACs)
final stage, 42–43, 43f–44f
initial stage, 42
second stage, 42
calcium hydroxide (CH), 21, 23
calcium silicate hydrate (C-S-H), 21–23, 23f
calcium sulfoaluminate cement (CSA), 38–39
capillary voids, 26–27, 27f
clinker minerals, 22, 22f
ettringite, 24
ferrite reaction, 22
heat evolution, 20, 21f
isothermal calorimetry, 20
magnesium oxychloride cements (MOCs), 52
portlandite, 23, 24f
setting and hardening, 25–27, 25f
Hydrodynamic forces, 121
Hydrogarnets, 24

I
Igneous rocks, 75
Immersion-type vibrators, 116–117
Impact toughness, 216–217, 216f–217f
Impermeability, 34
Industrial waste aggregate, 91–92
Inorganic cementing materials
alkali activated materials (AAM), 47–51
blended cement, 37
calcium aluminate cements (CACs), 41–47
calcium sulfoaluminate cement (CSA), 37–41
magnesium-based cements, 52–56
Portland cement, 17–37
Insoluble residue, 30
Interfacial transition zone (ITZ), 86
Internal vibrators, 116–117
Intrusive rocks, 75
Ionic bonding, 2
Isothermal calorimetry, 20

K
Kelvin's equation, 140
Killed steel, 219

L
Laplace pressure, 10
Lightweight aggregate, 70
Lignin, 105, 261
Lignite, 91
Lignosulfonates, 105
Limestone powder, 68, 193
Linear coefficient of thermal expansion, 9–10
Liquid asphalt, 287–288
Loading, 11, 11f
Loss of ignition (L.O.I), 30
Lumber, 239

M
Macro-defect-free cement (MDF), 362–363
Macrostructure, 5
Magma, 75
Magnesium, 237–238

Magnesium-based cements. *See also specific types*
application, 55–56
hydration, 52–53
manufacture and composition, 52
properties
abrasion resistance, 54–55
bonding strength, 54
fast setting and rapid strength gain, 54
flame retardant, 54
low electrical and thermal conductivity, 54
strength, 54
Magnesium oxychloride cement (MOC), 52, 55
Magnesium oxysulfate cement (MOS), 52–53, 55
Magnesium phosphate cement (MPC), 52, 56
Manufactured aggregate, 72
Marshall test, 310
Mass (m), 6
Mass fraction, 1
Material toughness, 13
Mechanical properties
brittleness, 12–13
durability, 15–16
dynamic, 14–15, 14f
elasticity, 11–12
hardness, 13
loading and strength, 11, 11f
plasticity, 11–12
toughness, 12–13
ultrahigh-performance concrete (UHPC), 371–374
Metakaolin (MK), 97t, 99, 193, 365
Metal atoms, 2
Metallic bonding, 3
Metals, 205
corrosion and prevention, 232–234
nonferrous metals, 234–238
standards and selection, 220–232
structural steel, 206–220
Metamorphic rocks, 76–77
Mica, 91
Micelles, 292
Microcracking, 143
Microsilica, 98
Microstructure, cement-based materials, 5, 5f
Minerals, 2
Mixing ratio, 1
Mixing water, for concrete
chemical compositions, 65, 67t
impurities, 66
requirements for, 65, 66t
Modulus of elasticity, 12, 245
Moisture conditions, aggregates, 86, 86f
Moisture content (MC), 242, 243f, 247, 249f
Molar fraction, 1
Molarity, 1
Monocarbonate, 24
Mortar press air meter, 340, 341f
Mylonites, 76–77

N

Natural aggregates, 71, 79
Natural polymers, 261
Natural pozzolans, 100–101, 100t
Newtonian fluids, 122–124, 295–296
Nonferrous metals, 205
aluminum, 236–237
copper, 234–236
magnesium, 237–238
Nonmetal atoms, 2
Non-Newtonian fluids, 295–296
Normality, 1
Normal weight aggregate, 70

O

Obsidian, 79
Off-site fabrication, 190–191
Ordinary Portland cement (OPC), 40
Oriented strand board (OSB), 254
Oven dry, 86

P

Particleboard, 254–256
Particle size distribution, 8, 28
Penetration test, 297–298, 298f
Phase composition, 1–2
Phenolic resin, 262

Physical properties, 5–10
Plastic cracking, 128, 128f
Plasticity, 11–12, 215–216, 215f–216f
Plasticizers, 103–105, 272
Plastics, 263f
  polymeric molecule, 264–266, 264f–265f
  properties of, 263–264
  settlement, 127
  shrinkage, 127
  thermoplastic, 266–271
  thermosetting polymers, 271
  viscosity, 121
Platy minerals, 76–77
Plies, 252
Plutonic rocks, 75
Plywood, 252–254, 255t
Poise, 295–296
Poisson's ratio, 129, 135, 245
Poly (ethylene oxide), 262
Polycarboxylates (PCs), 366
Polymelamine sulfonates (PMS), 103–105
Polymers, 14, 68, 282
  adhesives, 272–276
  cellulose, 261
  competitiveness, 262
  durability, 262
  economy, 262
  environmental advantages, 263
  fiber reinforced polymers (FRP), 276–284
  lightweighting, 262
  lignin, 261
  natural, 261
  phenolic resin, 262
  plastics, 263–271
  processability, 262
  recyclability, 263
  sealants, 271–272
  thatch, 261
  versatility, 262
  wood, 261
Polynaphthalene sulfonates (PNS), 103–105
Polypropylene fibers, 159
Polystyrene, 268
Polyurethane adhesive, 276
Polyvinyl acetal adhesives, 275
Polyvinyl acetate, 275
Polyvinyl alcohol, 275
Polyvinyl chloride, 268
Porosity, 5
Portland cement, 205
  application, 34
  blended cement, 36–37
  chemical properties
    alkali, 31
    insoluble residue, 30
    loss of ignition (L.O.I), 30
    total chloride content, 31
  composition of, 17–18, 19t
  concrete. *See* Concrete
  corrosion and prevention, 33–34
    acid, 32
    alkali, 32
    hardened cement, 31–33
    soft water, 31–32
    sulfate attack, 32–33
  expansive cement, 36
    AFm and AFt phases, 23
    calcium hydroxide (CH), 21, 23
    calcium silicate hydrate (C-S-H), 21–23, 23f
    capillary voids, 26–27, 27f
    clinker minerals, 22, 22f
    ettringite, 24
    ferrite reaction, 22
    heat evolution, 20, 21f
    isothermal calorimetry, 20
    portlandite, 23, 24f
    setting and hardening, 25–27, 25f
  manufacturing, 18–19, 18f
  physical properties
    bulk density, 30
    consistency, 28–29
    fineness, 28
    heat of hydration, 29–30
    setting time, 29
    soundness, 28
    specific gravity, 30
    strength, 29
  properties, 17–18

Portland cement (*Continued*)
sulfate resistance Portland cement, 35
white Portland cement, 35
Portlandite, 23, 24f
Portland-limestone cement, 37
Portland-pozzolan cement, 36–37
Portland-slag cement, 36
Prefabrication, 190
Pressure-aging vessel (PAV) procedure, 295, 304
Prestressed steel wires, 225
Primary bonds, 2
Pulverized fuel ash (PFA), 92–93
Pumice, 70, 79
Pumping concrete, 64, 64f, 114–115, 115f

Q

Quartzite, 76
Quenching, 211–212

R

Rail grinding, 230–232, 231f
Rail steel
grinding, 230–232, 231f
properties, 228–230
mechanical, 229–230, 230t
Rail wears, 230–231, 231f
Reactive powder concrete (RPC), 362–363
Ready-mixed concrete, 113
Recycled-concrete aggregate
absorption rates, 73
particle shape of, 73
side effect, 74–75
Recycled glass cullet (RGC), 366
Regional Metamorphism, 76–77
Reinforcement concrete, 134–135
Rheological behavior, 295–297, 296f–297f
Rheological parameters, 121, 122f–123f
Rice husk ash (RHA), 365
Rice husks, 254–256
Rigidity, 217
Rimmed steel, 219
Roller compacted concrete, 64, 64f
Rolling shear, 247
Rolling thin-film oven (RTFO) procedure, 294–295, 304
Rotational (Brookfield) viscometer test (ASTM D4402), 305
Rotational viscometer (RV) test, 298, 305
Rubber adhesives, 276

S

Sand, 71, 76
fineness modulus (FM), 82–83
particle gradation, 82, 83t
Sandstones, 76
Saturated-surface dry (SSD), 87, 170
SCC. *See* Self-consolidating concrete (SCC)
SCMs. *See* Supplementary cementitious materials (SCMs)
Sealants, 271–272
Sealed-space-filling self-compacting concrete (SSFSCC)
construction technology, 173f–174f, 178f, 180f, 185–190, 187f–188f
mix proportioning of, 178–179
parameters, 162f, 180–182
procedures, 182–185
typical mix for, 184–185, 185t–186t
property requirements of
filling ability, 176
fresh state, 176–178
hardened state, 175, 175t
passing ability, 176–177
stability, 161f, 177–178, 179t
Seawater, 66
Secondary bonds, 2
Sedimentary rocks, 75–76
Segregation, 125–127, 126f
Self-compacting concrete, 171
Self-consolidating concrete (SCC), 171
CRTS III ballastless slab track, 173–174
guidelines/specifications, 171–172
hydrostatic pressure hypothesis, 154f, 156f, 172–173
osmotic pressure hypothesis, 156f, 173–174

sealed-space-filling self-compacting concrete. *See* Sealed-space-filling self-compacting concrete (SSFSCC)
Semikilled steel, 219
Setting time, 25–27
alkali activated materials (AAM), 50
calcium aluminate cements (CACs), 44–45
final, 29
initial, 29
Sewage water, 66–68
Shallow-intrusive rocks, 75
Shear modulus, 245
Shear strength, 246–247, 246f
Shotcrete, 63, 64f
Shrinkage reducing admixture (SRA), 374
Silica fume (SF), 98–99, 364
Sinusoidal stress, 14, 14f
Skeletal volume, 6
Slag, 68
factors of, 165, 165t
Slump loss, 124
Slump flow, 120
Slump test, 118–120, 119f
Slurry infiltrated fiber concrete (SIFC), 362–363
Sodium chloride, 66
Sodium sulfate, 66
Soft water corrosion, 31–32
Solid solution, 1–2
Sorel cement, 52
Soundness, 28
aggregates, 88
Specific gravity, 6–8
Portland cement, 30
Specific surface area, 8
Split Hopkinson pressure bar (SHPB) test, 198–199
Stability, 310–311, 312f
Static strength test, 249
Steam-cured concrete, 64, 64f, 190
curing regime, 194–196, 194f–196f
durability, 197f–198f, 199–202
mechanical properties
compressive strength, 196–197, 196f
dynamic mechanical properties, 197–199
off-site fabrication, 190–191
raw materials, 192–193, 192t
Steel
corrosion, 149f, 151f, 160–162
chemical, 232, 233f
electrochemical, 232
prevention, 233–234
standards and selection, 220–232
structural, 206–220
Steel Q235, 221
Steel reinforcement, 61–62
Strain, 12
Strands, 254
Stress analysis, 14
Stress-strain behavior, 213–214, 214f
Stripping, 316
Strong alkali corrosion, 32
Structural steel, 205
chemical composition
aluminum, 206
carbon, 206, 207f
chromium, 208
manganese, 206
molybdenum, 208
nickel, 208
phosphorus, 207–208
silicon, 207
sulfur, 207–208
classifications
application, 219
composition, 217–219
deoxidation practice, 219
press-working modes, 220
shape, 220, 220f
mechanical properties
elasticity, 214–215, 214f
impact toughness, 216–217, 216f–217f
plasticity, 215–216, 215f–216f
rigidity, 217
stress-strain behavior, 213–214, 214f
strengthening mechanisms
alloying, 212–213
annealing, 212
grain size, 208, 209f

Structural steel (*Continued*)
heat treatment, 210–212, 211f
normalizing, 211
quenching, 212
strain hardening, 208–210, 209f–210f
tempering, 212
Sugar, 68
Sulfate attack, 150–152, 152t
corrosion, 32–33
Sulfate resistance Portland cement, 35
Superpave classification system, 300
Superplasticizers (SPs), 68, 103–107, 106f, 366
Supplementary cementitious materials (SCMs), 92
blast-furnace slag, 96–97
fly ash, 92–96
metakaolin (MK), 99
natural pozzolans, 100–101
silica fume, 98–99
steam-cured concrete, 192
Surface protective covering, 34
Surface texture, 79
Sustainability, 60
Synthetic resin adhesives
epoxy resin, 275
polyurethane, 276
polyvinyl acetal adhesives, 275
polyvinyl acetate, 275
polyvinyl alcohol, 275

T

Tars, 287
Temperature dependence, 299
Temperature susceptibility, 300–301, 301f
Tempering, 212
Tensile strength, 132–133, 133f
of aggregate, 89
Tensile test, 213–214, 214f
Tension strength, 246
Termites, 258
Ternary blended cement, 37
Thatch, 261
Thermal conductivity, 8–9
Thermal-cracking tests, 303, 316
Thermal expansion, 145–146
Thermoplastics, 263–264, 268f, 270f
composition and transition temperatures, 268, 269t
glass transition, 267
melting temperature, 267
van der Waals bond, 266
Thermosetting polymers, 263–264, 271
Thixotropy-enhancing agent (TEA), 370
Timber, 239
Time-of-loading, 299
Toughening agents, 272
Toughness, 12–13
Trinidad Lake asphalt, 287
True density, 6
True volume, 6

U

Ultrahigh-performance concrete (UHPC), 99
aggregates, 366
cementitious components
fly ash (FA), 365
granulated blast furnace slag (GGBFS), 365
limestone powder (LP), 365
metakaolin, 365
nanoparticles, 366
Portland cement, 364
rice husk ash (RHA), 365
silica fume (SF), 364
compressive strength, 363–364
dimensional stability, 374
durability, 362–363
carbonation, 375
fire resistance, 376
freeze-thaw resistance, 375
permeability, 375
fibers, 367, 367t
mechanical properties, 362–363
impact resistance, 373
strength, 371
stress-strain relationship, 372, 372f
toughness, 373
mix design
microstructure, 368
optimize pore structure, 368

toughness, 368–369
preparation and curing, 369–370
production of, 362–363
raw materials, 364–367
superplasticizers (SP), 366
workability/rheology, 370–371
Urbanization, 72

V

van der Waals bonding, 3, 266
Versatility, 59
Vertical shear, 246
Vibration, 63
consolidating concrete, 116
Vicat Test, 28–29
Vinyl copolymers, 103–105
Viscosity-modifying admixtures (VMAs), 111–112, 370
Viscosity-temperature susceptibility (VTS), 301
Viscous forces, 121
Volume fraction, 1

W

Waste materials, 91
Wastewaters, 66–68
Water reducers, 107
Water-reducing admixtures, 103–105
Weight (w), 6–8
Wetting, 10, 10f
White ants, 258
White Portland cement, 35
Wood, 240t, 261
composites, 251f
composition and manufacture, 240–241, 251–252, 253f
fiberboard, 256
flame-retardant composites, 256–257
oriented strand board (OSB), 254
particleboard, 254–256
plywood, 252–254, 255t
preservative-treated composites, 257
water-repellant composites, 256
dimensional stability, 242–243, 243f–244f
durability
decay, 257–258
moisture, 257
preservative treatments, 258–259
termites, 258
mechanical properties
anisotropy, 243–244
bending strength, 247, 248t
compression strength, 245–246, 245f
elastic, 244–245
shear strength, 246–247, 246f
straws, 243–244, 244f
tension strength, 246
moisture, 241–242, 243f
relative density, 241
strength, factors affecting
defects, 250
environment temperature, 249
moisture content (MC), 247, 249f
time under load, 249–250, 250f
structure, 240–241, 241f

Y

Yield stress, 121
Young–Laplace equation, 10
Young's modulus, 12

Kolibri 360 GmbH
Ferdinand-Jühlke-Straße 7
99095 Erfurt, Deutschland
produktsicherheit@kolibri360.de